Electrical Pre-Apprenticeship & Workforce Development Manual

"Developed by electricians for electricians"

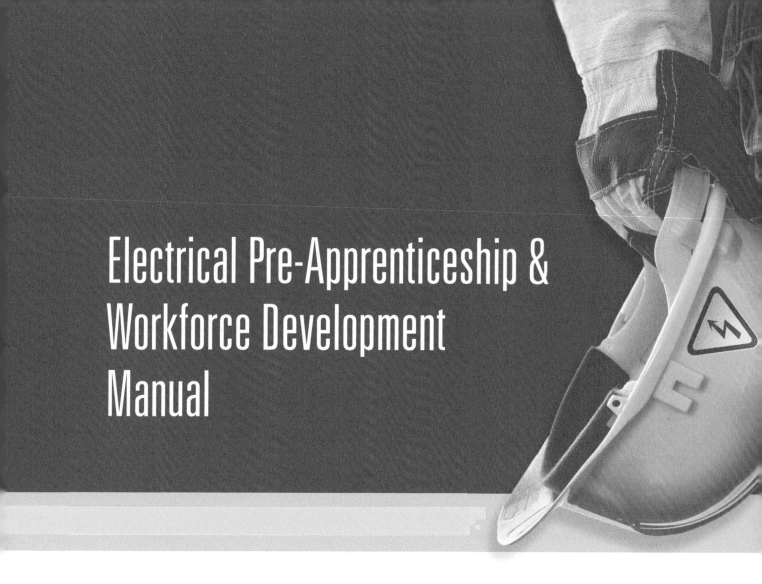

Electrical Pre-Apprenticeship & Workforce Development Manual

INDEPENDENT ELECTRICAL CONTRACTORS CHESAPEAKE
WESTERN ELECTRICAL CONTRACTORS ASSOCIATION INC.

Australia • Brazil • Canada • Mexico • Singapore • United Kingdom • United States

Electrical Pre-Apprenticeship & Workforce Development Manual
Author(s): Independent Electrical Contractors Chesapeake and Western Electrical Contractors Association Inc.

Vice President, Editorial: Dave Garza

Director of Learning Solutions: Sandy Clark

Acquisitions Editor: Stacy Masucci

Managing Editor: Larry Main

Senior Product Manager: John Fisher

Developmental Editor: Beth Jacobson, Ohlinger Publishing Services

Editorial Assistant: Kaitlin Murphy

Vice President, Marketing: Jennifer Baker

Marketing Director: Deborah Yarnell

Senior Marketing Manager: Erin Brennan

Marketing Coordinator: Jillian Borden

Senior Production Director: Wendy Troeger

Production Manager: Mark Bernard

Content Project Manager: Barbara LeFleur

Senior Art Director: David Arsenault

Technology Project Manager: Joe Pliss

For product information and technology assistance, contact us at
Cengage Customer & Sales Support, 1-800-354-9706 or support.cengage.com.

For permission to use material from this text or product, submit all requests online at **www.cengage.com/permissions.**

Library of Congress Control Number: 2011942330

ISBN-13: 978-1-111-31689-1
ISBN-10: 1-111-31689-9

Cengage
200 Pier 4 Boulevard
Boston, MA 02210
USA

Cengage is a leading provider of customized learning solutions with employees residing in nearly 40 different countries and sales in more than 125 countries around the world. Find your local representative at: **www.cengage.com.**

To learn more about Cengage platforms and services, register or access your online learning solution, or purchase materials for your course, visit **www.cengage.com.**

Notice to the Reader

Publisher does not warrant or guarantee any of the products described herein or perform any independent analysis in connection with any of the product information contained herein. Publisher does not assume, and expressly disclaims, any obligation to obtain and include information other than that provided to it by the manufacturer. The reader is expressly warned to consider and adopt all safety precautions that might be indicated by the activities described herein and to avoid all potential hazards. By following the instructions contained herein, the reader willingly assumes all risks in connection with such instructions. The publisher makes no representations or warranties of any kind, including but not limited to, the warranties of fitness for particular purpose or merchantability, nor are any such representations implied with respect to the material set forth herein, and the publisher takes no responsibility with respect to such material. The publisher shall not be liable for any special, consequential, or exemplary damages resulting, in whole or part, from the readers' use of, or reliance upon, this material.

Printed in Mexico
Print Number: 07   Print Year: 2020

Table of Contents

Preface • xiv
Acknowledgments • xx

1   A Career in the Electrical Trade • 2

The Electrical Industry..5
Training • 5

Career Paths..6
Electrical Construction Industry • 7
Electrical Maintenance • 13
Electrical, or Power, Distribution Electrician • 14
Electrical Distributor • 16
Electrical Manufacturer • 17
Electricians and Other Trades Workers • 18
Vertical Mobility in the Electrical Trade • 19

2   Professional Behavior and Study Skills   26

Professional Behavior in the Workplace..28
Active Workplace Participation • 28
On-the-Job Training and Workplace Experience • 29
Apprentice–Journeyman Workplace Interactions • 32
Workplace Practices and Procedures Regarding Alcohol and Drugs • 33
Punctuality and Attendance • 34

Building Personal Success..35
Active Course Participation • 35
Time Management Strategies • 37
    Time Management Tips • 37
Effective Study Techniques • 38
Benefits of Effective Study Techniques • 39
Comprehension Strategies for Reading Assignments • 40
Test-Taking Strategies • 41
    Before the Test Date • 41
    Taking the Test • 42

3 Safety • 50

Basic Safety Rules 53
Use the Buddy System • 54
Learn First Aid and Be Prepared to Administer It • 54

Understand How the Body Is Affected by Electric Current 55

The Jobsite and Proper Work Procedures 57
Occupational Safety and Health Administration (OSHA) • 57
    Employer Responsibilities • 57
    Trenches • 58
    Confined Spaces • 58
    Material Safety Data Sheet • 60
    Lockout/Tagout • 60

Personal Protective Equipment (PPE) 62
Head Injury Protection • 63
Foot and Leg Injury Protection • 63
Eye and Face Injury Protection • 64
Protection against Hearing Loss • 64
Hand Injury Protection • 64
Whole Body Protection • 65
Safety Harness • 66

Ladders and Scaffolding 67

4 Mathematics and the Metric System • 74

Whole Number Operations 76
Symbols • 76
Numbers • 77
Mathematical Terms • 77
Addition of Whole Numbers • 78
Subtraction of Whole Numbers • 79
    Borrowing • 79
    Borrow "1" Method • 80
Multiplication of Whole Numbers • 81
Division of Whole Numbers • 82

Fractions and Decimal Conversion 84
Definitions • 84
Fraction-to-Decimal Conversion • 84
Decimal Fractions • 85
Percent and Percentages • 86
    Changing a Decimal Fraction to a Percent • 86
    Changing a Common Fraction to a Percent • 86
    Changing a Percent to a Decimal Fraction • 87

Introduction to Algebra 87
Basic Algebra • 88
Numbers • 89
    Definite Numbers • 89
    General Numbers • 89
Signs of Operation and Grouping • 89
Algebraic Expressions • 89
    Coefficients • 90
    Powers and Exponents • 90
    Terms • 90

Translation into Algebraic Expressions • 91
Solving Equations • 91
Area and Calculations • 92
  Area Calculations • 93

Metric System and Units of Measurement 94
Metric System • 94
  Units of Measurement • 96
  The English System • 96
  Metric and English Systems Relationships • 96

The Tape Measure 97
The Tool • 98
  Reading the Tape Measure • 98
  Making a Measurement • 99

5   Basic Concepts of Electricity and Magnetism • 108

The Atom 111
Electric Charge 114
Electricity—Electrons in Motion 116
Valence Electrons • 116
Free Electrons and Electron Flow • 117
Ions • 117
Static Electricity • 117
  Practical Applications of Static Electricity • 118
  Nuisance Charges of Static Electricity • 119

Electric Sources 120
Direct Current • 120
Alternating Current • 120

Magnetism 121
Magnets • 122
Fields, Flux, and Poles • 122
Electromagnetism • 123
Magnetic Devices • 125

6   Electrical Theory • 130

Electrical Quantities and Units 132
Current • 132
Voltage • 133
Resistance • 134
Power • 135

Circuit Essentials 137
Basic Electrical Circuit • 137
Circuit Symbols and Diagrams • 138

Ohm's Law 139

Series Circuits 141
Series Multiple-Load Circuits and Subscripts • 142
Voltage Drops • 142
Resistance • 143
Calculating Series Circuit Values • 143
Ground as a Reference • 144

Parallel Circuits 146
Voltage • 146
Current • 147
Resistance • 148
Calculating Parallel Circuit Values • 149

Series-Parallel Circuits 152

7 Introduction to the *National Electrical Code (NEC)* • 160

The History of the *National Electrical Code (NEC)* 162

Changing and Writing *Code*, and Code-Making Panels 163
Changing and Writing *Code* • 163
Code-Making Panels • 165

The Purpose and Importance of the *National Electrical Code* 165

Code Book Chapters—Arrangement and Summary 166
Chapters • 166
Articles • 169
Parts • 169
Sections • 170
Tables • 170
Figures • 170
Exceptions • 171
Informational Notes • 171
Extractions • 171
Table of Contents • 171
Index • 172
Annexes • 172
Terms and Definitions • 172
Scope • 172
Boldface Type • 173
Gray Highlighting • 173
Bullets • 173

NEC Standards and Local Authorities 173
Standards • 174

Using the *Code* Book 176

8 Grounding • 182

Grounding and Bonding 185
Grounding • 185
Bonding • 186

Service Grounding 188
Grounding Requirements for a Residence Service Installation • 191
Bonding Requirements for a Residence Service Installation • 192
Grounding Electrode System • 193
    Grounding Electrode Conductor Connection • 193
Grounding and Bonding in a Single-Family Dwelling • 195
Grounding and Bonding in Commercial and Industrial Facilities • 197
    Concrete-Encased Electrode • 197
    Additional Ground Rod Requirements • 198
    Ground Clamps • 198

Incorrect Grounding or Lack of Grounding 200
Incorrect Grounding • 200
Lack of Grounding • 201

NEC Requirements—Bonding of Wiring Devices to Outlet Boxes 202
    NEC References and Interpretations • 202
    Equipment Bonding Jumper Installation • 205

Ground-Fault Circuit Interrupter (GFCI) 206
    Code Requirements for Ground-Fault Circuit Interrupters • 208

9   Tools • 212

The Importance of Caring for Hand Tools and Their Proper Use 214

Basic Hand Tools 215
    Wrenches • 216
        Open-End Wrenches • 216
        Adjustable Wrenches • 216
    Pliers • 217
        Lineman • 217
        Long-Nose Pliers • 218
        Diagonal Cutting Pliers • 218
    Wire Strippers • 219
        T®-Stripper Wire Stripper • 219
        Cable Ripper • 219
        Screwdrivers • 219
        Cordless Screwdriver • 221
    Knife • 222
    Hammer • 222
    Hacksaw • 222
    Tape Measure • 224
    Folding Rule • 224

Power Tools 225
    Power Drills • 226
        Pistol-Grip Drill • 227
        Hammer Drill • 228
        Cordless Drill • 229
    Power Saws • 229
        Circular Saw • 230
        Reciprocating Saw • 230
        Portable Bandsaw • 231

Specialty Tools 233
    Knockout Punch • 233
    Keyhole Saw • 233
    Fish Tape • 234
    EMT (Electrical Metallic Tubing) Bender • 236
    Level • 236
    Chisel • 237
    Cable Cutter • 238
    Hex Key Set • 238
    Fuse Puller • 239
    Rotary BX Cutter • 239

Electrician Tool Kits 240
    Standards • 240
    Tool Kits • 241
        12-Piece Electrician's Tool Set • 241
        13-Piece Journeyman Electrician's Kit • 242

10 **Wiring Overview • 246**

Specifications Used in Making Electrical Installations 248
Building Plan • 248
    Plot Plan • 249
    Floor Plans • 249
    Elevation Drawings • 250
    Sectional Drawings • 251
    Detail Drawings • 252
    Electrical Drawings • 252
    Schedules • 252
    Specifications • 253
NEC Requirements Regarding Services • 255
    Overhead Service • 255
    Underground Service • 256

Symbols and Notations Used in Electrical Drawings and Plans 259

Nationally Recognized Testing Laboratories 265
Purpose of an NRTL • 265
Example NRTLs • 265
    Underwriters Laboratories • 265
    CSA International • 266
    Intertek Testing Services • 267
National Electrical Manufacturers Association (NEMA) • 267

Fuses and Circuit Breakers and Current Ratings 268
Fuses • 268
    Plug Fuses • 269
    Cartridge Fuses • 270
Circuit Breakers • 271
Interrupting Rating • 272

11 **Wiring Devices • 278**

Receptacles—Markings and Operation 280
Conductor Identification • 280
    Grounded Conductor • 280
    Ungrounded Conductor • 281
    Three Continuous White Stripes • 281
    Grounded Neutral Conductor • 281
    Receptacles • 281
    Single Receptacle • 283
    Duplex Receptacle • 283

Switches—Markings and Operation 285
Single-Pole Switch • 285
Double-Pole Switch • 286
Three-Way Switch • 286
Four-Way Switch • 288

Dimmer Control Device 289
Electronic Dimmers • 289
Autotransformer Dimmers • 290
Fluorescent Lamp Dimming • 291

Fuses—Purpose and Operation 291

Circuit Breakers—Purpose and Operation 294

Ground-Fault Circuit Interrupter (GFCI) and Arc-Fault Circuit
    Interrupter—Installation and Operation 295
Ground-Fault Circuit Interrupter (GFCI) • 296
Arc-Fault Circuit Interrupter (AFCI) • 299

12 Wiring Methods • 308

NEC Requirements for the Installation of NMC 310

NEC Requirements for the Installation of MC Cable 312

NEC Requirements for the Installation of UF Cable 314

NEC Requirements for the Installation of EMT 315
Raceways • 315
Electrical Metallic Tubing (EMT) • 318

Switches—Wiring Methods 319
Types of Switches • 319
NEC Requirements for Three- and Four-Way Switches • 322
Installing Single-Pole Switches • 323
Installing Three-Way Switches • 326
Installing Four-Way Switches • 330

NEC Requirements for Replacing Existing Grounded
    and Ungrounded Receptacles 333
Replacing Existing 2-Wire Receptacles Where a Grounding Means Does Exist • 334
Replacing Existing 2-Wire Receptacles Where a Grounding Means Does Not Exist • 335

Five Types of Circuit Conditions 336
Normal • 336
Overload • 336
Short Circuit • 338
Ground Fault • 339
Open • 339

13 Wiring Calculations • 344

NEC Requirements for Calculating Branch-Circuit Sizing and Loading 346
General Lighting Circuits • 347
NEC Requirements and Calculations • 348
Small-Appliance Branch Circuits • 350
Laundry Branch Circuits • 351
Bathroom Branch Circuits • 352
Individual Branch Circuits • 352
Ampacity of a Conductor • 354

Conduit Fill Calculations as per NEC 356

Box Fill Calculations and Box Selection 358
Box Types and Selection • 358
Box Fill and Sizing Electrical Boxes • 361

Proper Conductor Size and Overcurrent Device for a Circuit 366
Conductors • 366
    American Wire Gauge (AWG) • 367
NEC Requirements • 367

14 Wiring Requirements • 376

NEC Requirements for Locating Receptacles, Switches,
and Luminaires for a Residential Dwelling 378
Circuit Layout • 378
NEC Requirements on Receptacle Locations • 381
NEC Requirements for Locating Lighting Outlets • 384
    Habitable Rooms • 384
    Additional Locations • 385
    Stairways • 385
    Basements, Attics, Storage, and Other Equipment Spaces • 385

NEC Requirements for GFCI Protection Locations
for a Residential Dwelling 386
NEC Requirements • 386
GFCI Receptacles and Circuit Breakers • 387

Cable Layout for a Master Bedroom in a Residential Dwelling 388
Cable Layout for a Master Bedroom • 389

15 Green Technology and the Electrical Industry • 394

Green Technology 396
Green Energy Solutions • 397
    Energy Conservation • 397
    Energy Efficiency • 397
    Responsible Energy Production • 398

Solar and Wind Technologies and Other Green Energy Sources 398
Solar Energy • 399
Wind Energy • 402
Other Alternative Green Energy Sources • 405
    Geothermal • 405
    Hydroelectric • 407
    Biomass • 408
    Hydrogen Energy and the Fuel Cell • 409

U.S. Green Building Council (USGBC) and Leadership in Energy
and Environmental Design (LEED) 410

Green Technology Employment Opportunities for Electricians 412
Green Areas of Opportunities • 413
Green Jobs • 413

16 The Job Search • 420

Understanding the Hiring Process 422
Traditional Job Hunting • 422
    Search the Want Ads • 423
Seek Out Apprenticeship Programs • 424
    Follow the Procedures of the Particular Program • 425
    Identify and Meet Certification Requirements • 425
    Find Job Placement and Sign an Indenture Agreement • 425

Preparing for the Job Search 426
Meet Basic Qualifications • 426
    Have a Clean Driving Record • 426
    Be Able to Pass a Drug Test • 426
Gather Your Occupational History • 427
    Create an Educational History and a Work History • 427

Create a List of Professional References • 428

Prepare a Skills Inventory • 429

Collect Needed Documents • 429

Research the Company • 429

Collect Information • 430

Prepare a List of Questions about the Company and the Position • 430

Completing the Job Application 432

Know How to Fill Out an Application Form • 432

Know How to Prepare an Electronic Résumé • 433

Section 1: Current Contact Information • 434

Section 2: Career Objective • 434

Sections 3 and 4: Education and Work History • 434

Section 5: Additional Skills and/or Accomplishments • 435

Section 6: References • 435

Format the Résumé • 436

Know How to Write a Cover Letter • 436

Sample Section 1 • 437

Sample Section 2 • 437

Sample Section 3 • 437

Interviewing Successfully 438

Exhibit Professional Behavior • 438

Write Down the Date, Time, and Location of the Interview • 438

Follow the Directions to the Interview Location • 439

Arrive on Time • 439

Greet Interviewers Politely • 439

Reflect a Positive Attitude • 440

Listen Carefully • 440

Answer Questions Briefly • 440

Show What You Know • 440

Dress Appropriately • 440

Demonstrate Job-Specific Knowledge • 440

Glossary • 448

Index • 460

Preface

Introduction

Workforce development for the electrical industry is a powerful mission for educators. Effective workforce development means that electricians are working competently and safely in a career that has the potential to provide them with financial security, and the public with functional, energy-saving, safe electrical systems. It is in this spirit that Independent Electrical Contractors Chesapeake, (IECC), Western Electrical Contractors Association, Inc. (WECA), and Cengage Learning joined together to develop a manual and educational guidelines for pre-apprenticeship programs that can make a contribution to workforce development.

Pre-apprenticeship programs are not for everyone. But in many cases, this type of program has the potential to set students up for success by enabling them to make more informed decisions about whether their capabilities and interests are a match with a career as an electrician, and to set them up for success by providing a solid foundation in core knowledge and skill sets related to a career in the electrical industry.

There is often confusion around what role pre-apprenticeship programs can play in the workforce development arena and about what learning objectives should be accomplished in this course of study. The goal of this publication is to clear up confusion on the roles that pre-apprenticeship programs can play and to provide guidance for implementation of an effective program when the need for pre-apprenticeship is evidenced.

Because the book's subject matter experts, Keith Chitwood and Jim Deal, have over 69 years of combined experience in the field and/or as instructors, it is clear that this book was designed "by electricians for electricians."[1] The structure of this book is intended to be user friendly, and the tone is designed to be down-to-earth with regard to factoring in the real world of working in the electrical industry. Program administrators, instructors, and students will be the judges of how effectively the goal is accomplished and how well these intentions are communicated.

Overview and Components

The contents of this book share much in common with a typical construction project. Its 16 chapters begin by establishing the foundation every electrician needs and build on that foundation until the student has all of the information needed to become an electrical apprentice or electrician trainee. Although it is recommended that instructors cover chapters in sequential order, all chapters are designed so that they can be discussed independently.

---

[1]"Training developed by electricians for electricians" is copyrighted by WECA, and permission is granted for use in this Preface.

An overview of the chapters' content follows:

Chapter 1: A Career in the Electrical Trade gives students an inside view of the inner workings within the electrical industry and details the many career opportunities that being an electrician can offer.

Chapter 2: Professional Behavior and Study Skills is designed to empower students with a set of skills that has the potential to set them up for success both in the classroom and on the job.

Chapter 3: Safety is one of the most important chapters in this book. Nothing means more to electricians and their families, employers, and coworkers than staying safe on the job, and in this chapter students learn foundational safety practices and procedures necessary to work safely.

Chapter 4: Mathematics and the Metric System simplifies subject matter that can seem daunting and difficult to understand to some students. This chapter lays out the material in a simple, step-by-step approach that makes mathematical concepts easy to understand and apply.

Chapter 5: Basic Concepts of Electricity and Magnetism gives students a platform from which to launch an understanding of the field's technical side. From the anatomy of an atom to how current flows through a conductor, this chapter covers the nuts and bolts of electricity.

Chapter 6: Electrical Theory introduces an array of basic terms, topics, and concepts, and relates them to on the job applications. Topics covered include Ohm's law, the components of a circuit, how to solve for circuit values, and much more.

Chapter 7: Introduction to the *National Electrical Code*® (*NEC*®*) serves as a blueprint for students to follow while navigating the most important textbook in the industry. Navigating the *Code* book is critical to the success of an electrician whether he or she is a rookie or a seasoned professional.

Chapter 8: Grounding covers the finer points of both grounding and bonding, giving students a comprehensive understanding of the concepts and of the *NEC* requirements that go along with grounding and bonding tasks

Chapter 9: Tools identifies and familiarizes students with everything they need to know about the equipment and tools they will use in the field. By the end of the chapter, students will have the know-how to build a tool box that would make a master electrician proud.

Chapter 10: Wiring Overview provides a foundation of the basics of wiring. Students are introduced to the building plan and the electrical specifications, drawings, and prints indicated on these plans.

Chapter 11: Wiring Devices describes some of the industry's most commonly used devices, including receptacles, switches, and dimmer controls. Students are also introduced to the purpose and operation of overcurrent, ground-fault, and arc-fault protection devices, including fuses, circuit breakers, ground-fault circuit interrupters, and arc-fault circuit interrupters.

Chapter 12: Wiring Methods prepares students for installing a wide variety of cables, including NMC, MC, UF, and EMT, with an emphasis on *NEC* standards.

Chapter 13: Wiring Calculations builds on the lessons of the previous chapter and covers procedures for calculating branch-circuit sizing and loading, how to determine maximum-size overcurrent devices, and how to compute box and conduit fill calculations.

Chapter 14: Wiring Requirements provides a comprehensive explanation of the *National Electrical Code (NEC)* requirements for locating receptacles, switches, and luminaires in residential wiring applications. It also touches on (*NEC*) requirements as they pertain to ground-fault protection.

Chapter 15: Green Technology and the Electrical Industry gives students an up-close look at emerging electrical and energy conservation technologies.

---

*National Electrical Code® and *NEC*® and NFPA 70® are registered trademarks of the National Fire Protection Association, Quincy, MA.

They will gain an appreciation for the value of these technologies, as well as for the career opportunities that these alternatives can provide.

Chapter 16: The Job Search provides a road map for the multistep process of finding an entry-level job in the electrical trade. Tips on building a résumé, writing cover letters, interviewing, and navigating the job boards will give students an edge as they head out into the real world.

Each chapter includes the following components, which tie the technical content to soft skills development while also incorporating adult learning reinforcement and motivational techniques:

- Career Profile
- Life Skills Covered
- Chapter Outline
- List of Key Terms
- Chapter Objectives
- Life Skill Goals
- Where You Are Headed
- Introduction
- Self-Check questions and Life Skills questions throughout the chapter
- Trade Tips
- Chapter Summary
- Review Questions
- Activity and/or Lab
- Glossary

Since its inception, the goal of this Manual has been more than simply providing first-class content. The development team's focus was on creating a complete learning experience. That's why each chapter of the *Electrical Pre-Apprenticeship and Workforce Development Manual* boasts a robust list of components and features that provide both student and instructor with added value before, during, and after the chapter has been read. The rationale for these components is presented below.

Career Profile

A Career Profile opens each chapter. Designed to inspire students, each profile tells the story of a person who started as an apprentice and rose through the ranks in the electrical industry. By reading these real-life examples, students learn that with hard work there is no limit to what they can accomplish.

Life Skills Covered

Specific life skills identified in each chapter of the book guide the student toward concepts and tools that can be useful to them in their life and career. The identified life skills correlate with the Life Skill Goals and are intended to provide the WIFM—what's in it for me motivation—to the student for consideration.

Chapter Outline

The chapter outline is a student's road map for navigating the text. It lists all of the main sections in the chapter, allowing a student to find the information that he or she needs quickly.

Chapter Objectives

If the Chapter Outline is the student's road map, the Chapter Objectives are the destination. These measurable goals state clearly what the student will know or be able to do at the end of each lesson.

Where You Are Headed

This chapter component addresses the age-old question: Am I ever going to use any of this in the real world? It provides students with specific examples of how what they are learning will help them to reach their career goals.

Key Terms

A list of key terms can be found at the beginning of each chapter. As each key term is introduced in the text, it is defined in the margin. This component ensures that each key term that may be new to the students is defined *before* it is used within the learning scenarios.

Life Skill Goals

Being an electrician is about more than knowing how to run wiring. The Life Skill Goals take practical concepts from the workplace, like cooperation, teamwork, and critical thinking, and apply them to the content and tasks covered in each chapter. Each section features an activity requiring students to practically apply one of these skills.

Self-Check Questions

Each main section of a chapter features at least one set of Self-Check questions that are designed to give students time to pause and reflect in order to increase comprehension on what they read. One of the most positively reviewed features of the text, these questions assist students to fully comprehend what they heard, saw, and read.

Trade Tips

Seamlessly integrated into the text are Trade Tips that offer real-life examples of how the student can apply chapter material in the field.

Review Questions

The chapter Review Questions offer the student an opportunity to gauge how well they understood the text. Targeted multiple choice, true-false, and fill-in-the-blank questions cover all of the chapter's objectives.

Labs and Activities

Each chapter concludes with a lab and/or an activity. These labs and activities take chapter concepts and apply them to real-world scenarios. All labs contain scoring rubrics based on observable or measurable criteria.

Glossary

The glossary includes all of the terms introduced in the chapters and makes them available in one convenient location for referencing.

Ancillary Package

Although the *Electrical Pre-Apprenticeship & Workforce Development Manual* is a complete and thorough text in its own right, we want to go beyond simply offering outstanding course material. That's why we have developed a comprehensive *Electrical Pre-Apprenticeship & Workforce Development Trainer Implementation Manual* that will help ease the preparation process for instructors. Some of the steps taken to ensure a simple, seamless teaching process are laid out here.

- An annotated **Chapter Objectives** list that correlates each objective with its related **Review Questions**
- A **Chapter Outline** that goes beyond the version included in the text. Broken up by unit, topic, and subtopic, each outline is annotated with helpful **instructional tips** from industry experts. Points of emphasis and potential discussion topics are included. Also, all **important figures** are identified.
- **Answers** to each unit's **Self-Check Questions**
- **Answers** to the chapter **Review Questions**, including the **unit heading** where the answer is found
- Suggested **answers** for end-of-chapter **Activities**
- **Perforated** copies of the chapter **Labs**, featuring space for instructor and student signatures

In addition to these manual-exclusive features, each chapter contains the content from the actual student manual.

An Instructor Resource CD (ISBN 978-1-1113-1691-4) is available for instructors to assist with class preparation, including teaching aids such as PowerPoint® presentations for each chapter, computerized test bank questions, answers to student manual questions, chapter outlines annotated with instructional tips, printable labs, and an image gallery.

About the Authors

Independent Electrical Contractors of Chesapeake (IEC Chesapeake), located in Maryland and servicing the Mid-Atlantic region, and the Western Electrical Contractors Association, Inc. (WECA), located in California, worked in cooperation with Delmar, Cengage Learning to create the *Electrical Pre-Apprenticeship & Workforce Development Manual* and *Trainer Implementation Manual*. IEC Chesapeake is a chapter of the Independent Electrical Contractors (IEC), and WECA is a California-based independent merit shop electrical contractors association. Having major contributors from each coast allows the text a perspective that truly stretches from sea to shining sea, covering everything in between. Much more than a regional handbook, the authors have focused on providing content that applies to aspiring electricians in all 50 states.

Contributing Author

Bill Hessmiller graduated magna cum laude from the Pennsylvania State University in Electrical Engineering Technology and acquired his electrical and electronics background in the U.S. Coast Guard. Currently Vice President of Publications with Ferguson Lynch, an information technology consultancy firm, Bill has also worked with several leading publishing companies through the years as a freelance writer and editor with a multidisciplinary background.

Bill holds a General (First Class) Radio/Telephone Operator License (Federal Communications Commission). He is a journeyman electrical equipment technician (International Union of Electrical, Radio, and Machine Workers) and serves as a curriculum advisor for Johnson College (Scranton, Pennsylvania). Bill holds the following credentials:

- Certification with the Pennsylvania Department of Education in Biomedical, Electrical, and Electronic Equipment Technology
- Certified Engineering Technician (National Institute of Certification in Engineering Technologies, Washington, D.C.)
- Electronics Technician and a Certification Administrator (Electronic Technicians Association, Greencastle, Indiana).

Editorial Advisory Board

Keith Chitwood

Keith Chitwood, an experienced electrician who ran his own electrical contract and service business for more than a decade, has devoted himself to training young electricians. He spent 17 years as an apprenticeship instructor for Associated Builders and Contractors (ABC). Chitwood's time with ABC also includes a 6-year stint as Apprenticeship Coordinator. Outside of his time at ABC, he has trained electricians for the Atlantic Technical Center, Helix Electric, and Rex Moore Electrical Contractors & Engineers. Today, Chitwood serves as manager of apprenticeship education programs for WECA.

Keith's degrees and certification/licenses attained and industry involvement include:

- Four-year apprenticeship program with ABC in Florida
- Journeyman's Electrician license
- Master's Electrician license
- Florida Department of Education part-time vocational certification
- Certified electrical craft instructor with NCCER

- Certified Master Trainer with NCCER
- First Aid/CPR/AED instructor/trainer
- California-certified general electrician
- California-certified voice data video technician
- California Private Post-Secondary and Vocational Training–credentialed instructor

Jim Deal

Jim Deal, a licensed master electrician, has dedicated himself to training young electricians. He has served as a pre-apprenticeship instructor and has been involved in the development and implementation of approved apprenticeship programs, a pre-apprenticeship program, and a continuing education program. He played a key role in the IEC Chesapeake Apprenticeship Chapter, being acknowledged three times as Apprenticeship Chapter of the Year with the IEC National Association. Deal also served as a plant manager for all mechanical and electrical systems at the University Hospital and Shock Trauma Unit in Baltimore, Maryland. He is currently the Director of Education at IEC Chesapeake.

Jim's degrees and certification/licenses attained and industry involvement include:

- Four years in military service in Vietnam— USAF
- Four-year electrical apprenticeship program completed with ABC in Maryland
- Electrical Master's license
- Chairman of Life Safety Workgroup of 100 nurses and faculty staff responsible for maintaining a safe hospital environment for patients and staff
- Seat on numerous career and technology committees in Maryland, including high school technical training advisory boards

Acknowledgments

There are several people that the authors would like to acknowledge for the role that they played in the production of this text.

Keith Chitwood, the Manager of Apprenticeship Education Programs at Western Electrical Contractors Association (WECA) and **Jim Deal**, Director of Education for Independent Electrical Contractors (IEC) Chesapeake have served as key subject matter experts for the project, providing invaluable industry knowledge and helping to shape the book's content at every stage of the development process.

Terry Seabury, WECA Executive Director, CEO, and **Grant Shmelzer**, IEC Chesapeake Executive Director, were key contributors to the project. Time and again, their guidance and willingness to provide thoughtful answers to difficult questions proved an invaluable resource.

The authors would like to acknowledge the following people who lent us their life experiences for the Career Profile component: Edmund T. "Ned" Johns, James W. Miller, Chris Ross (Construction Connect, Inc.), Henry Aden, Ray Shorkey, George Hockaday-Bey (G-11 Enterprises, Inc.), Jimmie Slemp, Sarah E. High, Patricia Brack, Trenton Johnston, Jim Taylor III, Gregory J. Anderson (Rex Moore Electrical Contractors & Engineers), Jeremy Grosser (Rex Moore Electrical Contractors & Engineers), Larry Carlyle, Jim Deal, and Keith Chitwood. Their stories offer concrete examples that will inspire students to believe that with hard work there is no limit to what they can accomplish.

The authors would like to thank **Ray C. Mullin**. Mr. Mullin is the author of many technical articles that have appeared in electrical trade publications and several books, including Delmar, Cengage Learning's *Electrical Wiring Residential,* 17e, and *Electrical Wiring Commercial,* 14e, which were co-authored with Phil Simmons. Mr. Mullin completed his apprenticeship training and worked as a journeyman and supervisor for residential, commercial, and industrial installations. He is a former electrical instructor for the Wisconsin Schools of Vocational, Technical, and Adult Education. Mr. Mullin's *Electrical Wiring Residential* served as a valuable resource throughout the process of writing this book.

The authors would also like to acknowledge Development Editor **Beth Jacobson** of Ohlinger Publishing Services. Beth skillfully managed the production of the text from concept to completion. Her vision and attention to detail was a crucial component of this project's success. **Dan Vest**, also from Ohlinger Publishing Services, assisted Beth with managing the project.

The authors and publisher would like to express thanks to those reviewers who provided insightful feedback throughout the development of this first edition.

Kevin Szol
College of the Rockies
Electrical Programs Instructor, Trades Programs
    Coordinator
Cranbrook, British Columbia

Jeff Llapitan
Bates Technical College
Electrical Construction Instructor
Tacoma, WA

Oscar Buschinelli
The Centre for Skills Development and Training
Lead Electrical Instructor
Burlington, Ontario

Katrina Cloud
Portland Community College
Apprenticeship & Training Specialist,
    Apprenticeship Administrator
Portland, OR

John Marks
IEC Chesapeake
Pre-Apprenticeship Instructor
Odenton, MD

Robert Palmer
IEC Chesapeake
Job Corp Instructor
Odenton, MD

Applicable tables and section references are reprinted with permission from NFPA 70®*-2011, *National Electrical Code*, copyright © 2010, National Fire Protection Association, Quincy, MA 02169. This reprinted material is not the complete and official position of the NFPA on the referenced subject, which is represented only by the standard in its entirety.

---

*National Electrical Code® and *NEC®* and NFPA 70® are registered trademarks of the National Fire Protection Association, Quincy, MA.*

Chapter 1

A Career in the Electrical Trade

Career Profile

Edmund T. "Ned" Johns' unusual career path proves that it is never too late to pursue the kind of work you truly want to do. Johns is currently an instructor in the Education Department of the Western Electrical Contractors Association (WECA) in Rancho Cordova, California, where he teaches classes for students entering the electrical trade. But before becoming a certified electrician, he worked full-time, first as a heavy-duty engine and transmission mechanic/machinist and later as a research assistant who helped to develop engine lubricating oil and fuel system deposit control additives.

About finding the courage to alter his work life in a radical way, Ned reveals, "It took me nearly four years to actually initiate the change in career fields. Before I finally chose the electrical field, a 25-year veteran electrician enabled me to assist him with some special electrical projects. With his direction and encouragement, I made the change and have never looked back."

Now, after completing Commercial Electrician apprenticeship training, earning certification as a California State "General" Electrician, and spending many years accumulating additional on-the-job training and in-class technical training, Johns instructs future electricians. He teaches such subjects as understanding and applying principles of the *National Electrical Code* (*NEC*), interpreting and using construction plans, and operating basic electrical motor controls.

When offering guidance to students embarking on their own careers, Johns says, "Every one of us has a particular skill or talent that he or she can do well—something that just seems to come 'naturally.' Our skills and talents can truly be of benefit to others (as a service that they are unable to perform for themselves) and to ourselves (as a livelihood). Therefore, my advice is 'Invest in yourself!' Seize the opportunities available in this country. Learn as much as you can to develop your skill (your talent) and refuse to let complacency or the negativity of others stop you from being the best you that only you can be! Others' lives may depend on it."

Chapter Outline

THE ELECTRICAL INDUSTRY

CAREER PATHS

Key Terms

Apprenticeship
Coaxial cable, or coax
Conduit
Construction electrician
Continuing education
Distribution electrician helper
Electrical distributors
Electrical inspector
Electrical distribution
Electrical product distribution
Electrical products manufacturer
Electrical supervisor
Electricity
Estimator
Fiber-optic cable
Foreman
Industrial electrician
Journeyman
Journeyman electrician
Maintenance electrician
Master electrician
National Electrical Code
Outside lineman
Plant electrician
Preventive maintenance
Project manager
Raceway
Senior construction electrician
Superintendent
Tradesman

Chapter Objectives

After completing this chapter, you will be able to:

1. Describe the electrical industry.
2. Identify various career paths available in the electrical trade.

Life Skills Covered

Seven life skills will be covered throughout this book. See the following list.

 Goal Setting To get anywhere, you have to know where you're headed. The Goal Setting life skill helps you learn the skills necessary to identify where you want your career to be headed and the best route to take to make that a reality.

 Cooperation and Teamwork As an electrician, you will always be working with other people. The Cooperation and Teamwork life skills teach you the finer points of building on-the-job relationships with coworkers, customers, supervisors, and others.

 Communication Skills Whether verbal or written, communication is what makes the business world go 'round. Communication life skills give you tips on how to hear others more clearly and how to make yourself heard in the classroom, on a jobsite, and elsewhere.

 Managing Stress The Managing Stress life skills assist you to perform under pressure. You'll learn useful tips to help you "keep your cool."

 Critical Thinking Thinking critically involves going beyond the *what* and understanding the *why* of how something works. These life skills teach you to ask the questions that will give you a deeper understanding of processes and why they are important.

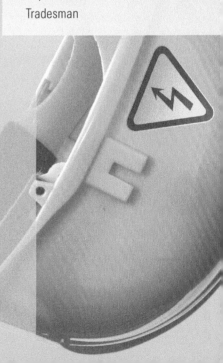

Self-Advocacy In the business world, there are no cheerleaders, and simply talking about how great you are won't get you far. The Self-Advocacy life skill goals are designed to provide you with suggestions for separating yourself from the crowd so that when you look for a job, you will have something real, unique, and valuable to offer an employer.

Take Action The Take Action life skills are all about taking your destiny into your own hands by making sure that you totally understand the material in these chapters. Pay extra attention to the Self-Check questions: Master them and you will master the content that goes with them.

We focus on one or more of these life skills in each chapter. At the beginning of each chapter, an icon or icons tell you which life skill(s) the chapter will focus on. Life skill activities for you to complete occur throughout the chapter. The life skills covered in Chapter 1 are:

 Goal Setting   Take Action

Life Skill Goals

Your primary life skill goal for this chapter is to create goals for yourself as you begin your career in the electrical industry. Your goals will serve as the map that leads you to your final destination—a career in the electrical field. The second life skill that this chapter focuses on is "take action." As you read through the chapter, complete the self-check questions. If you're unsure of an answer, ask a classmate.

Where You Are Headed

A career in the electrical trade is a major life decision, so it is important that you understand the five major areas of the electrical trade that most electricians work in. Having this information will give you the opportunity to think about what career path you might want to take.

Introduction

Electricity the theorized flow of electrons that cannot be seen, although its effects can be seen and measured.

Electricity is a basic requirement for a modern, comfortable life. Electricity is essential for light, power, air conditioning, and refrigeration. It is a general term that refers to the presence and flow of electric charges. The critical importance and diversity of the electrical trade has created unimaginable job opportunities for the electrical tradesman. Electricians install and maintain all of the electrical power systems, wiring, and control equipment for our homes, businesses, and factories. You can learn their trade through an apprenticeship or more informally, as a helper for an experienced electrician, giving you the opportunity to regularly interact with other skilled tradesmen on the jobsite.

1-1   The Electrical Industry

The U.S. Bureau of Labor expects an increase of nearly 12% in the number of job opportunities for electricians by 2018. As the population and economy grow, more electricians are needed to install and maintain electrical devices and wiring in homes, factories, offices, and other structures. New technologies such as computers, telecommunications equipment, and automated manufacturing systems require more complex wiring systems to be installed and maintained. Additional electrical jobs will be created as older structures are rehabilitated and retrofitted to meet existing electrical codes. At this time, about 79% of electricians work in the construction industry or are self-employed. There are also opportunities in the electrical distribution industries and for maintenance electricians. Electricians completing an electrical apprenticeship can make as much as someone completing a bachelor's degree.

Training

Apprenticeship a systematic method of training an individual in a trade or industry.

Most people learn the electrical trade by completing a 4- or 5-year **apprenticeship** program. An apprenticeship program is a systematic method of training an individual in a trade or industry. After indenture, the apprenticeship program ensures that you receive both the classroom instruction and the on-the-job training necessary for you to become proficient in your chosen field. Apprenticeship gives trainees a thorough knowledge of all aspects of the trade and generally improves their ability to find a job. Although more electricians are trained through apprenticeship than are workers in other construction trades, some still learn their skills informally, on the job. A candidate for the electrical apprenticeship program should have strong math and English skills. High school math is essential to solve mathematical problems in class and on the job. High school English skills and reading comprehension are key to a successful career in the electrical trade. English and reading skills are necessary to read and understand product documentation, material information, and project and job task descriptions, as well as to interpret the *NEC*.

A typical large apprenticeship program provides at least 144 hours of classroom instruction each year and 8000 hours of on-the-job training over the course of the apprenticeship. In the classroom, electrical apprentices learn blueprint reading, electrical theory, electronics, mathematics, electrical code requirements, and safety and first-aid practices. They also may receive specialized training in welding, communications, fire alarm systems, and cranes and elevators. On the job, under the supervision of experienced electricians, apprentices demonstrate mastery of the electrician's work. At first, they drill holes, set anchors, and set up conduit. Later, they measure, fabricate, and install conduit, as well as install, connect, and test wiring, outlets, and switches. They also learn to set up and draw diagrams for entire electrical systems.

Those who do not enter a formal apprenticeship program can begin to learn the trade informally by working as helpers for experienced electricians. Many helpers supplement this training with trade school or correspondence courses.

Journeyman a journeyman has the required skills and knowledge and has met the requirements of time in the field.

Continuing education the acquisition or improvement of work-related skills, generally referring to classes and seminars that focus on job-related skills and knowledge.

After completion of the apprenticeship program or after many years of working as an electrician's helper, you become a journeyman. A journeyman has the required skills and knowledge and has met the requirements of time in the field. Some electricians become licensed. However, licensing requirements vary from area to area. Licensed electricians usually must pass an examination that tests their knowledge of electrical theory, the *NEC*, and electric and building codes. Even journeyman electricians periodically take courses to keep abreast of changes in the *NEC*, materials, or methods of installation. With advancing technology and ever-changing trends in the electrical trade, continuing education has become less of an option and more of a necessity. Continuing education is the acquisition or improvement of work-related skills, generally referring to classes and seminars that focus on job-related skills and knowledge. In a field such as the electrical trade, whether you are interested in upgrading your work skills and knowledge in order to keep up with the latest trends affecting your industry or dealing with persistent technological advances, expect to enjoy lifelong learning. This learning comes with additional benefits; further training makes you a more valuable employee, and there may be potential for an increase in pay.

LIFE SKILLS

What questions do you have about the apprenticeship process?

Self-Check 1

1. What are some of the skills electrical apprentices learn in the classroom?
2. What are some of the skills electrical apprentices learn on the job?

1-2 Career Paths

More than half of all electricians are employed in the construction industry. Do you know where all other electricians work? We discuss here many electrician jobs that exist today and detail what is listed on typical electrician job descriptions.

As an electrician, your career path does not start when you complete your apprenticeship training. In fact, your career begins on the *first* day of your apprenticeship program. As an apprentice, you learn the trade in the classroom *and* on the job.

As a first-year apprentice or lead journeyman, you experience the diversity that exists in the electrical field. Electricians install and maintain all of the electrical power systems for our homes, businesses, and factories, installing such equipment as telephone systems, computers, street lights, intercom systems, and fire alarm and security systems. They may also specialize in wiring ships, airplanes, and other mobile platforms. Other specialized fields include marine electricians, research electricians, and hospital electricians.

Although there are hundreds of different jobs throughout the electrical industry, electricians work in one of the following fields:

- Electrical construction
- Electrical maintenance
- Electrical power distribution
- Electrical products distribution
- Electrical products manufacturing

National Electrical Code a United States standard for the safe installation of electrical wiring and equipment.

All electricians must follow state and local building codes and the *National Electrical Code* when performing their work. The *NEC* or NFPA 70 is a United States standard for the safe installation of electrical wiring and equipment.

Electrical Construction Industry

According to the U.S. Bureau of Labor, more than 79% of electricians are employed in the construction industry or are self-employed.

Conduit a protective passageway (pipe or tubing) for cables

Electricians specializing in *electrical construction* typically install wiring systems in factories, businesses, and new homes. In factories and offices, construction electricians first place conduit inside designated partitions, walls, or other concealed areas. A conduit is a protective pasageway (pipe or tubing) for cables. **Figure 1-1** shows electrical metallic tubing (EMT), and **Figure 1-2** illustrates installed conduit.

FIGURE 1-1   Electrical metallic tubing (EMT).

Photography courtesy of E.T. "Ned" Johns of WECA

FIGURE 1-2 Installed conduit through a basement wall.

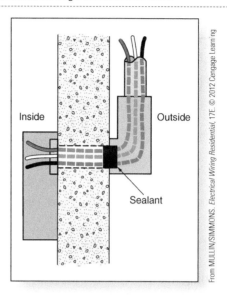

From MULLIN/SIMMONS. *Electrical Wiring Residential*, 17E. © 2012 Cengage Learning

Electrical construction specialists also fasten to walls small metal or plastic boxes that will house electrical switches and outlets. They then pull insulated wires or cables through the conduit to complete circuits among these boxes. In lighter construction, such as residential, plastic-covered wire is generally used instead of conduit.

Running wires through wall cavities can be a messy and cumbersome task, especially when remodeling. Electrical raceways make the job a little easier. An electrical raceway is an enclosed channel of metal or nonmetallic materials designed expressly for holding wires or cables. **Figure 1-3** shows an example of raceways and boxes securely fastened in place. *Note:* Insert points out that you may *NOT* use electrical raceways or cables, to support other raceways, cables, or nonelectrical equipment. See *300.11(B)* and *(C)*.

Figure 1-4 shows an electrician at work. Take note that whatever the job, *safety* precautions are always taken. This electrician is wearing a hard hat, safety goggles, a long-sleeve shirt, and gloves!

No matter what type of wire is used, electricians connect it to circuit breakers, transformers, or other components. They join the wires in boxes with various specially

Raceway
an enclosed channel of metal or nonmetallic materials designed expressly for holding wires or cables.

FIGURE 1-3 Raceways and boxes securely fastened in place. See circled X.

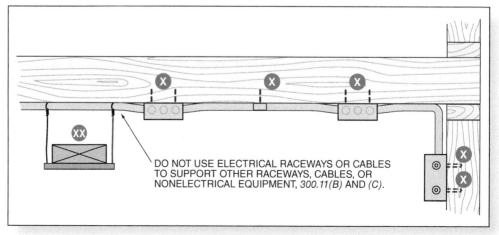

DO NOT USE ELECTRICAL RACEWAYS OR CABLES TO SUPPORT OTHER RACEWAYS, CABLES, OR NONELECTRICAL EQUIPMENT, *300.11(B)* AND *(C)*.

From MULLIN/SIMMONS. *Electrical Wiring Residential*, 17E. © 2012 Cengage Learning

FIGURE 1-4   An electrician at work.

© auremar/www.Shutterstock.com

designed connectors. Once the wiring is completed, they use test equipment such as ohmmeters, voltmeters, and oscilloscopes to check the circuits for proper connections, ensuring electrical compatibility and safety of components. **Figure 1-5** shows electricians using an analog multimeter to make a measurement.

In addition to wiring a building's electrical system, electricians may install coaxial or fiber-optic cable for computers or other telecommunications equipment.

FIGURE 1-5   Electricians making a measurement.

© Lisa F. Young/www.Shutterstock.com

FIGURE 1-6 Coaxial cable.

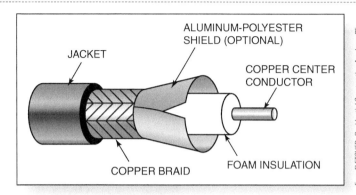

ALUMINUM-POLYESTER
SHIELD (OPTIONAL)

JACKET

COPPER CENTER
CONDUCTOR

COPPER BRAID

FOAM INSULATION

From FLETCHER. *Residential Construction Academy*, 3E.
© 2012 Cengage Learning

Coaxial cable, or **coax** an electrical cable with an inner conductor surrounded by a flexible, tubular insulating layer, surrounded by a tubular conducting shield.

Fiber-optic cable a thin, flexible, transparent fiber that acts as a wave guide or "light pipe," to transmit light between the two ends of the fiber.

A coaxial cable, or coax, is an electrical cable with an inner conductor encased in a flexible, tubular insulating layer, surrounded by a tubular conducting shield. **Figure 1-6** illustrates the construction of coaxial cable.

A fiber-optic cable is a thin, flexible, transparent fiber that acts as a wave guide, or "light pipe," to transmit light between the two ends of the fiber. **Figure 1-7** illustrates the construction of a fiber-optic cable.

Increasingly, electricians are called upon to install telephone systems, computer wiring and equipment, street lights, intercom systems, and fire alarm and security systems. Some electricians may connect motors to electrical power and install electronic controls for industrial equipment.

FIGURE 1-7 Fiber-optic cable.

From ZACHARIASON. *Electrical Materials*, 1E. © 2008 Cengage Learning

FIGURE 1-8 Fuse sockets and fuses.

© Jack Cobben/www.Shutterstock.com

Electricians who specialize in residential work may rewire a home and replace an old fuse box with a new circuit-breaker panel to accommodate additional appliances. **Figure 1-8** shows dated fuse sockets and fuses. Unfortunately, you may come across such dated and poorly wired boxes. The wiring shown is this figure is *not* up to *Code* and is never acceptable! **Figure 1-9** shows a circuit-breaker panel.

FIGURE 1-9 Circuit-breaker panel.

© Lisa F. Young/www.Shutterstock.com

There are many electrician jobs in the construction industry. Job titles sometimes vary from jobsite to jobsite. **Figure 1-10** shows a construction electrician.

The core electrical skills and duties are found in just about every electrician's job description. The following are a few electrical construction job titles and their associated job descriptions:

Construction electrician Basic responsibilities include installing and repairing telephone systems and high- and low-voltage electrical power distribution networks, both overhead and underground. Splice and lay cables, erect poles, string wires, and install transformers and distribution panels. Install, repair, and maintain street lighting, fire alarm, public address, and interoffice and telephone switchboard systems. Install, maintain, and repair interior wiring for lighting and electrical equipment. Work with batteries, electric motors, relays, solenoids, and switches. Operate electrical generators, read and interpret blueprints and prepare sketches for projects, and make estimates of material, labor, and equipment requirements.

Senior construction electrician Basic responsibilities include maintaining building electrical infrastructure, such as low-voltage ac (120V, 220V, 240V, 480V), medium ac (12,470V), low-voltage dc systems, LAN network and communication cabling, fire alarm systems, security surveillance systems, and programmable logic controller (PLC) systems. Specific duties include installation/construction of engineering equipment and test fixtures. Installation, construction, maintenance, and repair of facilities infrastructure (i.e., medium-voltage switch gear, main breakers, building lighting, security and safety systems, communication systems, etc.). Troubleshoot, diagnose, adjust, repair, construct, assemble, calibrate, install, and maintain all types of electrical and electronic high- and low-voltage circuitry systems and equipment. Read and interpret blueprints, schematics, and sketches. Handle customer relations and assist with quotations and project/job planning.

FIGURE 1-10 A construction electrician measuring to install a switch box.

Self-Check 2

1. List three basic responsibilities of a construction electrician.

2. List three basic responsibilities of a senior construction electrician.

Electrical Maintenance

Preventive maintenance
the care and service provided for the purpose of maintaining equipment and facilities in satisfactory operating condition by providing for systematic inspection, detection, and correction of failures before they occur.

Maintenance electricians generally spend much of their time providing **preventive maintenance**. Preventive maintenance involves maintaining equipment and keeping facilities in satisfactory operating condition by performing systematic inspections. These inspections can detect failures and correct them before they occur or before they develop into major defects. Maintenance electricians periodically inspect equipment, locate problems, and correct them before breakdowns occur. Electricians may also advise management on whether continued operation of equipment could be hazardous. Where required, they install new electrical equipment. Because maintenance electricians are needed in virtually every industry, from manufacturing facilities and hospitals, to railroads and airports, there are many jobs in electrical maintenance. **Figure 1-11** shows a maintenance electrician making a measurement.

The following are a few electrical maintenance positions and their job descriptions:

Plant electrician
basic duties include installing, troubleshooting, and maintaining lighting, electrical equipment, and power distribution systems.

Plant electrician Basic duties include installing, troubleshooting, and maintaining lighting, electrical equipment, and power distribution systems up to 4160 volts. Operate, monitor, and maintain all building support equipment, including electrical, HVAC systems, plumbing, and utility-related support equipment. Work with blueprints and schematics. Plan, coordinate, oversee, and participate in activities associated with electrical maintenance and construction projects. Communicate

FIGURE 1-11 A maintenance electrician making a measurement.

verbally, in writing, and electronically with all levels to assist in providing electrical and related services. The work involves using ladders or other elevated platforms and tools of the trade, heavy lifting, working in confined spaces, and using personal protective equipment. Maintenance electricians interface with maintenance personnel and contractors, and provide support for all trades.

Industrial electrician Basic duties include targeting and troubleshooting, electrical maintenance issues that arise in an industrial setting, and making necessary repairs. Complex wiring and re-routing of electricity is an inherent part of the job. Diagnose equipment problems and repair as necessary. Install electrical equipment and work with programmable logic controllers (PLCs).

Maintenance electrician Basic responsibilities include repairing, installing, replacing, and testing electrical circuits, equipment, and appliances. Use hand tools and testing instruments to isolate defects in wiring, switches, motors, and other electrical equipment. Inspect completed work for conformance with requirements of local building and safety codes. Mount motors, transformers, and luminaire into position, and complete circuits according to diagram specifications. Estimate time and material costs on electrical projects. Dismantle electrical machinery and replace defective electrical or mechanical parts.

Self-Check 3

1. What is one difference between an industrial electrician and a plant electrician?
2. List three basic responsibilities of a maintenance electrician.

Electrical, or Power, Distribution Electrician

Electrical distribution is the final stage in the delivery of electricity to end users. A distribution system's network carries electricity from the transmission system and delivers it to consumers. **Figure 1-12** provides a basic illustration of a

Industrial electrician basic duties are targeting and troubleshooting electrical maintenance issues that arise in an industrial setting, and making necessary repairs.

Maintenance electrician basic responsibilities include but are not limited to repairing, installing, replacing, and testing electrical circuits, equipment, and appliances.

Electrical distribution the final stage in the delivery of electricity to end users.

FIGURE 1-12 Electrical power grid.

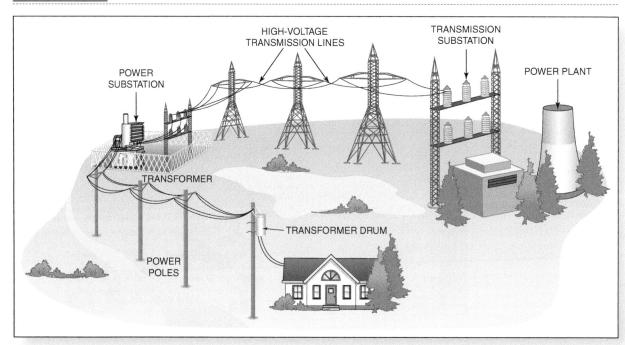

© Cengage Learning 2013

power grid. The electrical power is generated at the power plant and is transmitted by way of substations and transmission lines. Transformers on power poles change high voltage to usable voltage; in this figure, the transformer is connected to the home.

Electrical distribution electricians, also known as outside linemen, install and repair underground power lines, switch gears, transformers, and other hardware on the primary distribution system. Electrical or power distribution electricians also construct secondary service lines to buildings and make various electrical connections. They troubleshoot facilities and installations during power failures, connect and disconnect transformers and switching equipment, and make emergency repairs where necessary. Most power distribution electricians work on electric utility company distribution systems, often at high voltages. **Figure 1-13** shows a lineman working on power lines.

The following are power distribution electrician positions and their job descriptions:

Outside lineman Basic duties include planning and initiating projects. Establish OSHA and customer safety requirements, set towers and poles, and construct

Outside lineman typically installs and repairs underground power lines, switch gears, transformers, and other hardware on the primary distribution system.

FIGURE 1-13 Lineman working on power lines.

© Annette Shaff/www.Shutterstock.com

other devices to support transmission/distribution cables. Maintain and repair overhead distribution or transmission lines, string new wire or maintain old wire, and install and maintain insulators, transformers, and other equipment.

Distribution electrician helper Basic duties include assisting with electrical equipment, materials, and cables/conductors, and responding to trouble calls. Assist in installing electrical service to residential/commercial customers. Check equipment on truck for proper operation/inventory. Assist in setting up/operating hydraulic/aerial crane. Assist in repairing high-voltage wires. Assist in installing mast arms with light fixtures on poles. Assist in installing and servicing electrical vaults. Perform carpentry work in construction and repair of utility structures. Operate various power saws, drills, buffers, grinders, and sanders.

Distribution electrician helper basic duties include assisting with electrical equipment, materials, and cables/conductors, and responding to trouble calls.

LIFE SKILLS

Which electrical industry sounds most interesting to you? Why? What questions do you have?

Self-Check 4

1. What are electrical distribution electricians also known as?
2. List three basic duties of a distribution electrician helper.

Electrical Distributor

Electrical product distribution is the provision of electrical products to end users and electrical contractors. Markets for electrical products include the following:

- Residential and commercial electrical contractors
- Commercial and industrial companies
- Schools
- Hospitals
- Power companies
- Government and military installations

Electrical product distribution the provision of electrical products to end users and electrical contractors.

Electrical components are also purchased by manufacturing companies to use in all types of electrical products. Examples of electrical products distributed are listed here:

- Cable and wire
- Lighting and fan fixtures
- Circuit breakers
- Hand and power tools
- Automation and motor control
- Switchgear and apparatus

Those individuals who work in the electrical product distribution industry are considered **electrical distributors**. Electrical distributors are those workers, often considered *solutions providers,* who distribute materials to end users and contractors. Typical positions include salespeople, product experts, skilled and technical professionals, warehouse associates, and drivers.

The following are electrical product distributor positions and their job descriptions:

> **Electrical distributor** a worker, often considered a *solutions provider,* who distributes materials to end users and contractors.

- *Account manager*   Responsibilities include making sales calls to current and prospective customers. Demonstrate new products. Reporting industry updates to customers, and develop account plans and penetration strategies.

- *Quotations specialist*   Responsibilities include taking a Bill of Material for a construction project and creating a complete and professional quotation, using a personal computer. Interact with electrical contractors, architects, engineers, and owners to suggest products to be used on all types of construction projects. Take the Bill of Material from blueprints and specifications. Negotiate pricing with manufacturers and customers. Act as a source of technical expertise within the branch. Collect and analyze quotations data to determine patterns of success or failure. Measure these patterns by product type, manufacturer, bid amounts, and customers.

- *Outside sales*   Duties include expanding business by making visits to current and prospective customers. Promote products and services to customers. Maintain regular call coverage. Report all competitive activities and investigate new items. Prepare and execute business plans for assigned accounts. Develop and maintain strong customer relations while providing an excellent level of customer service.

Self-Check 5

1. What is electrical product distribution?
2. List three jobs that exist in the electrical product distribution industry.

Electrical Manufacturer

> **Electrical products manufacturer** produces those items that are needed by individuals working in the electrical trade.

Electrical products manufacturers produce those items that are needed by those individuals working in the electrical trade. Electrical hardware includes an extensive range of electrical hardware products, components, and accessories that are used for the distribution, supply, and usage of electricity. Electrical hardware includes everything from electrical cable, wire, and meters to luminaires, bulbs, and fans. Some examples of electrical hardware products manufactured are provided here:

- Panels, boxes, and box covers
- Circuit breakers and switchgear
- Conduit and pipe
- Cable, wire, and cord
- Sockets, plugs, and fixtures
- Earthing accessories
- Adapters

- Busbars
- Light bulbs
- Fans
- Meters

There are too many electrical products to list here, and the electrical products manufacturing industry continues to grow. Increasing construction activity continues to drive the demand for electrical hardware products, and increasing demand for products fuels the need for qualified individuals to work in this field. The following are electrical products manufacturing positions and their job descriptions:

- *Cost engineer (electrical)* Responsibilities include developing the means for subjective evaluation of systems or components under test. Analyze data to ensure that correct conclusions are reached. Develop major hardware proposals to carry out test activities. Direct a team of test engineers and technicians on specific programs. Coordinate the acquisition and deployment of engineering tools and human resources for major test or development programs. Originate unique approaches to solutions of experimental investigations. Integrates sophisticated measurement systems with test hardware.

- *Electro-Mechanical Assembly Technician* Duties include identifying and gathering necessary components and materials. Use hand and power tools in the fabrication process. Set up, run and assemble wire harnesses. Make mechanical and electrical (solder) connections on wired harnesses.

- *Product line sales manager* Responsibilities include working in all aspects of the company. Assist in making and implementing key decisions on capital allocation. Manage and record inquiries for products, and confirm that replies are made to inquiries.

Self-Check 6

1. Why are electrical product manufacturers important?

2. List three examples of electrical hardware products manufactured.

Electricians and Other Trades Workers

Tradesman a manual worker in a particular skill or craft.

An electrician is a tradesman specializing in electrical wiring of buildings, stationary machines, and related equipment. A tradesman is a manual worker in a particular skill or craft. Electricians may be employed in the construction of new buildings or maintenance of existing electrical infrastructure. **Figure 1-14** shows electricians reviewing prints on a construction site. In the construction industry, electricians are but one group of skilled tradesmen. Electricians work with just about every trade. Electricians work with bricklayers to run conduit inside block walls. Electricians work with HVAC and plumbers by supplying line voltage and control wires to some of their equipment such as flush valves, A/C units, heaters, and so on. Electricians are also responsible for providing temporary

FIGURE 1-14   Electricians at a construction site.

power and lighting for all trades, including electrical hook-up for job trailers. Other tradesmen that work in the construction industry include carpenters, heavy equipment operators, and pipefitters, to name just a few. Skilled tradesmen such as these work in the construction industry, and that requires working at ever-changing locations and work environments. The construction industry is the single most dangerous land-based civilian work sector. All tradesmen must be aware that it's not that the hazards and risks are unknown, but that they are difficult to control in constantly changing work environments. In the construction environment, tradesmen are often exposed to dirt, mud, inclement weather, heavy machinery on the move and workers all over the site.

Two of the biggest safety hazards on a work site are falls from heights and vehicles and electrical shocks. Some of the main health issues on site are asbestos, solvents, noise, cement dust, and manual handling activities.

Self-Check 7

1. What other tradesmen do electricians typically interact with on the jobsite?
2. What are the biggest safety hazards on a jobsite?

Vertical Mobility in the Electrical Trade

A **journeyman electrician** is an experienced electrician who may have completed training through an apprenticeship program and is able to work independently without supervision. To a first-year apprentice, becoming a journeyman electrician may seem a long way off, but before you know it, all courses have been completed,

Journeyman electrician an experienced electrician who may have completed training through an electrical apprenticeship program and is able to work independently without supervision.

Foreman a tradesman who oversees and manages the building, maintenance, and troubleshooting of electrical and electronic systems.

Electrical inspector examines the installation of electrical systems and equipment to ensure that they function properly and comply with electrical codes, standards, and regulations.

Estimator a person who often performs in a supervisory capacity for a general construction company and accurately assesses the amount and cost of electricity required for a building or process.

Superintendent a person who oversees the implementation, maintenance, and safety of an electrical system.

Project manager a person who works on site and interprets technical statements of work and design documentation as it relates to project planning, budgeting, procurements, implementation, testing, training, and project completion.

and you are a journeyman. The electrical tasks you have been performing are becoming second nature. With *safety* always in mind, your ability to accomplish job tasks is taking less time. With each passing year of apprenticeship, your paycheck grows. You are a journeyman now, but it may not end there. You can be a "tool-carrying" electrician for as long as you want, but other opportunities may present themselves along the way.

The next step up is to become a foreman, then a lead foreman, a general foreman, and, finally, an area foreman. An electrical foreman is a tradesman who oversees and manages the building, maintenance, and troubleshooting of electrical and electronic systems. Some experienced electricians go through extensive on-site and classroom training in colleges or vocational schools to become master electricians. Most states license master electricians based on examinations, accrued experience, and on-the-job training as a journeyman or apprentice electrician. The designation "master" is only given to electricians who can display extensive job knowledge and are tested to have an extensive understanding of the electrical safety code.

Another electrician with an extensive understanding of electrical codes and standards is the electrical inspector. Electrical inspectors examine the installation of electrical systems and equipment to ensure that they function properly and comply with electrical codes, standards, and regulations. Electrical inspectors examine everything from the connection of the main line to the wiring of individual outlets. These inspectors also inspect appliances that are hardwired into a structure, and they may examine appliances that are simply plugged in, to confirm that they will work properly with the electrical system. Training to become an electrical inspector often includes a college degree in electrical engineering and/or extensive experience with practical real-world electrical installation experience.

Other supervisory positions include these:

- **Estimator** A person who works for a general construction company that accurately assesses the amount and cost of electricity required for a building or process.

- **Superintendent** A person who oversees the implementation, maintenance, and safety of an electrical system.

- **Project manager** A person who works on site and interprets technical statements of work and design documentation as it relates to project planning, budgeting, procurements, implementation, testing, training, and project completion.

These types of positions are not for everyone. Hard work and the desire to "work and manage" must be there. Often, electricians will take college courses leading to degrees that will help them move onto this vertical career path and these opportunities. The following are typical supervisory positions and their job descriptions:

Electrical supervisor Responsibilities include supervising electricians and trades helpers in the repair, maintenance, and installation of electrical systems. Here are a few examples of duties:

- Supervise two or more full-time employees on a regular basis.
- Make decisions regarding hiring, evaluation, promotion, and termination of employees, or make related recommendations.
- Schedule, assign, plan, and monitor the work of electrical shop personnel performing a variety of electrical projects.
- Plan and develop work schedules, equipment usage schedules, priority of electrical expenditures, and procedures for equipment and appliance maintenance.
- Train or inform employees of new work methods or procedures, changes in safety regulations, work policies, and electrical codes and standards.
- Ensure budgeting guidelines are met; prioritize shop expenditures to stay within budget, and assist in preparation of capital and operations budgets.
- Inspect new construction and remodels.
- Assist engineers in design requirements.

Master electrician Electrical contractors must have a Master Electricians license in order to pull a permit for electrical work. Master electricians provide hands-on troubleshooting, construction installation, and maintenance and repair of the electrical components consisting of low and high voltage. Some specific qualities and skills of a good master electrician are listed here:

- Ability to perform inspections and preventive maintenance on all electrical equipment as required
- Attention to detail, ensuring that all safeguards of equipment are installed as needed to conform to governmental and state regulations.
- Excellent verbal and written communication skills with the ability to follow and provide detailed instructions
- Working knowledge and comprehension of state and federal safety regulations, as well as the local, state, and federal electric code standards.
- Experience working with 3-phase motors and controls.
- Working exposure to PLC (inputs, outputs, and logic).
- Experience in reading prints and troubleshooting circuitry problems.
- Ability to communicate well with others, both verbally and written.
- Ability to work successfully and safely under time-sensitive, deadline-oriented situations.

Electrical supervisor responsibilities include supervising electricians and trades helpers in the repair, maintenance, and installation of electrical systems.

Master electrician an electrician who can display extensive job knowledge and has been tested to have an extensive understanding of the electrical safety code.

LIFE SKILLS

How do you feel about the different opportunities available to you after completing your apprenticeship?

Self-Check 8

1. List four possible supervisory positions available to journeyman electricians.
2. List four job responsibilities for an electrical supervisor.
3. What qualifies an electrician to attain the designation "master" electrician?

Summary

- Electricians install and maintain all of the electrical power systems for our homes, businesses, and factories.
- Electricians specializing in construction typically install wiring systems into factories, businesses, and new homes.
- Electricians specializing in maintenance generally repair and upgrade existing electrical systems and service electrical equipment.
- Electricians specializing in electrical, or power distribution, typically install and repair underground power lines, switch gears, transformers and other hardware, as well as construct secondary service lines to buildings.
- Electrical distributors are those workers, often considered solutions providers, who distribute materials to end users and contractors.
- Electrical products manufacturers produce those items that are needed by individuals working in the electrical trade.
- Construction industry tradesmen include electricians, bricklayers, carpenters, plumbers, heavy equipment operators, and pipefitters.
- Construction site hazards and risks are difficult to control because of the constantly changing work environment.
- After you become a journeyman, possible supervisory positions include foreman, estimator, superintendent, and project manager.
- To become a master electrician, you must be able to display extensive job knowledge and are tested to have an extensive understanding of the electrical safety code.

Review Questions

True/False

1. Electricians only work in the construction industry.   (True, False)

2. Electricians who specialize in electrical maintenance generally spend much of their time providing preventive maintenance.   (True, False)

3. Basic duties of an industrial electrician are splicing and laying cables, erecting poles, stringing wires, and installing transformers and distribution panels.   (True, False)

4. Electrical distribution electricians are also known as inside wiremen.   (True, False)

5. A journeyman is an experienced electrician who has completed training through an apprenticeship program and can work without supervision.   (True, False)

6. The U.S. Bureau of Labor projects a decrease in job opportunities for electricians by 2018.   (True, False)

7. Electricians who specialize in residential work generally spend much of their time providing preventive maintenance.   (True, False)

8. Many apprenticeship programs provide over 144 hours of classroom instruction each year and 8000 hours of on-the-job training during the entire program.   (True, False)

9. After the successful completion of an apprenticeship program, the apprentice becomes a journeyman.   (True, False)

10. Electrical product distribution is the provision of electrical products to end users and electrical contractors.   (True, False)

11. Those individuals who work in the electrical product distribution industry are also called distribution, or power, electricians.   (True, False)

12. Electrical products manufacturers produce those items that are needed by those individuals working in the electrical trade.   (True, False)

13. There are only about 200 electrical products manufactured in the United States each year.   (True, False)

14. Electrical inspectors examine the installation of electrical systems and equipment to ensure that they function properly and comply with electrical codes, standards, and regulations.   (True, False)

Multiple Choice

15. What is the designation given to an electrician who displays extensive job knowledge and has passed a test, signifying an extensive understanding of the electrical safety code?
   A. Tradesman
   B. Lead journeyman
   C. Master
   D. Foreman

16. A person who often performs in a supervisory capacity for a general construction company and accurately assesses the amount and cost of electricity required for a building is a(n) _____.
   A. Foreman
   B. Estimator
   C. Superintendent
   D. Project manager

17. An electrician who typically installs and repairs underground power lines is a(n) _____.
   A. Electrical distribution electrician
   B. Construction electrician
   C. Maintenance electrician
   D. Residential electrician

18. At a construction work site, the person who provides job duties and also evaluates work performance is a(n) _____.
   A. Foreman
   B. Instructor
   C. Apprentice
   D. Journeyman

19. Which of the following workers is generally not found at a construction site?
   A. Pipefitter
   B. Heavy equipment operator
   C. Cartographer
   D. Electrician

20. Construction site hazards are difficult to control because of _____.
   A. The lack of OSHA regulations
   B. The lack of *NEC* and *NFPA* construction standards
   C. The large number of workers at a particular site
   D. Constantly changing work environments

ACTIVITY 1-1   Why a Career in the Electrical Trade?

Write a description of your electrical experiences and why you desire a career in the electrical trade.

ACTIVITY 1-2   Setting Goals for Yourself

Write your expectations and goals for the next 4 to 5 years (by no coincidence, this happens to be the length of an apprenticeship).

ACTIVITY 1-3   From Electrical Generation to Your Home

Describe how electricity is generated by the utility and becomes available at electrical outlets in your house.

ACTIVITY 1-4   Guest Speaker

An electrician who works in the electrical construction, maintenance, or distribution industry will visit your class. Create a list of three questions to ask the guest. Write a summary of what you learned from the guest speaker.

Chapter 2

Professional Behavior and Study Skills

Career Profile

James W. Miller is a State of Maryland Master Electrician and currently works as Foreman Electrician at F.B. Harding, Inc. Explaining his choice of career, he says, "Although I was always interested about how mechanical things work, I never really understood how electrical devices functioned. I thought this would be a good opportunity to learn something new. Eleven years later, I am still learning new electrical concepts. I never go through a day without new challenges." And looking to the future, he declares, "I want to keep sharpening my leadership skills. There is no limit for a person who can get others to perform with success."

On a day-to-day basis, Miller's work involves installing "gear, cable, and all sorts of devices." Even as he jokes about his work, James stresses the importance of cooperating with others on the jobsite: "I work with every trade. Generally, you talk to the electrician if you want to know something about the job!"

Indeed, in working continually to develop both his technical knowledge and his teamwork skills, Miller has determined that jobs well done are the result of effective teamwork and preparation: "My activities consist of planning for the successful installation of electrical equipment for our customers. I focus on efficiency and safety. I work with other electricians to answer questions." After consulting with his fellow electricians to draft a plan, Miller then seeks "to interact with other trades to achieve overall on time completion of our project." This process can be challenging, he notes: "There are always time and resource shortages that affect a job. The idea is to foresee some of these situations. It doesn't matter if it is a small task or a big job—you need a plan to be successful."

To students who are beginning electrical apprenticeships and careers, Miller asserts again the merits of preparation and collaboration: "Be the guy that every boss wants. Know how to do your job, ask questions, and listen. Be organized and helpful to your fellow workers. Don't be afraid to show extra effort without promise of praise—it will be noticed if it becomes a habit."

Chapter Outline

PROFESSIONAL BEHAVIOR IN THE WORKPLACE

BUILDING PERSONAL SUCCESS

Key Terms

Absenteeism

Active learning

ASAP

Attendance

Cram

Flashcards

Main idea

Mnemonic

On-the-Job Training (OJT)

Paraphrasing

Passive learning

Personal Protective Equipment (PPE)

Punctuality

Reading comprehension

Substance abuse

Tardiness

Time management

Zero tolerance

Chapter Objectives

After completing this chapter, you will be able to:

1. Provide examples of behaviors that reflect active participation in the workplace and in this course.
2. List a variety of study techniques and describe the benefits of using them.
3. Explain the value of on-the-job training.

Life Skills Covered

 Cooperation and Teamwork

 Communication Skills

 Managing Stress

 Critical Thinking

 Take Action

Life Skill Goals

For the second chapter of this book, your life skill goals focus on the big picture—on Cooperation and Teamwork, Communication Skills, Managing Stress, Critical Thinking, and Taking Action. Your first life skill goal for this chapter is to understand that the ability to complete a task within a team environment equates to improved performance on the job and potential increases in wages. You are to begin to think about how you can demonstrate professional behavior as well as teamwork and cooperation in the classroom and on the jobsite. You should also begin to think about strategies to manage your time and express yourself in a positive way. You have the opportunity to review a variety of Critical Thinking strategies that you can apply to this

course as well as to the workplace. Finally, you continue to work on the Take Action life skill as you answer questions throughout the reading.

Where You Are Headed

You learn about workplace interactions and expectations so that you can begin your career in the electrical industry as a positive professional. Strategies that help you succeed in the classroom can help you succeed on the job too. You learn the tips that other people use to have successful lifelong careers in the electrical trades. Study skills and thinking critically are discussed, as well as time management, reading comprehension skills, and test-taking tips.

Introduction

Learn how important professional behavior is *before* you get to the workplace. To succeed on the job and manage the diverse work environment that electricians regularly encounter day to day, you need to be an active participant in the workplace. This chapter provides some tips to help you achieve success. Be prepared mentally and physically, be on time, and know that you are part of a team at the workplace. Also, with safety in mind, zero tolerance for alcohol and drugs *must* be practiced.

2-1 Professional Behavior in the Workplace

When you are finally ready to choose your career path in the field, on-the-job training at an actual jobsite is an important part of your learning process as an electrician. This training helps you acquire hands-on professional skills, as well as appropriate job behavior, and allows you and your coworkers to perform productively. You are supervised by a journeyman electrician, who helps you learn to get the job done correctly and safely.

Drug-free workplace policies are also discussed in this section; given the potentially dangerous jobsite environment electricians must face, zero tolerance for alcohol and drugs is the rule in all work situations.

Finally, attendance and punctuality are addressed. These positive work habits are among the most important factors in your future job retention and advancement.

Active Workplace Participation

Each and every one of us can certainly use advice, reminders, or examples of specific traits that make us more efficient, productive, and valuable members of the workforce. As a skilled tradesman, your most valuable asset is yourself, and how you manage yourself determines just how well you do in the workforce.

Electricians rarely go to an office to start their day. Diverse jobsites require the electrician to travel to different locations to complete electrical work. The nature of electrical work requires you to work with many different people and new jobsites on a regular basis. Getting to a jobsite on time is *not* enough. It is very important to start a new job, or any workday, with a good attitude and thorough preparation.

Here are a few reminders to help you actively participate on the job:

- With "safety first" in mind, arrive with the proper protective equipment, and don't just be on time—be early. This is not only fair to your employer but also to your coworkers.

- When you arrive, don't waste time. Work diligently. Due diligence is a necessary characteristic of any great team member and ensures that assigned jobs get accomplished in a timely, effective manner.

- Always be willing to take on additional responsibilities. Be productive at all times. If your task has been completed, clean and organize your workspace and get prepared for your next task; by doing so, you set a good image. Try your best to do something extra when you are asked, and volunteer for tasks that allow you to learn something new.

All employers want their employees to succeed. Your foreman is one of the most important people at your work site. This person helps you learn your job duties and evaluates how well you do them. Keep a positive attitude toward your foreman. Treat your supervisors well, and they should treat you well.

Very few people work entirely alone. As an electrician, you often interact with other skilled tradesmen and others at the work site. Remember to treat these people with the respect and good attitude you would like people to show you. Cooperation and courtesy are important in getting a job done well.

LIFE SKILLS

How do you already demonstrate active workplace participation? How could you improve?

Self-Check 1

1. What are two ways you can actively participate on the job?
2. Why is it important?

On-the-Job Training and Workplace Experience

You learned the material, passed your tests, and developed electrical skills; now you have been assigned to a jobsite. You have been told what to expect, but until you get there, it isn't the same. Your first day at the site is much like your first day on a new job. Rest assured, you are still in training. Your on-the-job training is every bit as

FIGURE 2-1   A journeyman and an apprentice electrician.

© Lisa F. Young/www.Shutterstock.com

challenging, exciting, and perhaps as tiring as *working* as an electrician. **Figure 2-1** shows a journeyman electrician overseeing an apprentice on the job.

Don't wait until your first day to get directions to the work site. Know where to go and whom you report to. Be on time and ready to work. With "safety first" in mind, have the proper personal protective equipment (PPE) for the job. Personal protective equipment refers to equipment worn to minimize exposure to a variety of hazards (e.g. gloves, foot and eye protection, earplugs, hard hats). Always look and act professionally. Maintain a positive attitude at all times. Although you are in training, you are a member of a team. Your professional attitude helps your team function productively. Remember, the work site can be a dangerous place. There are many different trades working together to complete a project. You may have to work in poor weather conditions, in dirt and mud, and where there are heavy machinery and people all over the site. Stay focused. **Figure 2-2** shows a construction site.

Training that is provided at the jobsite is often referred to as on-the-job training (OJT), job "shadow" training, or hands-on training. Whatever it's titled, studies indicate that it is the most effective form of job training. On-the-job training focuses on the acquisition of skills within the work environment under normal working conditions. Through on-the-job training, workers acquire both general skills that they can transfer from one job to another and specific skills that are unique to a particular job. On-the-job training typically includes verbal and written instruction, demonstration and observation, and hands-on practice and imitation. In addition, the

Personal Protective Equipment (PPE) equipment worn to minimize exposure to a variety of hazards (e.g. gloves, foot and eye protection, earplugs, hard hats).

On-the-Job Training (OJT) the acquisition of skills within the work environment under normal working conditions.

FIGURE 2-2 Construction site.

on-the-job training process involves one worker (a journeyman) passing on knowledge and skills to a trainee (apprentice).

You have been trained to do electrical work; now you follow the day-to-day activities of a senior journeyman. You learn by watching and "doing with supervision and direction." For example, you learn more about bending conduit, and *this* conduit is used for an actual construction project. Under the supervision of experienced electricians, you get hands-on experience. Your first tasks may include drilling holes, setting anchors, and attaching conduit. Later, you measure, bend, and install conduit, as well as install, connect, and test wiring, outlets, and switches. You are also called on to clean up, organize material, and "go for" materials, tools, and (of course) coffee. You learn by watching and asking questions.

LIFE SKILLS

What are some ways you can show that you are acting professionally during on-the-job training?

Self-Check 2

1. List three practical job tasks that are typically performed during the on-the-job segment of electrician training.
2. List three typical learning components of on-the-job training.

Apprentice–Journeyman Workplace Interactions

Everyone, from first-year apprentice and journeymen electricians to foremen, knows why thcy are at the jobsite and what they have to do. As an apprentice, you should be told what is expected of you. If you aren't told, *you should ask* what is expected of you. You work under the supervision of a journeyman electrician. You often hear the phrase "time is money." It is important to get the job done in a timely and safe manner, but you should *take the time* to ask a job-related question. **Figure 2-3** shows a journeyman and an electrician planning for a job.

During the course of a day, you hear journeymen talk about the jobsite, the project at hand, what tasks need to be performed right away, and what work lies ahead. Listen carefully. You are all on the same team. Remember, those journeymen were apprentices once. They want to help you attain those skills that get the job done properly and on time.

Often, journeymen ask what you are learning in class. They remind you how important those classes are. They are speaking from experience. Talk to them. Ask them what areas they would have liked to study more. They will give you real-life examples of job tasks that, up to this point, you have only heard about in class. Some of these journeymen have years of experience. Many of them are willing to share job experiences, skills learned over time, and practical methods of doing certain tasks. Again, ask questions and listen carefully. After several conversations with a few journeyman electricians, you soon realize that learning doesn't stop once you become a journeyman.

FIGURE 2-3 Journeyman and apprentice electricians.

© Lisa F. Young/www.Shutterstock.com

How does building a good relationship with more experienced electricians help you accelerate your own career?

Self-Check 3

1. What does everyone know at the jobsite?

2. List two examples of information or insights that a journeyman can share with an apprentice electrician at the work site.

Workplace Practices and Procedures Regarding Alcohol and Drugs

Zero tolerance
the strict policy of enforcing all the laws of a state, or the rules of an institution, and allowing no toleration or compromise for first-time offenders or petty violations.

Why do you think that your student handbook and just about any other school or company handbook spells out, in no uncertain terms, that there is zero tolerance for "possession, use, or selling of alcoholic beverages or non-prescribed drugs?" Why have companies and businesses everywhere developed policies that adhere to "drug-free workplace laws?" Safety, safety, safety.

Zero tolerance is the strict policy of enforcing all the laws of a state, or the rules of an institution, and allowing *no tolerance* or compromise for first-time offenders or petty violations. Zero tolerance imposes automatic punishment for infractions of a stated rule, with the intention of eliminating undesirable conduct. *Fact:* The majority (over 65%) of current (past month) unlawful drug users 18 years or older are working either full-time or part-time.

Alcohol and other drug abuse in the workplace can create problems that may jeopardize the safety of you and your coworkers. There is not a career field anywhere that is exempt from the devastating problems that alcohol and drug abuse can cause. As an electrician, safety must always be the number 1 priority. Think about the job tasks that electricians perform. Electricians work with tools, make measurements, wire circuits, and work with dangerous voltages. and they *must* understand the *NEC.* The slightest impairment from alcohol or drug abuse could be catastrophic. An impaired person puts his or her own personal safety at risk, as well as the safety of coworkers. Think about the circuits that could be wired incorrectly while under the influence; a mistake could cause massive damage, drive up costs, and put lives at risk.

Substance abuse
a maladaptive pattern of use of a substance (or drug) that is not considered dependent.

Remember, it's not your responsibility to diagnose substance abuse. However, it is your job to observe behavior that might be typical of someone abusing substances and take measures to protect yourself and your coworkers. Substance abuse, also known as drug abuse, refers to the excessive use of and dependence on a substance. If you suspect someone has a substance abuse problem and/or you feel that your life is being put into danger, you should immediately contact your supervisor or the training director of your pre-apprenticeship program.

How can drug or alcohol use negatively impact your career prospects?

Self-Check 4

1. Explain zero tolerance regarding alcohol and drugs and why you think zero tolerance is important.

Punctuality and Attendance

When you accept a position, you agree to follow the job description and workplace guidelines provided by your employer. Many job descriptions clearly list as a *requirement* "to maintain an acceptable record of attendance, punctuality, and meeting deadlines." It may also be written as follows: "attendance/punctuality: is consistently at work and on time, uses paid leave within policy requirements, and informs supervisor about

Attendance the act of being present.

necessary absences in a timely manner." Attendance is being present at work every day;

Punctuality being on time.

punctuality is being on time. Why are punctuality and attendance so important? When employers schedule their workday and plan what is to be accomplished, they need to know they can depend on the people they have scheduled. It is not just about the employer. Being habitually late or missing work is not a personal or private thing; it affects others. When one member of a team is not there to do his or her assigned tasks, others suffer.

Employers do not expect you to work if you are really sick. They also understand

Tardiness the quality or habit of not adhering to a correct or usual or expected time.

that emergencies occasionally come up that result in you being tardy. Tardiness is the habit of not adhering to a correct, usual, or expected time. Simply put, tardiness is being repeatedly late for work. In these cases, it is vitally important to let your employer know as early as possible so appropriate action can be taken to cover for you while you are out. Most employers have a specific policy regarding when, how, and to whom you must report your tardiness or absence. If your employer does not have a policy in place, listed here are some acceptable steps to follow:

1. Have a good reason why you are absent.

ASAP means as soon as possible.

2. Call your employer ASAP (as soon as possible).

3. Make sure you talk to the person in charge.

4. Explain your reason for missing work clearly, completely, and respectfully.

Absenteeism a habitual absence from work.

Being constantly late for work or absenteeism, not going to work at all, are easy habits to fall into. Absenteeism is the habitual absence from work. Each time it happens, it gets a little easier to do. These are bad habits that could cost you your job and a chance at future employment. Statistics show that the two biggest reasons why people fail to keep a job are being late for work and missing days from work. Punctuality and regular attendance are important for job advancement and retention. If you are repeatedly late for work, you jeopardize future pay increases, and you may be subject to dismissal.

Employers need dependable people. In fact, most employers would rather have an employee who is less capable on a job but at work every day than an employee who is very capable but misses work a lot. Why? It's called dependability. Employers don't like to guess at who is coming in today and who is not. There is money and reputation at stake. Employers need to know that a job is covered and completed on time.

Employers need dependable people to get the job done. Good attendance and punctuality also show an employer that you have a good attitude toward your job. Dependability and attitude go a long way in helping you to get pay raises, promotions, and better references.

Remember, punctuality not only applies to arrival time in the morning, but also to returning from lunch or break. When you work for someone and accept pay from him or her, you are forming a work contract. It is your responsibility to be at work consistently and on time. Give your employer "eight hours of work for eight hours of pay." Success is often built on teamwork. If *you* don't do your part, you cause the team to fail. When you do your part, the company succeeds and makes money and *you* make money.

LIFE SKILLS

On the jobsite, how do you show that you are a part of a team?

Self-Check 5

1. Why is regular attendance and consistent punctuality important in the work place?
2. List recommended steps you should take when reporting lateness or absence.

2-2 Building Personal Success

The most common barrier to success encountered by students in any program of study is a lack of effective techniques for study and exam preparation. If you are asked how you prepare for your classes and tests and answer, "I just go over my notes," then you need to learn about effective study techniques and their benefits. Your success in this or any course depends on your *active participation* in the course. Read on and learn how effective time management, better reading comprehension, and useful test-taking skills aid in successful learning.

Active Course Participation

If you think that *class participation* is just saying something when called upon or simply "being there," then you want to learn what researchers say about *active* class participation, behavior, and active learning.

Passive learning
acquiring
knowledge
without active
effort, such as
listening to lectures
and watching
experienced
professionals
in the working
environment.

Active learning
involves learning
by being engaged
in the educational
process.

Learning is often accomplished in a passive manner by having instructors transmit content to the learners for them to absorb. Passive learning is acquiring knowledge without active effort, such as listening to lectures and watching experienced professionals in the working environment.

Active learning involves learning by being engaged in the educational process. Activities such as exploring, asking questions, preparing for class, and actually using new information or experiences in real-world situations can help improve your ability to retain new knowledge. Active learning refers to methods of instruction that focus the responsibility of learning on learners.

Research shows that learning is an active process, not a passive one. Students learn best when they take an active part in the learning process. Some aspects of active class participation are listed here:

- Asking questions
- Contributing ideas and providing new insights in the form of supporting arguments, personal views, opinions, and experiences
- Clarifying materials presented
- Exploring new perspectives

In active class participation, you need to use Critical Thinking skills, such as observing, analyzing, and evaluating information to understand concepts and solve problems. These skills require you to go beyond the basic recall of information. Passive class participation happens when you think on your own and keep everything to yourself. Active participation requires you to adopt an open mind and share what you think with your classmates.

You, as a student, should be aware of the behaviors your instructors want to see and those that do not reflect active class participation. Here are several desirable behaviors:

- Hardworking
- Positive attitude
- Motivated
- Interested

You should try to avoid any negative behavior, attend every class, and participate in class discussions.

Here are some classroom behaviors to avoid:

- Long-winded comments—Students should keep their answers, questions, and support for classmates as short, specific, and relevant as possible.
- Repetitive responses—Students should pay attention and keep up with the class.
- Participation monopolizers—Students should try not to carry the entire class load on their shoulders.
- Responses that discourage others from contributing—Signs of impatience, boredom, or superiority can be shown by gestures and facial expressions. Show classmates that you are interested in what they are saying.

Discouraging behaviors also include cutting someone off when the person is trying to make a point, one upmanship, and unnecessary argument (there is no need to challenge everything that is said). These behaviors *take away* from the learning process and make for a less than ideal learning environment.

Student participation can help a class move forward effectively if it is entered into willingly, enthusiastically, and purposefully.

What are your participation strengths? What is something you would like to work on in the area of active participation? Why is it important that you participate in class discussions and activities?

Self-Check 6

1. List two of the most common detracting classroom behaviors.
2. List two desirable classroom behaviors.

Time Management Strategies

Does it seem like there's never enough time in the day to get everything done? Are you feeling overwhelmed by your classes, assignments, and work schedule? Good time management is a key element to successful learning. Learn about the goals of time management and how you can apply them to your own schedule.

Time management is a range of skills, tools, and techniques used to manage time when accomplishing tasks, projects, and goals. When you are a student, conflicting responsibilities pile up quickly—classes, homework, job, and so on. Learning how to best use your time to get the most out of your education can sometimes be a challenge. Here are a few time management goals to help you see the benefits of these skills:

Time management a range of skills, tools, and techniques used to manage time when accomplishing tasks, projects, and goals.

- Help yourself become aware of how you use your time to organize, prioritize, and succeed in your studies.
- Enable yourself to take control of your life. Manage your time; don't let it manage you.
- Improve the quality of your life—be healthier and happier with much less stress.

Time Management Tips

The following are some suggestions for taking control of your time and organizing your life:

- Make a "to do" list every day.   Prioritize by putting things that are most important at the top, and do them first.

- Be organized.   Have an organized workplace, and don't waste time constantly looking for your work.

- Create good study habits.   Study at the same time each day, and that practice will become a habit.

- Plan out your time.   Figure out how much time you have for each assignment, and plot this out in your calendar. (When you plot out your time, be sure to schedule in breaks. Working straight through without a break can make you *less* efficient and cause stress. Avoid overload.)

- Avoid procrastinations and distractions.   Scheduled breaks are good (and necessary) but procrastination is not. A detailed schedule helps you to keep on track and avoid wasting time.

- Be flexible.   The unexpected happens (sickness, car troubles, etc.). You need to know how to rearrange your schedule when necessary.

- Take responsibility for time management.   While in training, no one is there to manage your time but yourself. You must take the initiative to get yourself organized.

Remember, time management is a learned skill, and it may be a new skill for you. If you attempt to organize your time and it doesn't go well at first, don't give up. The more you manage your time, the easier this habit becomes!

Which of the time management tips listed above do you plan on using? Why?

Self-Check 7

1. What is time management?
2. List two time management goals and two tips that can help you achieve them.

Effective Study Techniques

Just about everyone has something to offer regarding study techniques. Keep in mind that no two people study exactly the same way. What works for one person may not work for another. However, there are some general techniques that seem to produce good results. The following list contains suggestions that enable you to increase your effectiveness as a student.

- Develop a study schedule. A good schedule keeps you from wandering off course.

- Set up your study times. A good rule of thumb is that studying should be carried out only when you are rested, alert, and have planned for it.

- Limit your study sessions and take breaks. It has been proven that short bursts of concentration repeated frequently are much more effective than one long session.

- Select a place to study. You can study anywhere. However, places like libraries, study lounges, or private rooms are typically better choices. Wherever you choose to study, there should be a minimum of distractions.

- Develop thinking skills. Effective thinking skills cannot be studied, but must be built up over a period of time. Good thinkers see possibilities where others see only dead ends.

- Take good notes. If an instructor writes it on the board, you should write it in your notes. Learn to keep notes logically and legibly.

- Outline textbooks by using a highlighter. Experience has shown that important text passages highlighted are generally more easily remembered.

- Set reasonable goals. If you set reasonable goals, you get into the habit of accomplishing them; you can set higher goals over time.

- Maintain a positive and optimistic attitude. Whether you are working on a project, completing a homework assignment, or preparing for a test, maintaining a positive attitude and being optimistic goes a long way.

Applying these study techniques makes a huge difference, not only in remembering subject matter but also in taking quizzes or exams or working on projects. Remember, effective study skills must be practiced in order for you to improve. It is not enough to simply "think about" studying; you have to actually do it.

Self-Check 8

1. Which of the study techniques is the most helpful for you? Why?
2. Which of the study techniques is the hardest to incorporate into your routine?

Benefits of Effective Study Techniques

The main objective for using effective study techniques is to *increase your effectiveness* as a student. If you practice effective study techniques, you begin to see improvements in information retention, resulting in academic success and excellence.

Effective study techniques enhance learning *how* to learn, memorization, and studying speed. These techniques become tools or guides for self-improvement and better grades!

Students who are serious about applying effective study techniques become motivated, efficient learners. These students develop critical organizational and academic skills and improve their test scores and grades. By applying these techniques, you will realize the transfer of your learned skills to other aspects of your life and will notice that your overall abilities are improving as well.

Keep in mind however, effective study skills depend on one thing: your willingness to want to improve and do well. If you really don't want to make the effort and investment, no amount of suggestions, tips, or techniques will help!

Comprehension Strategies for Reading Assignments

Reading comprehension understanding the information that you read.

Reading about 200 words per minute is considered average for most students. For normal reading rates, 75% is an acceptable level of comprehension. Reading comprehension is understanding the information that you read. Reading rates vary; some of us read slower and some faster. What's more important is your comprehension, or ability to understand what it is you are reading. So, read on and learn how you can increase your level of comprehension.

Most reading comprehension strategies have the same basic components: pre-reading and active reading.

Pre-reading strategies include a first look at material and making predictions about what is covered. This list includes some helpful previewing information:

- Book and chapter titles
- Chapter headings and subheadings
- Chapter objectives
- Pictures, graphs, and photo captions
- End-of-chapter Review Questions

Active reading strategies include highlighting important text information and taking notes. This active involvement requires the reader to *think* about what he or she is reading. It's a good method to use so you can review and study information later without re-reading it all.

Here are some other strategies that improve your reading comprehension and enable you to read more effectively:

Main idea words that tell the reader what the paragraph is mostly about.

- Skimming and scanning—using a quick survey of the text to recognize the main idea, identify text structure, and confirm question predictions.
- Reasoning from context—using prior knowledge of the subject and ideas in the text as clues to the meanings of unknown words instead of stopping to look them up.

Paraphrasing stopping at the end of a section to check comprehension by summarizing and restating the information and ideas in the text.

- Paraphrasing—stopping at the end of a section to check comprehension by summarizing and restating the information and ideas in the text.
- Knowing the structure of paragraphs—often, the first sentence provides a framework for adding details. Also, look for words, phrases, or paragraphs that change the topic.

See Table 2-1 for an example of a formal strategy to improve your reading comprehension.

Like any study method, reading comprehension strategies must be practiced. Reading comprehension is an active process that can be improved if you are serious about using effective strategies.

TABLE 2-1 The SQR3 Study Reading System

SQR3 Study/ Reading System	This system helps you learn and remember. Here are the highlights of SQR3:
Survey	Gather the information necessary to focus and formulate goals.
Question	What questions do you have? Help your mind to engage and concentrate.
Read	Read actively to answer your questions and to fulfill your purpose.
Recite	Recite what you read. Retrain your mind to concentrate and learn as it reads.
Review	Review what you read. Refine your mental organization and begin building memory.

© Cengage Learning 2013

LIFE SKILLS

Which reading strategy do you think is the most useful? Why?

Self-Check 9

1. What are the two basic components of most reading strategies?
2. List three strategies that improve reading comprehension.

Test-Taking Strategies

The most obvious sign that you are *not* prepared for a test is when you have to stay up all night to "cram." Cram is a slang term for last-minute studying. The only thing cramming does is make you so tired you won't be able to think clearly enough to answer the questions that you do know. Did you know that the most common test-taking mistake is *not reading (or not following) the directions*? Read on and learn how you can be better prepared to take your next test.

Cram a slang term for last-minute studying.

Before the Test Date

Preparing for a test should start long before the test date. Start preparing for your first test on the very first day of class. Here are test-taking strategies that should be practiced *before* the test date:

- Make sure that you understand the material as you are learning it.
- Plan reviews as part of a weekly study schedule.

- While reviewing, ask yourself questions on the material. Turn the main points of each topic into questions, and be sure answers come to you quickly and correctly.
- Schedule reviews into several manageable time segments rather than one long period.
- It may seem elementary, but flashcards can help. Flashcards are a set of cards with words or numbers on either side, which can be used in classroom drills or in private study.
- Use mnemonic techniques to memorize lists, definitions, and other specific kinds of information. A mnemonic is a memory, or learning, aid (e.g., "one **t**ea, two **s**ugars" to help us remember how to spell po**t**a**ss**ium).
- Form study groups with other class members to discuss and quiz each other on important information.
- Get a good night's sleep before the test.

Flashcards cards with words or numbers on either side, which can be used in classroom drills or in private study.

Mnemonic a memory or learning aid.

A very good way to prepare is to analyze how you performed on previous tests. Review tests taken in the past, especially when studying for a final examination. The more tests you take, the better your test-taking strategies will become.

Taking the Test

Always arrive at the test site early. This gives you enough time to select a seat, organize your materials and get yourself relaxed. Be sure you have pencils, paper, calculator, and other needed materials. The following is a list of strategies that help to improve your test-taking skills:

- Dump your brain. Write down things like formulas, dates, and keywords you used in learning the material that might help you remember.
- Read the directions carefully. Can more than one answer be correct? Will you be penalized for guessing? Don't lose points because you assumed you knew what to do.
- Preview the test *before* you answer anything. Mark questions you know and ones that you don't. Note the point value of each question. This helps you to budget your time.
- Answer the easy questions first. This builds confidence and keeps you from wasting time on questions you are unsure of. Mark those questions, and come back to them.
- Ask for clarification. If you are unsure of the wording or the meaning of a question, ask the instructor.
- Keep an eye on the clock, and pace yourself. Make sure you have time to finish the questions with the highest point values. Don't get anxious if you hear others finishing the test early.

- Stay calm, relaxed, and confident. Regardless of the pressure to succeed, stressing out works against you. Remind yourself that you are well prepared.

- Don't second-guess. More times than not, you will change a right answer to a wrong one.

- Double-check your work. Make the time to re-check your answers, and be sure to go back to any questions that you marked as more difficult.

Most examinations are in a multiple-choice format. Multiple-choice tests ask a student to recognize a correct answer among a set of options (choices) that include three or four wrong answers, referred to as distracters. The following are some tips on answering multiple-choice questions.

- Cover the choices and try to answer the question.

- Eliminate answers you know are not right.

- Treat each choice as a true/false question, and choose the "truest" choice.

- Read all choices before choosing your answer.

- If there is no guessing penalty, take an educated guess and select an answer.

- Do not keep changing your answer; generally your first choice is correct.

- In "All of the above" and "None of the above" choices, if you are sure that one of the statements is true, then don't choose "None of the above." If one of the statements is false, then don't choose "All of the above."

- In a question with an "All of the above" choice, if you see at least two correct statements, then choose "All of the above."

- More often than not, the correct answer is the choice with the most information.

Test-taking strategies, like all study techniques, are works in progress. Remember to review your test preparation methods. Identify and maintain those habits that work well and replace those that don't.

Self-Check 10

1. List three strategies that improve test-taking skills.

2. Why do you think effective study techniques are important?

Summary

- The nature of electrical work requires you to work with many different people and new jobsites on a regular basis.

- Behaviors of active, productive members of the workforce include minimal absenteeism, punctuality, positive attitude, preparedness, and diligence.

- The *NEC* is a United States standard for the safe installation of electrical wiring and equipment.

- On-the-job training (OJT) provides practical, "hands-on" training at a work site.
- Apprentice–journeyman interactions provide invaluable technical insights and practical job skills to apprentice electricians at the jobsite.
- Alcohol and other substance abuse in the workplace can create problems that may jeopardize the safety of you and your coworkers.
- It is the responsibility of all workers, supervisors, and employers to be aware of their surroundings and do what they can to make the work environment safe.
- Employer guidelines and job descriptions often list regular attendance and consistent punctuality as a requirement.
- Students learn best when they take an active part in the learning process.
- Time management refers to a range of skills, tools, and techniques used to manage time when accomplishing tasks, projects, and goals.
- Effective study techniques improve subject matter retention and thereby also grades on quizzes, tests, and projects.
- Pre-reading and active reading are strategies that improve reading comprehension.
- Students should continuously review test-taking strategies, identifying and using those that work.

Review Questions

True/False

1. Cramming is an effective study technique. (True, False)

2. An effective strategy for taking a test is to begin preparing on the first day of class. (True, False)

3. Analyzing how you performed on previous tests is a good way to prepare for a test. (True, False)

4. Don't lose points by wasting time reading the directions. (True, False)

5. If you are unsure of the meaning of a question, ask the instructor. (True, False)

Multiple Choice

6. Protective clothing, helmets, goggles, or other garments designed to protect the wearer's body from injury are referred to as _____.
   A. Foul weather gear
   B. Hazardous duty equipment
   C. Protective personnel equipment
   D. Personal protective equipment

7. Which of the following is not a typical task performed while an electrician receives on-the-job training?
   A. Setting anchors
   B. Drilling holes
   C. Wiring boxes
   D. Gathering tools

8. Why have companies and businesses developed policies that encourage zero tolerance for possession, use, or selling of alcoholic beverages or non-prescribed drugs?
   A. To help stimulate the economy
   B. To minimize problems and maintain a safe work environment
   C. To receive government grants
   D. To provide more opportunities for substance abuse workers

9. Which of the following is not generally listed as a job requirement on job descriptions?
   A. Acceptable record of prior paid leave
   B. Acceptable record of attendance
   C. Acceptable record of punctuality
   D. Acceptable record of meeting deadlines

LIFE SKILLS

What test-taking strategy or strategies did you use to answer the multiple-choice and true-false questions above?

Fill in the Blank

10. _____ learning refers to those methods of instruction that focus the responsibility of learning on students.

11. Hardworking, motivated, and interested are all _____ employee behaviors.

12. Have a(n) _____ workplace and don't waste time constantly looking for your work.

13. A good study _____ keeps you from wandering off course.

14. Effective _____ skills cannot be studied, but must be built up over a period of time.

15. Practicing effective study techniques results in improved information _____.

16. Reading _____ is understanding the information that you read.

17. _____ reading strategies include highlighting important text information and taking notes.

18. The _____ sentence is a term used to describe the sentence that summarizes the main idea of the paragraph.

19. When preparing for a test, schedule reviews into _____ time segments.

20. Zero tolerance imposes automatic _____ for infractions of a stated rule.

21. As an employee, it is your job to observe _____ that is typical of someone abusing substances and protect yourself and your coworkers.

22. Statistics show that one of the main reasons why people fail to keep a job is _____.

ACTIVITY 2-1   To Do List

Make a "to do" list every day. Prioritize by putting things that are most important at the top and do them first.

ACTIVITY 2-2   Create a Schedule

Plan out your time. Figure out how much time you have for each assignment, and plot this out in your calendar. (When you plot out your time, be sure to schedule in breaks. Working straight through without a break can make you *less* efficient and cause stress. Avoid overload.)

ACTIVITY 2-3   Most Useful Technique

Describe the technique you have learned in this chapter that you think will be most useful to you on the job. Which will be most useful in the classroom?

ACTIVITY 2-4   Setting Goals for Professional Behavior

Write down three goals you have for exhibiting professional behavior in the classroom and the workplace.

ACTIVITY 2-5   Scenario Reflection

You are going to participate in an active workplace participation scenario. After the scenario, write a paragraph describing what you learned from this scenario.

ACTIVITY 2-6   Using the Five Steps

You are going to participate in a scenario where you apply five steps to solving a workplace problem. After the scenario, describe how using the five steps helped you to identify a variety of solutions.

(Continues)

(CONTINUED)

▪ ACTIVITY 2-7 ▪ Active Workplace Participation

Active participation requires that the sender of information, in this case the foreman, ensures that the receiver, the apprentice, clearly understands the direction he has been given. As for the apprentice, this person's duty is to pay close attention to details provided by the foreman so that he or she can accurately restate the information when asked to do so. If the apprentice is unsure of a specific point or piece of information, once the sender is done speaking, he or she should ask questions to get information relative to that specific piece of information so that there is a clear understanding of what to do.

For the following scenarios, two students will be chosen: One acts as the Foreman, and the second student acts as the Electrical Apprentice.

Scenario 1: Incomplete Information

Foreman (*looking directly at the apprentice*): I want you to go to the gangbox and get the chipping hammer. From there, you need to go up to the second floor, to room 210, and in the northeast corner of the room you'll see there's a PVC conduit that didn't fall into the block wall as was intended. I need for you to chip it out so we can fit the conduit back in the cells of the block wall. Do you understand?

Apprentice (*looks all around, pays attention to something happening in the distance while the foreman is providing instructions*): Yeah, sure! I get to go to the gangbox and then go up to the second floor and play with the chipping hammer.

Scenario 2: Complete Information

Foreman (*looking directly at the apprentice*): I want you to go to the gangbox in the trailer and get the AEG chipping hammer. From there, you need to go up to the second floor, to room 210, and in the northeast corner of the room you'll see that there is a PVC conduit that didn't fall into the block wall as was intended. I can show it to you right here on the plans. I need for you to chip it out so we can fit the conduit back in the cells of the block wall. Keep in mind that you need to make the trench deep enough so that we can get 2 inches of concrete cover over the pipe when it's finished. If you have any questions, please ask me, so we can be sure to do this correctly the first time.

Apprentice (*looks directly at the foreman while he is receiving direction*): Yes I understand. I need to go to the trailer and get the AEG chipping hammer from the gangbox and then go to room 210 on the second floor. In the northeast corner of the room, I should see a conduit that is not in the wall; I can refer to the plans to make sure that I am working on the right conduit. I am to chip the concrete so that we can get the conduit to lay back in the cells of the block wall, and I also need to make sure that I chip the trench deep enough so that we can cover the conduit with 2 inches of concrete when I am done.

(Continues)

(CONTINUED)

Discussion Questions

1. What is different about the foreman in each scenario?

2. What is different about the apprentice in each scenario?

3. If you were the apprentice in the first scenario, what could you say to the foreman so that he/she will provide you with additional information?

4. What feedback would you give to the apprentice in each scenario?

ACTIVITY 2-8 Critical Thinking Skills

A problem is defined as a situation that arises when there is a difference between the way something is and the way that you want it to be. Not only can it be frustrating when this happens, but if not handled correctly, it causes delays on the jobsite.

Critical thinking is a method that uses five steps to solve problems. They are as follows:

1. Define the problem.

2. Analyze and explore alternatives.

3. Choose a solution and plan how to implement the solution.

4. Put the solution into effect and monitor the results.

5. Evaluate the final results.

Scenario

You are the project foreman of a downtown high-rise apartment building. The parking garage for all of the workers on the project is located 1 mile away from the project. Your crews have to carry their heavy toolboxes from their cars to the jobsite every day. Your start time is 7:00 a.m. each morning, but many members of your crew arrive late on various days or tired from carrying their tools from the parking garage. This is negatively impacting your schedule.

Together with the other members of the class, use the steps above to help develop solutions to this problem.

Chapter 3

Safety

In the view of **Chris Ross, Construction Connect, Inc.**, the construction industry is a diverse field, rich with opportunities for observant, interested workers who recognize and value frequent chances to gain different types of experience and take on new challenges.

Ross entered the industry by training to perform basic electrical rough-ins and assist in the installation of wiring. He was then accepted into the Western Electrical Contractors Association (WECA) program and successfully completed the WECA course in 2001. He soon found work as an estimator/electrician at ZSI, Inc. (Sacramento, California), specializing in automation and motor controls for water treatment plants and pump stations.

After working on projects from start to finish at ZSI, Ross was qualified to move into his current position as vice president of Advanced Electric, Inc. (Newcastle, California), a commercial construction company that has specialized in public works and private construction for over 28 years. Ross is also founder and chief executive officer of Construction Connect, Inc. (Folsom, California), a company that creates and sells web-based software solutions for the construction industry.

To explain his rapid career advancements, Chris says, "One of the largest factors that contributed to my success is the desire and confidence to take on as many new responsibilities as possible. I would always ask to be part of the most challenging and difficult tasks on a job."

Ross also credits the unique nature of education through apprenticeship with his success in the industry: "Learning a trade was the most rewarding school experience I ever had. It is extremely motivating when something you learned in class can be used in the real world the next day. Unlike studying something you may do in the future, apprentices and trainees have the satisfaction of putting their education to use immediately."

Chris believes that student electricians should follow his example in seeking to understand their field from a variety of perspectives: "Stay active in industry events, groups and any networking opportunities. Being a good Electrician, Foreman, or Project Manager is always valuable, but the people who really excel are interested in more than just a paycheck. One of the biggest assets you can develop is a solid reputation and a network that helps you expand your business and generate meaningful opportunities."

Chapter Outline

BASIC SAFETY RULES

UNDERSTAND HOW THE BODY IS AFFECTED BY ELECTRIC CURRENT

THE JOBSITE AND PROPER WORK PROCEDURES

PERSONAL PROTECTIVE EQUIPMENT (PPE)

LADDERS AND SCAFFOLDING

Key Terms

Artificial respiration

Buddy system

Cardiopulmonary Resuscitation (CPR)

De-energized

Doff

Don

Electrical shock

Energized

Lockout/tagout (LOTO)

Material Safety Data Sheets (MSDS)

Microamperes (μA)

Milliamperes (mA)

Occupational Safety and Health Administration (OSHA)

Scaffolding

Voltmeter

Chapter Objectives

After completing this chapter, you will be able to:

1. List general safety rules.
2. List levels of current and recognize how electric current affects the human body.
3. Describe OSHA regulations as listed in Standard 1910, Subpart S, that relate to safety and the prevention of work-related injuries.
4. Give two examples of when employees use material data safety sheets.
5. Use lockout and tagout procedures as per OSHA and the NFPA.
6. List three types of personal protective equipment (PPE) and how each provides protection.
7. Describe proper procedures when using ladders and scaffolding.

Life Skills Covered

 Critical Thinking

 Take Action

Life Skills Goals

The life skill goals covered in this chapter are Critical Thinking and Take Action. Your goal for this chapter is to practice the reading comprehension strategies you learned in Chapter 2. To become the best possible electrician, you must develop critical thinking skills to help you take in all of the information presented, process it, ask questions about it in class, and then make an educated decision about the culture that you desire for your personal safety and the safety of others working with and around you. Critical thinking is an

invaluable skill, providing a way of looking at information and deciding what is useful, prioritizing that information, and finally making decisions based on all available pertinent information (e.g., if a part of the circuit is not working, deciding where you should begin to look).

You also continue to work on the Take Action life skill as you answer questions throughout the reading.

Where You Are Headed

The construction industry is a dangerous place. Your safety and the safety of your coworkers is vital. In this chapter, you learn about OSHA, a federal government agency, and become familiar with state and local agencies. OSHA is a government agency developed to ensure safe and healthful working conditions for working men and women by setting and enforcing standards and by providing training, outreach, education, and assistance. You also learn about a variety of techniques to keep yourself and your coworkers safe.

 Practice Pre-reading. Preview the chapter headings and subheadings, pictures, key terms, Self-Check questions, summary, and end-of-chapter Review Questions. What topics are covered in this chapter?

Introduction

OSHA, or the Occupational Safety and Health Administration, estimates that the lockout/tagout standard saves 122 lives and prevents 28,000 lost workday injuries each year. This standard represents specific practices and procedures to safeguard you and your coworkers from unexpected startup of machinery and equipment, or the release of hazardous energy during service or maintenance activities. It's likely that well over 800 lives have been saved since the standard went into effect in 1989. This means that more than 800 people still go home to their families, friends, and loved ones.

This chapter covers proper work procedures, the importance of OSHA, why we need personal protective equipment, and the proper use of ladders and scaffolding. Learn how to *work safely*, and use the lockout/tagout procedure, a standard that saves lives.

Can you think of a few safety rules that we follow in our everyday lives?

Knowing and following safe practices and procedures protects you and your coworkers. Read and understand them. Remember them. Practice them.

 What does the first sentence in the next paragraph tell you about what the focus of the paragraph will be?

3-1 Basic Safety Rules

The biggest safety hazard associated with working as an electrician is electrical shock, a reflex response to the passage of electric current through the body. Whether you are working in commercial, industrial, or residential wiring, electric shock hazards exist.

Before you work on a circuit, always disconnect power!

A circuit that has power (electricity) connected to it is considered energized. When power is removed, the circuit is de-energized. Use the following procedure to make certain that the circuit you are working on is de-energized:

1. NEVER assume that a circuit is off (de-energized).

2. Following the appropriate National Fire Protection Association (NFPA) 70E guidelines, secure and don appropriate personal protective equipment (PPE) *prior* to conducting any voltage testing. *Don* means "to put on" or "to dress oneself in." In contrast, doff means "to remove" (as to remove an article of clothing).

TRADE TIP   NFPA 70E, entitled Standard for Electrical Safety in the Workplace, is a standard of the NFPA.

TRADE TIP   PPE refers to protective clothing, helmets, goggles, or other garments designed to protect the wearer's body from serious workplace injuries or illnesses resulting from contact with chemical, radiological, physical, electrical, mechanical, or other workplace hazards.

3. Verify that your test instrument is operational by first testing a known working source, and then test the circuit.

4. When *you* establish that the circuit is de-energized, affix a warning tag at the point of disconnection (using such a tag keeps someone from restoring power). Use a lockout device to prevent anyone from restoring power.

TRADE TIP   If lockout is not possible, use a buddy system to make sure that power is not restored while someone is still working on a circuit. The buddy system is a procedure in which two people operate together as a single unit so they are able to monitor and help each other.

Use the Buddy System

When it's necessary to work in potentially dangerous locations, never work alone. Another worker can assist on the job and, if necessary, administer first aid such as artificial respiration or cardiopulmonary resuscitation (CPR). On the jobsite, one person certified in CPR is required to be part of the crew. Artificial respiration is the act of simulating respiration, which provides for the overall exchanges of gases in the body by pulmonary ventilation, external respiration, and internal respiration. CPR is an emergency procedure for people in cardiac arrest or, in some circumstances, respiratory arrest.

Some voltages can force enough current through the skin to produce an electrical shock. Electrical shock occurs upon contact of a human body with any source of voltage high enough to cause sufficient current through the skin, muscles, or hair. If the current is 50 milliamperes (mA) or more, the shock can be fatal.

TRADE TIP When possible, work with your right hand only. A shock passing from hand to hand is often fatal because the current passes directly through the heart.

Self-Check 1

1. Why is utilizing the buddy system when working in potentially dangerous situations important?
2. What are the four steps to make certain the circuit you are working on is de-energized?

Learn First Aid and Be Prepared to Administer It

Read the first sentence of the paragraph below. What will you learn in this paragraph?

Learning first aid is very useful when working in any field. However, it is invaluable in the electrical field, especially when working with voltages of 50 volts or more. The main goals of first aid are listed here:

- Preserve life
- Prevent further harm or prevent the condition from worsening
- Promote recovery

Every electrician should know how to administer first aid and be trained in CPR. Here are two conditions every electrician should be very familiar with:

- Electrical burns
- Electrical shock

An electrical burn may appear minor or not show on the skin at all, but the damage can extend deep into the tissues beneath the skin. If the electrical current passing through the body is strong enough, internal damage such as heart rhythm disturbance or cardiac arrest can occur.

The danger from an electrical shock depends on the type of current, the level of voltage, the path current traveled through the body, the person's overall health, and how quickly the person is treated.

In any case, while waiting for medical help, follow these steps:

- Look first. Don't touch. The person may still be in contact with the electrical source. Touching the person may pass the current through you.

- Turn off the source of electricity.

Reading about CPR and learning when it's needed will give you a basic understanding of the concept and procedure. It is strongly recommended that you learn the details of how to perform CPR by taking a course. Nearby hospitals and local chapters of the American Heart Association (AHA) and the American Red Cross are good sources for finding a CPR course in your area. The following websites provide additional information: www.americanheart.org and www.redcross.org. Taking a CPR course could mean the difference between life and death.

Self-Check 2

1. What are the two health hazards every electrician should be familiar with?

2. What is CPR?

3-2 Understand How the Body Is Affected by Electric Current

It is important that you understand the dangers and consequences of electrical shock. Use the process outlined below to improve your understanding of this section.

- Survey through Section 3-2, *Understand How the Body Is Affected by Electric Current*.

- Skim through the section. Take a look at the pictures. Read the Self-Check questions at the end of the section.

- Create the questions you are going to answer by reading this section.

- Read this section. Take some notes. Find the answers to your questions.

- Recite these answers.

- Review. Go back and highlight the main points in the section. Add them to your notes.

Electric circuits can be dangerous and *electric shock* can kill. Voltage is the force that causes current to flow through a circuit. The effect of this force is similar to the pressure that pushes water through a pipe. If the pressure is high, then more water flows through the pipe.

Milliamperes (mA) a unit of current equal to one thousandth (10^{-3}) of an ampere.

Microamperes (μA) a unit of current equal to one-millionth (10^{-6}) of an ampere. (Ampere is the base unit used to define the rate of current flow.)

A current through the human body in excess of 10 **milliamperes (mA)** can paralyze a victim and make it impossible to let go of a live circuit. A milliampere is a unit of current equal to one-thousandth (10^{-3}) of an ampere. Ten milliamperes is only 10 one-thousandths of an ampere. A flashlight can provide more than 40 times that amount of current. Flashlight batteries are safe to handle because human skin has a high enough resistance to keep the current level very low. The current produced by a 1.5V battery produces only **microamperes (μA)**. A microampere is one-millionth of an ampere. Such a low level of current is unnoticeable.

A shock can be produced (and felt) when a voltage causes current to flow through the skin. Even though **Figure 3-1** shows that a current as low as 50 milliamperes can be life threatening, ANY SHOCK can be fatal if it interrupts the natural rhythm of the heart.

TRADE TIP Electricians who work with high voltage must be properly trained and use required safety equipment.

Figure 3-1 shows how different levels of current affect the body. The information shown is typical, but body (skin) resistance varies from person to person. Skin resistance is generally low when wet and even lower if cut. If the length of the current path (shock) is longer (i.e., hand to foot versus hand to hand), then resistance is higher. Some people have a high tolerance to electricity, and some have a very low tolerance. Figure 3-1 is provided to show that even low levels of current *can be lethal*!

FIGURE 3-1 The effects of electric current on the body.

Current Flow	Effect on the Human Body
Less than 1 milliampere	No sensation.
1 milliampere	Possibly a tingling sensation.
5 milliamperes	Slight shock felt; not painful, but disturbing; most people can let go; strong involuntary reactions may lead to injuries.
6 to 30 milliamperes	Can definitely feel the shock; it may be painful and you could experience muscular contraction (which could cause you to hold on).
50 to 150 milliamperes	Painful shock, breathing could stop, severe muscle contractions; death is possible.
1000 to 4300 milliamperes	Heart convulsions (**ventricular fibrillation**), paralysis of breathing; usually means death.
10,000 milliamperes	Cardiac arrest and severe burns; death is probable.

Self-Check 3

1. Why is body (skin) resistance important as it relates to electrical shock?

2. What precautions can be taken to increase body (skin) resistance?

3-3 ■ The Jobsite and Proper Work Procedures

Since 1970, work-related fatalities are down 50%, and occupational injuries have declined by 40%, due in large part to the creation of OSHA in that year and its health and safety standards for the workplace. OSHA mandates employer standards, construction industry requirements, confined space regulations, and material safety data sheets that all specify procedures to help protect electricians on the jobsite. In addition, procedures known as lockout/tagout can help safeguard you and your coworkers from electrical malfunction during service or maintenance of equipment.

Occupational Safety and Health Administration (OSHA)

Occupational Safety and Health Administration (OSHA) an agency of the US Department of Labor whose mission is to prevent work-related injuries, illnesses, and occupational fatalities by issuing and enforcing standards for workplace safety.

The Occupational Safety and Health Administration (OSHA) is an agency of the United States Department of Labor. It was created by Congress under the Occupational Safety and Health Act, signed by President Richard Nixon on December 29, 1970. Its mission is to prevent work-related injuries, illnesses, and occupational fatalities by issuing and enforcing standards for workplace safety and health. The agency is headed by a Deputy Assistant Secretary of Labor.

Electricity has long been recognized as a serious workplace hazard. OSHA's electrical standards are designed to protect employees exposed to dangers such as electric shock, electrocution, fires, and explosives. OSHA standards cover personal protective equipment, hazardous locations, training, and use of equipment. Standard 1910, Subpart S, details regulations that relate to electrical safety. OSHA's website, http://www.osha.org, is a valuable resource, listing standards for many areas.

NFPA 70E, Electrical Safety for Employee Workplaces, also contains safety regulations that an electrician should be aware of. The NFPA 70E standards are incorporated into OSHA regulations.

Employer Responsibilities

Employers have some real responsibilities to their employees, as mandated by the standard listed in Section 5(a)(1) of the Occupational Safety and Health Act. This section is often referred to as the General Duty Clause; it requires employers to "furnish to each of his employees employment and a place of employment which are free from recognized hazards that are causing or are likely to cause death or serious physical harm to his employees." Section 5(a)(2) requires employers to "comply with occupational safety and health standards promulgated under this act."

OSHA regulations (Standards—29 CFR) are the law (CFR stands for Code of Federal Regulations). This entire standard details safety in the workplace for general industry.

Part 1926 in the OSHA regulation (Standards—29 CFR) specifies safety and health regulations for construction. This standard describes rules for anyone in the construction industry, not just electricians. Topics include medical services and first aid, safety training and education, personal protective equipment, hand and power tools, electrical requirements, and required signs and tags.

LIFE SKILLS

Paraphrase what you just read about employer responsibilities. Paraphrasing is summarizing and restating the information you just read in your own words.

Trenches

OSHA Standard 1926, Subpart P, App C, titled "Timber Shoring for Trenches," contains information that can be used when timber shoring is provided as a method of protection from cave-ins in trenches that do not exceed 20 feet in depth. In the electrical construction field, installing underground conduit is common practice. Measures must be taken to prevent potential injuries should a cave-in occur. The OSHA standard and its procedures must be understood and followed. The following is a list of safety precautions that should be followed in and around trenches:

1. Always maintain a distance of no less than 2 feet from sides of trenches or ditches. Pedestrian traffic can cause dirt to loosen, and a possible cave-in may result.

2. Erect a fence or a barricade that is at least 6 feet tall, with openings no greater than 4 inches between vertical supports.

3. When working in and around a trench, be certain that there is a safe access into and out of the trench.

TRADE TIP   Use ladders with adequate load capacity, and always use proper placement.

Confined Spaces

Many workplaces contain spaces that are considered "confined" because their configurations hinder the activities of employees who must enter, work in, and exit them. A confined space has limited or restricted means for entry or exit and is not designed for continuous employee occupancy. Following are some examples of confined spaces:

- Underground vaults
- Tanks
- Storage bins
- Manholes
- Pits
- Silos
- Pipelines

The hazards encountered when entering and working in confined spaces are capable of causing bodily injury, illness, and death to the worker. Accidents can occur to workers who fail to recognize that a confined space is a potential hazard. Danger of explosion, poisoning, and asphyxiation are present at every entrance to a confined area.

Ventilation is often a problem because of an area's configuration and limited openings. Protective personal equipment should include safety goggles, hard hats, special clothing, and often a separate supply of air. OSHA Section 12, "Confined Space Hazards," details those regulations associated with working in confined spaces.

Workers must be specially trained before entering a confined space. Anyone working in a confined space must be constantly alert for changing conditions. In the event of an alarm from monitoring equipment or any other indication of danger, workers should immediately leave the confined space. Anyone working in a confined space *must* wear appropriate protective equipment and clothing. The OSHA-mandated safety practice is to always wear a harness with a rope that extends to an outside person. The outside person, referred to as the safety watch or standby, can pull the worker to safety. **Figure 3-2** illustrates a confined area. A standby or safety watch is in position. The safety watch has certain responsibilities, detailed here:

- Understand the hazards that may be found inside the particular confined space and recognize signs, symptoms, and behavioral effects that workers in a confined space could experience.

- Monitor the confined space and surrounding area; be on the lookout for dangerous conditions.

- Remain outside the confined space; do no other work that may interfere with the primary duty of monitoring the workers inside the confined space.

- Maintain constant communication with the workers in the confined space.

- Order immediate evacuation if a potential hazard, not already controlled, is detected.

- Provide entry rescue only after the most stringent precautions are taken and another safety watch is in place.

FIGURE 3-2 Working in a confined area.

Paraphrase what you just read about confined spaces.

Material Safety Data Sheet

The material safety data sheet (MSDS) is a detailed document that provides product users and emergency personnel with information and procedures needed for handling and working with chemicals. This sheet dictates the type and style of protective equipment needed when using a particular chemical. An MSDS provides information about proper storage of a substance, first aid, spill response, safe disposal, toxicity, flammability, and other useful material. MSDSs are not limited to reagents used for chemistry; common household products such as cleaners, gasoline, pesticides, certain foods, drugs, and office supplies are also included. An MSDS would be needed if an electrician accidentally got some threading oil in the eye, spilled chemical on the skin, or breathed chemical fumes. Everyone at the jobsite needs to know where to access the MSDS and how to interpret the information. *If you do not know where the material safety data sheets for your work area are kept, check with your supervisor and find out.*

Information on an MSDS aids in the selection of safe products and helps everyone respond effectively to daily exposure situations as well as emergency situations. Employers must ensure that each employee knows how to find information on an MSDS and how to make use of that information. Being familiar with MSDSs allows for precautions to be taken for potentially dangerous products. Sometimes, seemingly safe products may contain unforeseen hazards. An employee's workplace is required to have MSDSs available for every hazardous chemical or substance used or encountered on the job. MSDSs must be readily available for employee review at all times in the workplace. In other words, they cannot be locked in an office or filing cabinet where access is not available to everyone. OSHA's Hazard Communication Standard (HCS) specifies certain information that must be included on MSDSs.

Paraphrase what you just read about material data sheets.

Lockout/Tagout

Lockout/tagout (LOTO) refers to specific practices and procedures to safeguard you and your coworkers from unexpected energization or startup of machinery and equipment, or the release of hazardous energy during service or maintenance activities.

Approximately 3 million workers service equipment; they face the greatest risk of injury if LOTO is not properly implemented. Compliance with the LOTO standard, OSHA 29, CFR 1910.147, prevents an estimated 120 fatalities and 50,000 injuries each year. Generally, industries implement their own policies and procedures. Some examples of LOTO devices are shown in **Figure 3-3**. Many industries require a padlock be used to lock out the equipment. A padlock is one of the devices shown in Figure 3-3. **Figure 3-4** illustrates the use of multiple padlocks and a safety tag.

Once a piece of equipment is locked out and tagged, it should be tested to make sure that no electrical power is present. Ensure that no one is exposed to the equipment when testing. The following procedure should be used:

1. Verify that the equipment is shut down (i.e., identify and lock out all potential sources of energy).

2. Isolate and dissipate any hazardous potential energy sources (such as charged capacitors).

3. Place power switch to ON, and test the equipment to ensure that it will not run or start.

FIGURE 3-3 Lockout/tagout devices.

From FLETCHER, *Residential Construction Academy, 3E.* © 2012 Cengage Learning

FIGURE 3-4 Multiple locking device shows that the equipment can be locked out by several different people.

From HERMAN, *Delmar's Standard Textbook of Electricity, 5E.* © 2011 Cengage Learning

4. Following the appropriate NFPA 70E guidelines, secure and don appropriate personal protective equipment (PPE) *prior to* conducting any voltage testing.

5. Verify that your voltage tester is operational by first testing a known working source.

Voltmeter an instrument used to measure the electrical potential difference between two points in an electric circuit.

6. Use a properly operating voltage tester such as a voltmeter, an instrument used for measuring the electrical potential difference between two points in an electric circuit, to verify that the power is absolutely de-energized.

7. After testing, return the power switch to the OFF position.

What is a voltmeter? Use the clues in the text to help you figure out the meaning of the word.

Self-Check 4

1. What is OSHA and why is it important?
2. List four categories that OSHA addresses in its electrical standards.
3. List three examples of confined spaces.
4. What is a material safety data sheet?
5. What is the importance of lockout/tagout?

3-4 Personal Protective Equipment (PPE)

Practice the SQR3 reading system.

- Survey through Section 3-4, *Personal Protective Equipment (PPE).*
- Skim through the headings. Take a look at the pictures. Read the Self-Check questions at the end of the section.
- Create the questions you are going to answer by reading this section.
- Read this section. Take some notes. Find the answers to your questions.
- Recite these answers.
- Review. Go back and highlight the main points in the section. Add to your notes.

OSHA and NFPA require the use of PPE to reduce employee exposure to hazards when engineering and administrative controls are not feasible or effective in reducing these exposures to acceptable levels. *Employees must be knowledgeable and determine whether personal protective equipment is needed.*

PPE garments are designed to protect the wearer's body from serious workplace injuries or illnesses resulting from contact with chemical, radiological, physical, electrical, mechanical, or other workplace hazards. Besides face shields, safety glasses, hard hats, and safety shoes, PPE includes a variety of devices and garments such as goggles, coveralls, gloves, vests, earplugs, respirators, and safety harnesses.

Head Injury Protection

Hard hats can protect your head from impact, penetration injuries, and electrical injuries caused by falling or flying objects, fixed objects, or contact with electrical conductors. **Figure 3-5** illustrates a typical electrician's hard hat. This type of hard hat is made of nonconductive plastic and has a pair of safety goggles attached that can be used when necessary.

TRADE TIP Although there is no OSHA standard specifying the service life (cycle) of hard hats, OSHA does recommend that users adhere to the replacement guidelines established by the manufacturer. As a general guideline, it is recommended that the hard hat be replaced at least every *3* years; if your hard hat takes a heavy blow or an electrical shock, replace it with a new one.

Foot and Leg Injury Protection

In addition to foot guards and safety shoes, leggings (e.g., leather) can prevent injuries by protecting from hazards such as falling or rolling objects, sharp objects, wet and slippery surfaces, molten metals, hot surfaces, and electrical hazards.

FIGURE 3-5 Electrician's hard hat with goggles attached.

From HERMAN, *Delmar's Standard Textbook of Electricity*, 5E. © 2011 Cengage Learning.

Eye and Face Injury Protection

Spectacles, goggles, special helmets or shields, and spectacles with side shields and face shields can protect against the hazards of flying fragments, large chips, hot sparks, radiation, and splashes from molten metals. They also offer protection from particles, sand, dirt, mists, dusts, and glare. **Figure 3-6** shows a pair of safety glasses with side shields.

Protection against Hearing Loss

Wearing earplugs or earmuffs can help prevent damage to hearing. Exposure to high noise levels can cause irreversible hearing loss or impairment, as well as physical and psychological stress. Earplugs made from foam, waxed cotton, or fiberglass wool are self-forming and usually fit well. A professional should fit workers individually for molded or preformed earplugs.

OSHA Standard 1926, Subpart D, titled "Occupational Noise Exposure," details noise levels and safe exposure limits. The American National Standards Institute (ANSI) also lists their mandate (A10.46-2007), which helps employers prevent occupational hearing loss to construction workers with noise exposures of 85 decibels and above.

Hand Injury Protection

Workers exposed to harmful substances through skin absorption, severe cuts, or thermal burns will benefit from hand protection. Gloves are a common article of protective clothing. Electricians often wear leather gloves with rubber inserts when it is necessary to work on energized circuits. **Figure 3-7** shows a pair of leather gloves with rubber inserts.

Kevlar® gloves are used to protect against cuts when stripping cable with a sharp blade. **Figure 3-8** shows a picture of Kevlar® gloves.

FIGURE 3-6 Safety glasses with side protection.

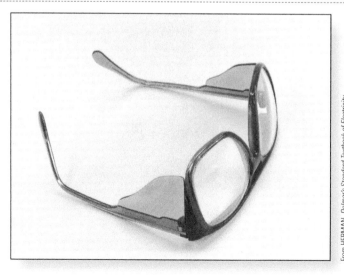

from HERMAN. Delmar's Standard Textbook of Electricity, 5E. © 2011 Cengage Learning

FIGURE 3-7 Leather gloves with rubber inserts.

From HERMAN. *Delmar's Standard Textbook of Electricity,* 5E. © 2011 Cengage Learning

FIGURE 3-8 Kevlar® gloves protect against cuts.

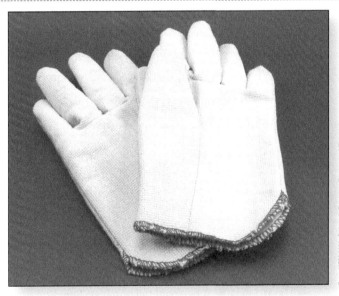

From HERMAN. *Delmar's Standard Textbook of Electricity,* 5E. © 2011 Cengage Learning

Whole Body Protection

In some cases, workers must shield their entire bodies against such hazards as exposure to heat and radiation. In addition to fire-retardant wool and cotton, materials used in whole-body PPE include rubber, leather, synthetics, and plastic. Fire-retardant clothing is often required for maintenance personnel who work with high-power sources such as transformer installations and motor-control centers.

TRADE TIP The **PPE** worn by a worker is dictated by the function the worker is performing and the policies of the worker's employer. In certain instances, all of the previously mentioned **PPE** may be necessary to perform a given task.

Safety Harness

Safety (body) harnesses are designed to minimize stress forces on a body in a fall, while providing freedom of movement to allow work to be performed. OSHA Standard 1926, Subpart M, titled "Fall Protection," contains information that addresses falls and harnesses.

> **TRADE TIP** As of January 1, 1998, body belts are not acceptable as part of a personal fall arrest system. These belts impose a danger of internal injuries when stopping a fall.

Figure 3-9 illustrates a personal fall arrest system. Note how this harness buckles around the upper body with leg, shoulder, and chest straps. The back of the harness has a heavy D-ring.

A *fall arrest system* is attached to the D-ring and should be secured to a stable structure above the worker. This system must be connected to a point of attachment that is capable of sustaining 5000 pounds. The lanyard limits the distance the worker drops in the event of a fall. The following conditions dictate wearing a safety harness:

- When working at distances greater than 6 feet above the ground or floor.
- When working near a hole or drop-off.
- When working on aerial platform lifts.

Figure 3-10 shows a safety harness.

Self-Check 5

1. What is the purpose of personal protective equipment?

2. Which OSHA standard addresses fall protection?

FIGURE 3-9 A personal fall arrest system.

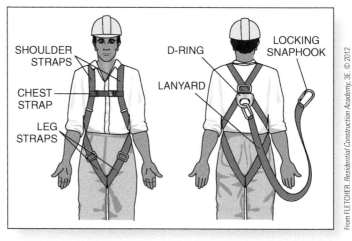

SHOULDER STRAPS

CHEST STRAP

LEG STRAPS

D-RING

LANYARD

LOCKING SNAPHOOK

From FLETCHER. *Residential Construction Academy*, 3E. © 2012 Cengage Learning

FIGURE 3-10 Safety harness.

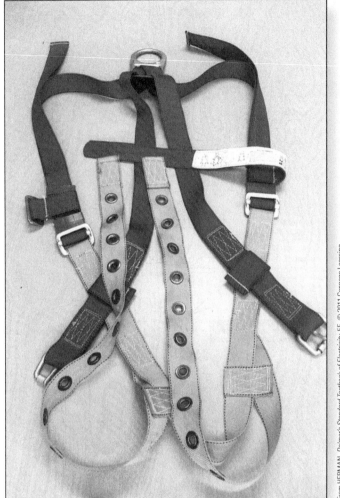

From HERMAN, *Delmar's Standard Textbook of Electricity*, 5E. © 2011 Cengage Learning

3-5 Ladders and Scaffolding

Scaffolding a temporary structure used to support people and material.

When it becomes necessary to work in an elevated location, ladders or scaffolding should be used. Scaffolding, or staging, is a temporary structure used to support people and material. Scaffolds typically provide the safest elevated working platforms. According to OSHA Standard 29, CFR 1926, Subpart L, titled "Safety Standards for Scaffolds Used in the Construction Industry," scaffolds shall be erected for persons performing work that cannot be done safely from the ground or from solid construction.

Scaffolds are generally constructed on the work site from standard sections. Figure 3-11 shows a section of scaffolding and Figure 3-12 illustrates how two end sections are connected by X braces that form a rigid platform. Sections of scaffolding are stacked on top of each other to attain the desired height.

FIGURE 3-11 Section of scaffolding.

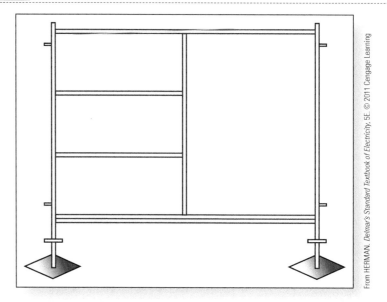

From HERMAN. *Delmar's Standard Textbook of Electricity, 5E.* © 2011 Cengage Learning

FIGURE 3-12 X braces connect scaffolding together.

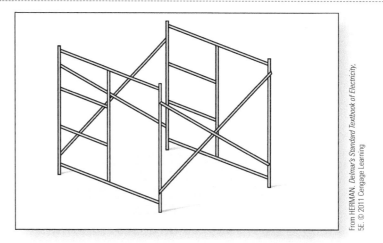

From HERMAN. *Delmar's Standard Textbook of Electricity, 5E.* © 2011 Cengage Learning

Scaffolding with wheels attached to the bottom section can be moved from one position to another. This type of scaffolding is called rolling scaffolding.

LIFE SKILLS

Paraphrase what you just read about scaffolding.

TRADE TIP The wheels are generally built onto the scaffolding with a mechanism that allows them to be locked, once the scaffold is placed in the desired location.

There are two main types of ladders:

- Straight
- Step

A straight ladder is shown in **Figure 3-13**; a stepladder is shown in **Figure 3-14.**

FIGURE 3-13 Straight ladder.

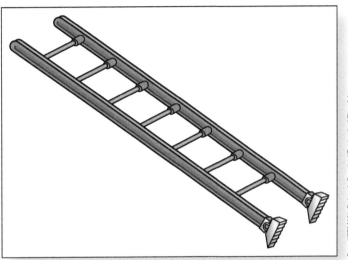

FIGURE 3-14 Stepladder.

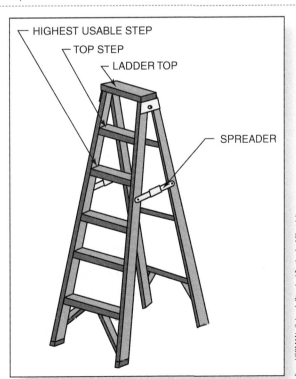

HIGHEST USABLE STEP

TOP STEP

LADDER TOP

SPREADER

Here are the guidelines for ladder safety:

- Ladders, including stepladders, shall be placed so that each side rail (or stile) is on a level and firm footing, making the ladder rigid, stable, and secure.
- Ladders shall be used at such an angle that the horizontal distance from the foot of the ladder to the structure the ladder rests against is one-quarter (¼) of the length of the ladder.
- Ladders shall rise to a height of at least 1 meter (or 3 feet) above any landing place for people using those ladders.
- Only one person at a time may use or work from a single ladder.
- Always face the ladder when ascending or descending it.
- Always maintain three points of contact with a ladder (i.e., both hands and one foot, or both feet and one hand.)
- Carry tools in a tool belt, pouch, or holster, not in your hands, so you can continue to grip the ladder.
- Wear fully enclosed slip-resistant footwear when using a ladder.
- Do not climb higher than the third rung from the top of the ladder.
- When working from a ladder, try to have the work area directly in front of the ladder. Reaching causes accidents.
- No metal ladder and no ladder reinforced with wire shall be used in the vicinity of any electrical conductor or any electrified equipment or apparatus that may result in a person receiving an electric shock.

Self-Check 6

1. What is scaffolding?
2. List four guidelines for ladder safety.

Summary

- Knowing and following safe practices and procedures will protect you and your coworkers.
- Always disconnect power before you work on a circuit.
- Learn first aid and CPR and be prepared to administer it.
- A current of 50 milliamperes or more through the skin can be life-threatening.
- Occupational Safety and Health Administration (OSHA) is an agency whose mission is to prevent work-related injuries, illnesses, and occupational fatalities by issuing and enforcing standards for workplace safety and health.
- Always maintain a safe distance of no less than 2 feet from trenches or ditches.
- The hazards associated with working in confined spaces are capable of causing bodily injury, illness, and death to the worker.

- The material safety data sheet (MSDS) is a detailed document that provides product users and emergency personnel with information and procedures needed for handling and working with chemicals.

- Lockout/tagout refers to specific practices and procedures to safeguard you and your coworkers from unexpected energization or startup of machinery and equipment, or the release of hazardous energy during service or maintenance activities.

- Personal protective equipment (PPE) is designed to protect you and your coworkers from serious workplace injuries or illnesses.

- Safety (body) harnesses are designed to minimize stress forces on your body in the event of a fall, while providing sufficient freedom to allow work to be performed.

- When it becomes necessary to work in an elevated location, use ladders and scaffolding in accordance with OSHA standards.

- The two main types of ladders are straight and step.

- No metal ladder and no ladder reinforced with wire shall be used in the vicinity of any electrical conductor or any electrified equipment or apparatus that may result in a person receiving an electric shock.

Review Questions

True/False

1. A circuit that has power connected to it is referred to as energized. (True, False)

2. A current through the human body in excess of 10 milliamperes can paralyze a victim. (True, False)

3. OSHA's electrical standards are designed to protect employees exposed to dangers such as shock, electrocution, and fires. (True, False)

4. A confined space has an unlimited means for entry or exit. (True, False)

5. Material safety data sheets are not easily accessed, but are properly filed off site in the Safety Director's corporate office. (True, False)

Multiple Choice

6. Once a piece of equipment is locked out and tagged, which of the following is not part of the procedure to verify that no electrical power is present?
   A. Isolate and dissipate capacitors.
   B. Verify that your test instrument is operational by first testing a known working source.
   C. Use a properly operating power meter that measures the level of electrical energy.
   D. After testing, return the power switch to the OFF position.

7. The OSHA standard titled "Fall Protection" states that the fall arrest system attached to a safety harness must be connected to a point of attachment that is capable of sustaining
   A. 500 pounds
   B. 5000 pounds
   C. 50,000 pounds
   D. 5000 kilograms

8. What type of clothing is often required for maintenance personnel who work with high-power sources such as transformer installations and motor-control centers?
   A. Water-resistant
   B. Fire-retardant
   C. All-weather
   D. Cotton-fiber

9. What generally provides the safest elevated working platforms?
   A. Scaffolds
   B. Stepladders
   C. Building ledges
   D. Straight ladders

10. Regarding safety guidelines when using ladders, you should not climb higher than which rung from the top of the ladder?
    A. First
    B. Second
    C. Third
    D. Fourth

Fill in the Blank

11. Always maintain _____ points of contact with a ladder.

12. Workers must be _____ in confined spaces prior to entering the confined space.

13. Every electrician should know how to _____ first aid and be trained in CPR.

14. _____ shock can be fatal if it interrupts the natural rhythm of the heart.

15. When working with trenches, always erect a _____ that is at least 6 feet tall.

ACTIVITY 3-1   Most Helpful Reading Strategy

What reading strategy did you find to be the most helpful as you read this chapter? Describe this strategy and explain why you find it helpful. Be prepared to share what you wrote with the class.

ACTIVITY 3-2   Summarize CPR Training

You will be going through CPR training. After you complete your CPR training, summarize what you learned.

ACTIVITY 3-3   Confirm That No Electrical Power Is Present

Once a piece of equipment is locked out and tagged, it should be tested to make sure that no electrical power is present. Find a partner. Explain the procedure that should be used.

ACTIVITY 3-4   Explain Ladder Safety

You learned guidelines for ladder safety. Find a partner. Take turns explaining the guidelines for ladder safety.

Chapter 4

Mathematics and the Metric System

Chapter Outline

WHOLE NUMBER OPERATIONS

FRACTIONS AND DECIMAL CONVERSION

INTRODUCTION TO ALGEBRA

METRIC SYSTEM AND UNITS OF MEASUREMENT

THE TAPE MEASURE

Key Terms

Algebra

Algebraic expression

Area

Coefficient

Constants

Digit

Dividend

Divisor

Equation

Even number

Factor

Formula

Fraction

Graduations

Metric system

Odd number

Percent

Percentage

Power

Prime number

Product

Quotient

Signs of operation

Term

Variables

Chapter Objectives

After completing this chapter, you will be able to:

1. Perform basic math operations, using addition, subtraction, multiplication, and division.
2. Convert fractions to decimals.
3. Perform basic operations of algebra, such as solving equations and calculating area.
4. Define the basis of the metric system of measurement.
5. Demonstrate how to read a tape measure.

Life Skills Covered

 Critical Thinking

 Communication Skills

 Take Action

Life Skills Goals

In this chapter, the life skills goals include Critical Thinking, Communication Skills, and Take Action. Your first life skill activity is to practice paraphrasing. Paraphrasing is summarizing and restating the information you just read in your own words. You are going to learn how to perform a variety of mathematical operations. Try to put the steps into your own words. Your second life skill goal is to work on talking about what you are learning with the other students in the class. Find someone in the class to talk to about the steps for a variety of the mathematical operations you will learn in this chapter. Finally, you continue to work on the Take Action life skill as you answer questions throughout the reading.

Can you think of a career that doesn't use mathematics? As a matter of fact, we use some form of math every day of our lives. As an electrician, you are expected to use mathematics on the job. Whether you are installing a raceway, determining a voltage drop, or making sure your paycheck is right, you have to use your math skills. This chapter covers whole number operations, fractions and decimal conversion, an introduction to algebra, the metric system, and using the tape measure. By developing math skills, you increase your chances for finding a successful apprenticeship program and becoming a qualified journeyman.

Introduction

Mathematics is, by definition, the science of numbers. Mathematics is a universal language. You will be successful in mathematics as an electrician, or in any chosen field, *only* if you learn and understand mathematics. Although calculators have become commonplace, everyone should know how to perform the fundamental operations of arithmetic. To be a successful electrician, you must be competent in mathematics! This chapter helps you learn the language of mathematics, starting with whole numbers, then describing essential definitions and technical terms, and ending with practical applications of algebra. Learn the language of mathematics to be successful!

4-1　Whole Number Operations

What does an electrician who is installing outlets and determining lengths of conduit have in common with another electrician who is trying to determine how many 145-foot-long pieces of wire he can get from three 500-foot reels? ANSWER: They both need to use whole number operations. Make these operations part of your mathematics skill set.

Symbols

The Arabic numerals are 0, 1, 2, 3, 4, 5, 6, 7, 8, and 9. They are known as the digits of arithmetic. A digit is a single character in a numbering system.

Digit a single character in a numbering system.

Here are the basic operations of mathematics:

- Addition   +
- Subtraction   −
- Multiplication   ×
- Division   ÷

Signs of grouping are as follows:

- Parentheses ( )
- Brackets [ ]
- Braces { }
- Vinculum ——

Signs that denote order include these:

- Less than <, as 3 < 8 (note that the sign points to the smaller quantity)
- Greater than >, as 7 > 5 (note that the sign points to the smaller quantity)
- Equal to =, as 4 = 2 × 2

Numbers

Arithmetic uses combinations of numbers and symbols for mathematical problems. Counting is used daily by almost everyone. Numbers simply represent things counted. The hours on a clock, speed limit signs, and the horsepower listed on a motor are expressed by numbers. **Figure 4-1** shows an inline ammeter used to measure direct current. Note that numbers are used to represent the level of current.

Mathematical Terms

When you understand the terms associated with mathematics, you are able to perform required operations. You may have heard that "mathematics is an exact science," and

FIGURE 4-1 Inline ammeter.

it is. When you know the terms and perform operations accurately, mathematical problems are easily solved. Here are some basic, but important terms:

Factor any of the numbers (or symbols) that form a product when multiplied together.

Prime number a number that has no factors except itself and 1.

Even number a number exactly divisible by 2.

Odd number a number that is not evenly divisible by 2.

Product the result of multiplying two or more numbers together.

Quotient the result of dividing one number by another.

Dividend a number to be divided.

Divisor a number that divides.

- **Factor** is any of the numbers (or symbols) that form a product when multiplied together. A factor, or a divisor, of a whole number is any other whole number that divides evenly into the number. Therefore, 6 and 3 are factors of 18.
- **Prime number** is a number that has no factors except itself and 1. Examples of prime numbers are 2, 3, 5, 7, and 11.
- **Even number** is a number exactly divisible by 2. Therefore, 2, 4, and 6 are examples of even numbers.
- **Odd number** is a number that is not evenly divisible by 2. Therefore, 1, 3, and 5 are examples of odd numbers.
- **Product** is the result of multiplying two or more numbers together.
- **Quotient** is the result of dividing one number by another.
- **Dividend** is a number to be divided.
- **Divisor** is a number that divides.

LIFE SKILLS

It's time to practice your teamwork skills. Choose a partner. Quiz each other on the mathematical terms and their definitions.

Addition of Whole Numbers

Numerical units with no fractional parts are referred to as whole numbers. Addition is the mathematical operation that represents combining collections of objects together into a larger collection, or the process of finding the sum (total) of two or more numbers. Whole numbers are added by placing them in a column with the numbers aligned on the right side of the column. The column farthest to the right is added first. The last digit of the sum is written in the answer. The remaining digit is carried to the next column and added. This procedure is followed until all columns have been added.

Example
What is the sum of the following numbers: 15 + 6 + 217 + 339 + 171?

2	**12**	**12**
15	15	15
6	6	6
217	217	217
339	339	339
+171	+171	+171
8	48	748

Practical Example

You are asked to install outlets in a TV repair shop. The first outlet is 36 inches from the panel. The second outlet is 96 inches from the first and the third outlet is 72 inches from the second. How much conduit will you need?

$$\begin{array}{r} 36 \\ 96 \\ +72 \\ \hline 204 \end{array}$$ *inches of conduit*

LIFE SKILLS

Find a partner, and explain to your partner what you learned about adding whole numbers.

Subtraction of Whole Numbers

Finding or calculating the difference between two numbers is referred to as subtraction. The smaller of the two numbers is placed below the larger, keeping the right column of numbers aligned.

Example

Subtract 241 from 673.

$$\begin{array}{r} 673 \\ -241 \\ \hline 432 \end{array}$$

Borrowing

In subtracting whole numbers, it is sometimes necessary to "borrow" from the number in the adjacent column. When this is done, the amount borrowed must be in increments of value of the column borrowed from. What this means is that starting from the right, the first column represents units or 1s, the second column represents 10s, the third column represents 100s, the fourth column represents 1000s, and so on. The number 7642 could actually be rewritten as 1000 seven times, 100 six times, 10 four times, and 1 two times.

Example

$$\begin{array}{r} 7642 \\ -5595 \\ \hline \end{array}$$

In this example, 5 cannot be subtracted from 2. Therefore, the 2 must borrow from the 4 in the column adjacent to it. Because the 4 is in the 10s column, 10 is borrowed, leaving 30 in that column. The borrowed 10 is added to the original 2, making 12 (10 + 2 = 12). Now, 5 can be subtracted from 12, leaving a difference of 7.

$$
\begin{array}{r}
763 \ \textbf{(12)} \\
-559 \ \ \textbf{(5)} \\
\hline
7
\end{array}
$$

In the next column, 90 must be subtracted from 30. Because this is not possible, 100 is borrowed from 600 in the adjacent column and added to the existing 30, making 130. (This now leaves 5 in the 100s column.) The difference is 40.

$$
\begin{array}{r}
75 \ \textbf{(130)} \ 2 \\
-55 \ \ \textbf{(90)} \ \ 5 \\
\hline
4 \ \ \ 7
\end{array}
$$

In the third column, 500 is subtracted from 500, leaving a difference of 0.

$$
\begin{array}{r}
7642 \\
-5595 \\
\hline
047
\end{array}
$$

In the fourth column, 5000 is subtracted from 7000, leaving a difference of 2000.

$$
\begin{array}{r}
7642 \\
-5595 \\
\hline
2047
\end{array}
$$

Practical Example

You are on a jobsite and have a 500-foot reel of wire. You use 270 feet from the reel. How much wire is still on the reel?

$$
\begin{array}{r}
500 \\
-270 \\
\hline
230
\end{array}
$$
feet of wire is still on the reel

Borrow "1" Method

Borrow "1" is another borrowing procedure used in the subtraction of whole numbers. This method's name is a bit misleading because 1 can be borrowed only from the units column. Many people, however, use this method and find it easier to understand.

Example

$$
\begin{array}{r}
721 \\
-46 \\
\hline
\end{array}
$$

Because 6 cannot be subtracted from 1, 1 is borrowed from the adjacent 2. The 1 now becomes 11, and the 2 becomes 1.

$$
\begin{array}{r}
\textbf{71 (11)} \\
-4 \ \ 6 \\
\hline
5
\end{array}
$$

The 4 must now be subtracted from the 1 in the second column. Because 4 cannot be subtracted from 1, 1 is borrowed from the adjacent 7, and 1 becomes 11. The 7 now becomes a 6.

$$\begin{array}{r} \textbf{6 (11) 1} \\ -4\ \ 6 \\ \hline 6\ \ \ 7\ \ \ 5 \end{array}$$

LIFE SKILLS

Paraphrase what you just read about subtracting whole numbers.

Multiplication of Whole Numbers

The mathematical operation of multiplication is a method of addition used when like numbers are added. If three 4s are added, the answer is 12. If the number 4 is multiplied by 3, the answer (known as the product) is equal to 12. Therefore, 3×4 is the same as adding three 4s.

$$\begin{array}{r} 4 \\ 4 \\ +4 \\ \hline 12 \end{array} \qquad \begin{array}{r} 4 \\ \times 3 \\ \hline 12 \end{array}$$

When multiplying larger numbers, first write the number to be multiplied. Then underneath it, write the number of times it is to be multiplied.

Example

$$153\ \times\ \ 25$$

$$\begin{array}{r} \textbf{1} \\ 153 \\ \times 25 \\ \hline 5 \end{array}$$

(*Note:* $5 \times 3 = 15$; 5 is listed below the 5 and the 1 is carried to the next column. Also note that the 1 carried over is actually 10 carried to the next column, which is the 10s column.)

Then multiply ($5 \times 5 = 25$) and add the 1; ($25 + 1 = 26$). Because the 25 is in the 10s column, the answer is actually ($250 + 10 = 260$).

$$\begin{array}{r} \textbf{2} \\ 153 \\ \times 25 \\ \hline 65 \end{array}$$

(*Note:* List 6 below the 2, and carry the 2 to the next column.)

Then multiply (5 × 1 = 5). The 1 is located in the 100s column (5 × 100 = 500); then add the 2 (5 + 2 = 7). Because the 7 is in the 100s column, the answer is actually 500 + 200 = 700. Place the 7 beside the 6. Next, multiply each digit of 153 by 2 in 25. The answer is brought down in the same manner except that one space is skipped (*Note:* The reason for skipping the space is because the 2 in 25 is actually 20, not 2).

Once each number is multiplied, the final step is to add the two sets of products together to obtain the total. The *final answer* after all numbers are multiplied and (products) added is 3825.

Practical Example

Your supervisor told you that you and your crew should be able to wire a two-bedroom house in 4 hours and a three-bedroom house in 6 hours. If his estimates are accurate, how long will it take you and your crew to wire 6 two-bedroom houses and 4 three-bedroom houses?

First: 6 × 4 = 24
       4 × 6 = 24
Then: 24 × 24 = 48 *hours*

LIFE SKILLS

Explain to a student in the class what you learned about multiplying whole numbers.

Division of Whole Numbers

Multiplication is adding a number to itself a specified amount of times. Division is the inverse, or opposite, of that process. Division is the process of subtracting a smaller number from a larger number a specified amount of times. The number to be divided is larger and referred to as the dividend. The number used to indicate the number of times the dividend is to be divided is called the divisor. The answer is referred to as the quotient. The process of division begins with the dividend being placed inside the division bracket; the divisor is placed to the left of the dividend. The answer, or quotient, is placed above the dividend.

$$\text{Divisor } \overline{)\text{Dividend}}^{\text{Quotient}}$$

Example
Divide 675 by 14.

$$14\overline{)675}^{4}$$
$$\underline{-56}$$
$$11$$

Note that the number 675 is placed under the division bracket and the number 14 to the left of it. The number 14 cannot be divided into a number that is smaller than itself. Therefore, 14 is divided into 67 first. The number multiplied by 14 that is closest to 67 must be determined (*Note:* This product must not exceed 67). Four is that number (4 × 14 = 56). Place 4 in the quotient directly over 7 in the dividend. The number 56 is placed below 67 and subtracted from it. This leaves 11. Because 14 cannot be divided into 11, the next number in the dividend is brought down to the right of 11. When the 5 is placed beside 11, 14 is then divided into 115.

$$
\begin{array}{r}
48 \\
14\overline{)675} \\
-56 \\
\hline
115
\end{array}
$$

The nearest number that 14 can be multiplied by and not go over 115 is 8 (8 × 14 = 112). Place 8 in the quotient directly over 5 in the dividend. Then place 112 below 115 and subtract from it. This leaves a remainder of 3. Because there are no more numbers in the dividend, the 3 is taken to the quotient and shown as R3. This signifies a remainder or 3. The *final answer* is 48 R3.

Practical Example

You ordered 2000 feet of BX cable. The cable is shipped in 250-foot coils. How many coils are shipped?

$$
\begin{array}{r}
8 \ coils \\
250\overline{)2000} \\
-2000 \\
\hline
0
\end{array}
$$

Paraphrase what you just read about subtracting whole numbers.

Self-Check 1

1. What is the sum of the following numbers: 12 + 294 + 528?

2. Subtract 289 from 734.

3. Multiply 234 by 57.

4. Divide 438 by 16.

4-2 Fractions and Decimal Conversion

Calculations with fractions are essential to the solution of many problems. Accurate measurements often require us to be more precise than whole numbers allow. Today, electronic calculators are used for many types of calculations. However, they should not replace a thorough understanding of the use of fractions. There are different ways to work with fractions. For an electrician, the easiest way to work with fractions is to change them into decimals.

Definitions

The number 8 divided by 4 gives an exact quotient of 2. This can also be written as 8/4 = 2. However, if you attempt to divide 3 by 4, you are unable to calculate an exact quotient. Dividing 3 by 4 can also be listed as ¾ and read as three-fourths. This is a

Fraction a number
that can represent
part of a whole.

fraction. A fraction is a number that can represent part of a whole. The fraction ¾ represents a number, but not a whole number.

Fourths signify that a unit is divided into four equal parts. The fraction ¾ indicates that three of the four parts are included in the number.

The divisor, or the number below the line in a fraction, is called the denominator. The dividend, or the number above the line in a fraction, is called the numerator. The denominator specifies how many parts the unit is divided into. The numerator tells us how many of these parts are included.

In **Figure 4-2**, a ruler is shown to have four equal parts per inch. Three-fourths of an inch is three of these equal fourths. Three-fourths of an inch is therefore marked as ¾. Counting those equal parts from the left of the ruler, the third part (3/4) is three-fourths of an inch. A measurement is illustrated by the arrow above the left side of the ruler. Note that the length measured is 3¼ inches. That is, 3 inches and one part (¼-inch long).

Fraction-to-Decimal Conversion

Like most mathematical operations, the simplest method to convert a fraction to its equivalent decimal is to use a calculator. Simply divide the top (numerator) of the fraction by the bottom (denominator), and read off the answer.

FIGURE 4-2 Ruler.

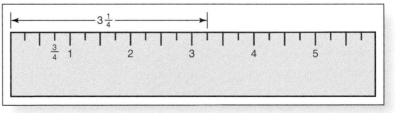

© Cengage Learning 2013

Example 1

Convert 5/8 to its decimal equivalent.

With calculator in hand, type in "5 / 8 ="

The displayed answer should be 0.625.

When it is necessary to convert a fraction to a decimal manually, follow these steps:

1. Determine a number that can be multiplied by the bottom of the fraction to make it 10, 100, 1000, or any 1 followed by 0s.
2. Multiply both top (numerator) and bottom (denominator) by that number.
3. Write down just the top number. Be sure to put the decimal place in the correct spot (one space from the right for every zero in the bottom number).

Example 2

Convert ¾ to its decimal equivalent.

1. Multiply 4 by 25 to produce 100.
2. Multiply top (numerator) and bottom (denominator) by 25:

   $3 \times 25 = 75$

   $4 \times 25 = 100$

3. Write down 75 with the decimal place two spaces from the right because 100 has TWO zeros. ANSWER = 0.75

Example 3

Convert 1/3 to its decimal equivalent.

1. Because there is no way to multiply 3 to become 10, 100, or any 1 followed by zeros, we must calculate and approximate the decimal by selecting, for example, 333.
2. Multiply top (numerator) and bottom (denominator) by 333.

   $1 \times 333 = 333$

   $3 \times 333 = 999$

3. Because 999 is approximately 1000, write down 333, with the decimal place three spaces from the right because 1000 has THREE zeros. ANSWER = 0.333

   (*Note:* This answer is only accurate to three decimal places.)

LIFE SKILLS

Find a partner. Explain to your partner how to convert a fraction to a decimal.

Decimal Fractions

Calculations using fractions often take considerable time and can be cumbersome. Decimals greatly reduce the time required for calculations. Decimals streamline calculations and are essential in both the metric system and the use of electronic calculators.

A decimal fraction is a fraction written with a denominator of 10 or a multiple, or power of 10. Therefore, 3/10, 19/100, 64/1000, 3521/1000, and 2782/10,000 are examples of decimal fractions.

When writing a decimal fraction, omit the denominator and indicate the denominator by placing a period (.), called a decimal point, in the numerator; there are as many digits to the right of this point as there are zeros in the denominator. For example, 3/10 is written 0.3, 19/100 is written 0.19, 64/1000 is 0.064, 3521/1000 is 3.521, and 2782/10,000 is 0.2782.

When working with decimals, most texts place a zero to the left of the decimal point when there is no whole number figure. So, .32 may be written 0.32. This notation is used in this text.

When there are fewer figures in the numerator than there are zeros in the denominator, zeros are added to the left of the figures to make up the required number. So, 57/1000 = 0.057 and 2/10,000 = 0.0002.

Percent and Percentages

Percentage a way of expressing a number or value as a fraction of 100.

Percent hundredths, or number per 100.

A **percentage** is a way of expressing a number or value as a fraction of 100. **Percent** means hundredths, or number per 100. One hundred percent (100%) means *all* of a number or value. For example, if an electrician has 100% of the 100 receptacle boxes that he just purchased, he has 100 boxes. If, on the other hand, he has 50% of the boxes, then he has one-half, or 50, receptacle boxes. If this same electrician has 25% of the boxes, he has one-fourth (or one-quarter) of the receptacle boxes, or 25 boxes.

Electricians use percentages in many instances, such as application to electrical circuits. For example, the current increased by 15% because a load decreased in value; or a motor may be delivering only 75% of its rated capacity. Examples of percentages when using the *NEC* could mean specifying the number of wires allowed in conduits or the overload sizes for equipment protection.

Changing a Decimal Fraction to a Percent

In the previous section, we worked with decimal fractions. To change a decimal fraction to a percent, move the decimal point two places to the right (multiply by 100), and add a percent sign.

Example
Express 0.20 as a percent.
Move the decimal point two places to the right, and add a percent sign.
0.20 = 20%.

Changing a Common Fraction to a Percent

Before we change a common fraction to a percent, the common fraction must be changed to a decimal fraction by dividing the numerator by the denominator. Once we have a decimal fraction, we move the decimal point two places to the right, and add a percent sign, just as we did in the previous example.

> *Example*
> Express ¾ as a percent.
> Divide the numerator by the denominator. 3 divided by 4 = 0.75.
> Move the decimal point two places to the right, and add a percent sign.
> 0.75 = 75%.

Changing a Percent to a Decimal Fraction

To change percent to a decimal fraction, move the decimal point two places to the left (divide by 100), and drop the percent sign.

> *Example*
> Express 8% as a decimal fraction.
> Move the decimal point two places to the left, and drop the percent sign.
> 8% = 0.08.

LIFE SKILLS

Find a partner. Explain to your partner how to change a fraction to a percent. Then have your partner explain to you how to change a percent to a decimal fraction.

Self-Check 2

1. Convert 7/8 to its decimal equivalent.
2. List the decimal equivalent for 3/10,000.
3. Express 0.65 as a percent.

4-3 Introduction to Algebra

Electricians need to know how to work with and rearrange formulas in order to solve problems involving electrical quantities. Many times, the relationship between these quantities is expressed in an algebraic equation. To solve these problems, you must be familiar with problems expressed in algebraic form. Algebra requires you to systematically apply a few simple rules. There is a lot of repetitive drill in math classes to establish the habit of *thinking systematically* about problems. After working through simple algebra problems, you will be able to apply that same sort of reasoning to real-life problems. This section covers basic algebra, types of numbers, signs of grouping, expressions, solving equations, and calculating areas.

Basic Algebra

Algebra a branch of mathematics that uses letters or other symbols to represent numbers, with rules for manipulating these symbols.

Algebra is a branch of mathematics that uses letters or other symbols to represent numbers, with rules for manipulating these symbols. Algebra enables us to solve problems by simple operations; it is more difficult to use arithmetic with only numbers in certain situations. Algebra is the arithmetic of literal expressions. The symbols that make up these literal expressions and formulas should not be new to you. An example of a formula that you used in elementary school is the formula for the area of a rectangle:

$$\text{Area} = \text{length} \times \text{width}$$

Written as a formula it becomes:

$$A = lw$$

However, when using algebra, the multiplication sign ($\times$) is not used. The multiplication sign could be confused with the letter x. Letters are written next to each other to denote multiplication. The letters in the above formula for finding the area of a rectangle assume different values.

> *Example 1*
> Find the area of a rectangle if its length is 15 cm and its width is 7 cm.
> $$\text{Length} = 15 \text{ cm}$$
> $$\text{Width} = 7 \text{ cm}$$
> $$A = lw$$
> $$A = (15)(7)$$
> $$A = 105 \text{ cm}^2$$

Variables those letters or literal symbols that can be assigned different values.

The letters A, l, and w in the formula above are referred to as variables. **Variables** are those letters or literal symbols that can be assigned different values. Letters representing numbers that do not vary are called **constants**. In the following example, a constant is part of the formula.

Constants letters representing numbers that do not vary.

> *Example 2*
> The formula for calculating current when power and voltage are given is
> $$I = P/E$$
> What is the current when a 2500-watt electric iron is connected to a 240 voltage source?
> $$I = P/E$$
> $$I = 2500/240$$
> $$I = 10.4 \text{ Amperes, or } 10.4 \text{ A}$$

LIFE SKILLS

Paraphrase what you just read about algebra. What are the most important points that were made?

Numbers

Generally, numbers are obtained by counting a group of objects or by measuring something. Therefore, the measure of an object, like the width of a room, is the number of times it is contained in the unit of measure. If we use 1 foot as the unit of measure, the measure of the width of a room 15 feet wide is the number 15.

Definite Numbers

The numerals 0, 1, 2, 3, and so on, have definite meanings. For example, the symbol "5" represents the idea "five," which means a quantity of five of something. This five might be 5 ft, 5 cm², $5, or 5 of any units. Five is a definite number.

General Numbers

General numbers are represented by letters. For example, in the formula used to determine the area of a rectangle A = bh, A represents the area (in square units), b the length of the base, and h the height of the rectangle. The letters A, b, and h are referred to as general numbers. The formula A = bh is for calculating the area of any rectangle.

Signs of Operation and Grouping

Signs of operation in algebra, the same signs and meanings as used in arithmetic (+, −, ×, and ÷).

The **signs of operation** in algebra are the same signs and meanings used in arithmetic. These signs are +, −, ×, and ÷. However, when using algebra, the multiplication sign (×) is not used; a multiplication is indicated between letters. In algebra, multiplication is signified by using the center dot (·) or parentheses (b)(h), or no sign is used between the letters (e.g., bh).

Grouping symbols organize an algebra expression that contains multiple groups; they help establish the order used to apply math operations. Grouping symbols include the following:

- Parentheses ( )
- Brackets [ ]
- Braces { }
- Vinculum ——

The first three symbols of grouping are placed around the parts grouped. The vinculum is usually placed over what is grouped. They all signify that the parts enclosed are to be taken as a single quantity. When performing operations in a problem containing the symbols of grouping, the operations within the grouping signs must be performed first.

Algebraic expression uses signs and symbols to represent numbers or quantities.

Algebraic Expressions

An **algebraic expression** uses signs and symbols to represent numbers or quantities. A numerical algebraic expression is made up entirely of numerals and signs. A literal algebraic expression includes both letters and numbers separated by signs.

Here's an example of a numerical algebraic expression:

$12 + 11 - (2 + 7)$

Here's an example of a literal algebraic expression:

$7xy + 3z$

Coefficients

Coefficient a multiplier of a term in a formula.

In the expression 5xyz, the 5, x, y, and z are factors of this expression. Any one of these factors or the product of any two or more of them is the coefficient of the remaining part. A coefficient is a multiplier of a term in a formula. What this means is that 5xy may be considered the coefficient of z or 5x, the coefficient of yz. Usually, though, when working with algebraic expressions, only the numerical portion is considered the coefficient. (*Note:* If no numerical quantity is expressed, 1 is understood to be the numerical coefficient.)

Powers and Exponents

Power a mathematical notation indicating the number of times a quantity is multiplied by itself.

When all of the factors in a product are equal, for example $x \cdot x \cdot x$, the product of the factors is referred to as a power of one of them. A power is a mathematical notation indicating the number of times a quantity is multiplied by itself. The form $x \cdot x \cdot x$ is written x^3. The small number (superscript) above and to the right denotes how many times x is used as a factor. In the expression x^3, x is referred to as the base, and the superscript 3 is the exponent, which is sometimes called the power.

Terms

Term the part of an algebraic expression not separated by a plus or minus sign.

In an algebraic expression, the part of the expression not separated by a plus or minus sign is called a term. The + or − sign that precedes a term is considered a part of that term. For example, in the algebraic expression $2xy + 5ab - c$, the terms are $+2xy$, $+5ab$, and $-c$. Algebraic expressions having various numbers of terms are defined as follows:

- A *monomial* is an algebraic expression of one term. 2xyz is a monomial expression.

- A *binomial* is an algebraic expression of two terms. $5a + 4x$ is a binomial expression.

- A *trinomial* is an algebraic expression that consists of three terms. $2a - 5b + 3c$ is a trinomial expression.

- A *polynomial* is any algebraic expression with two or more terms. A polynomial is sometimes referred to as a multinomial.

Like, or similar, terms are exactly the same or differ only in their numerical coefficients. Examples of expressions with like terms are $5x^2z$, $-8x^2z$, and $11x^2z$. Terms that differ other than in their coefficients are unlike or dissimilar terms. Examples of dissimilar terms are $3a^2b^3$ and $-9ab^3$.

LIFE SKILLS

Choose a partner. Quiz each other on the different types of algebraic expressions.

Translation into Algebraic Expressions

The importance of algebra and its properties is realized when solving practical, real-life problems. To put the process of algebra to work, the ability to translate from sentences and phrases to mathematical expressions must be mastered. Very often in practical problems, formulas are stated in words and then translated into algebraic equations. For example, the formula for finding the voltage dropped across a load (i.e., resistance) is voltage equal to current times resistance. This statement is translated to $E = I \times R$ or $E = IR$, where E = voltage, I = current, and R = resistance. Here is another example of translating a statement into an algebraic expression.

Example
The sum of two numbers doubled equals 40.
Solution: Let a and b represent the two numbers. Then, $2(a + b) = 40$.

Solving Equations

Equation a mathematical statement that asserts the equality of two expressions.

Electricians often need to solve problems involving electrical quantities. Generally, the relationship between these quantities is expressed in an algebraic **equation**. An equation is a mathematical statement that asserts the equality of two expressions. To solve such a problem, you must be familiar with algebraic equations. You must also be able to identify the factors in these equations and apply them accordingly.

You will use the basic algebraic equation throughout your career in the electrical industry. To perform successfully as an electrician, you must develop fundamental skills in this area.

Example 1
$A = 4$
$B = 6$
$C = A + B$
$C = 10$

Example 2
$X = 4$
$Y = 2X + 6$
$Y = 14$

Example 3

$E = I \times R$

$E = 240$ V

$R = 80\ \Omega$

$I = ?$

$E = I \times R$

$I = E/R$

$I = 240/80$

$I = 3$ A

Area and Calculations

Area the amount of surface of an object or the amount of material required to cover the surface.

Area is the amount of surface of an object or the amount of material required to cover the surface. Area is measured in square units. The basic example of area is a square whose sides are each one linear unit. Thus, a square foot is the area of a square that is 1 foot on a side. A square centimeter is a square that is 1 cm on a side, and so on. Table 4-1 below lists area measurements for the English system.

Table 4-2 lists area measurements for the Metric system; Table 4-3 shows English–Metric equivalents of length measurements.

TABLE 4-1 English Area Measure

144 square inches (sq. in.)	= 1 square foot (sq. ft)
9 square feet (sq. ft)	= 1 square yard (sq. yd)

© Cengage Learning 2013

TABLE 4-2 Metric Area Measure

100 square millimeters (mm²)	= 1 square centimeter (cm²)
100 square centimeters (cm²)	= 1 square decimeter (dm²)
100 square decimeters (dm²)	= 1 square meter (m²)
100 square meters (m²)	= 1 square dekameter (dam²)
100 square dekameters (dam²)	= 1 square hectometer (hm²)
100 square hectometers (hm²)	= 1 square kilometer (km²)

© Cengage Learning 2013

TABLE 4-3 English–Metric Equivalents Length Measure

1 square inch (sq. in.)	= 645.16 square millimeters (mm²)
1 square inch (sq. in.)	= 6.4516 square centimeters (cm²)
1 square foot (sq. ft)	= 0.092903 square meter (m²)
1 square yard (sq. yd)	= 0.836127 square meter (m²)
1 square millimeter (mm²)	= 0.001550 square inch (sq. in.)
1 square centimeter (cm²)	= 0.15500 square inch (sq. in.)
1 square meter (m²)	= 10.763910 (sq. ft)
1 square meter (m²)	= 1.119599 (sq. yd)

© Cengage Learning 2013

Area Calculations

Formula a rule
expressed in
letters, symbols,
and constant terms.

The measurement of area is a two-dimensional quantity. Area is the product of its length and its width. Formulas are used to determine exact measurements. A **formula** is a rule expressed in letters, symbols, and constant terms. The measurement of area when analyzing a rectangle is found by multiplying length and width. The formula is:

$$A = l \times w$$

Figure 4-3 lists the formulas for determining the area of a circle.

Figure 4-4 lists the formulas used to find the area of a rectangle.

Figure 4-5 lists the formulas used to find the area of a square.

Figure 4-6 lists the formulas used to find the area of a trapezoid.

Figure 4-7 lists the formulas used to find the area of a triangle.

FIGURE 4-3   Formulas used to find the area of a circle.

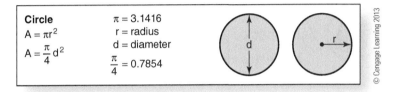

FIGURE 4-4   Formulas used to find the area of a rectangle.

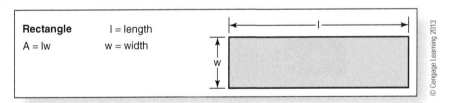

FIGURE 4-5   Formulas used to find the area of a square.

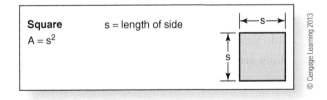

FIGURE 4-6   Formulas used to find the area of a trapezoid.

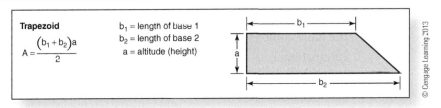

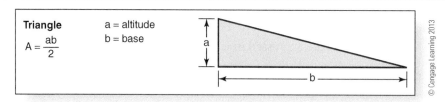

FIGURE 4-7 Formulas used to find the area of a triangle.

Triangle

$A = \dfrac{ab}{2}$

a = altitude
b = base

© Cengage Learning 2013

LIFE SKILLS

Paraphrase what you just read about area calculations.

Self-Check 3

1. What is algebra?
2. If A = 6 and B = 3A + 5, what does B equal?
3. If P = I × E, and I = 5A and E = 120V, what does P equal (in W)?
4. What is the definition of area?
5. What formula is used to determine the area of a rectangle?
6. What is the area of a square if one side of the square is 12 feet?

4-4 Metric System and Units of Measurement

The simplicity of the metric system is based on the fact that there is only one unit of measurement (or base unit) for each type of quantity measured. The most common base units in the metric system are the meter, gram, second, and liter. Each type of measurement has a base unit to which prefixes are added to indicate multiples of ten. For example, instead of saying 440,000 watts of power, we can say 440 thousand watts or 440 kilowatts (the prefix *kilo* means 1000). Since the metric system is based on multiples of ten, converting within the system is simple. Metric and English system values are also compared so conversion can be understood.

Metric System

Metric system
an international decimalized system of weights and measures.

The metric system is an international decimalized system of weights and measures used throughout the scientific community. The United States is moving toward using the metric system. In 1960, the metric system, started in France in 1790, became the more complete International System (abbreviated SI, which stands for Système International).

The metric system is based on units of 10. This is the same base as our number system, the decimal system. The U.S. system is based on numbers such as 12 and 16. The old English system was based on these numbers; they were used because they had several divisors and were easy to work with as fractions. With the proliferation of electronic calculators, the division of fractions is no longer important.

The most commonly used metric units are deka, hector, kilo, deci, centi, and milli. Deka means 10, hector means 100, and kilo means 1000. Deci means 1/10, centi means 1/100, and milli means 1/1000. Standard units of metric measure are shown in Table 4-4 below. A kilovolt, for instance, is 1000 volts and a milliampere is one-thousandth (1/1000) of an ampere.

Metric units progress in steps of ten; electrical measurements progress in steps of 1000. These units of measure are typically called engineering units. The most commonly used engineering units above the base unit are kilo, mega, and terra. Engineering units below the base unit are milli, micro, nano, and pico. These standard engineering units are shown in Table 4-5 below.

In the metric system, the fundamental unit of length is the meter. From the meter, the unit of capacity derived is the liter; from the unit of mass, the gram; and from the unit of area in measuring land, the are. All other units are decimal subdivisions or

TABLE 4-4 Standard Units of Metric Measure

Kilo	1000
Hecto	100
Deka	10
Deci	1/10 or 0.1
Centi	1/100 or 0.01
Milli	1/1000 or 0.001
Base unit 1	

© Cengage Learning 2013

TABLE 4-5 Standard Engineering Units

Engineering	Symbol	Multiply by Unit
Tera	T	$1,000,000,000,000 \times 10^{12}$
Giga	G	$1,000,000,000 \times 10^{9}$
Mega	M	$1,000,000 \times 10^{6}$
Kilo	K	$1,000 \times 10^{2}$
Base Unit		1
Milli	m	0.001×10^{-2}
Micro	µ	0.000001×10^{-6}
Nano	n	$0.000000001 \times 10^{-9}$
Pico	p	$0.00000000001 \times 10^{-12}$

© Cengage Learning 2013

multiples of these measures. For example, 1 cubic decimeter (cu dm) = 1 liter; 1 liter of water = 1 kilogram (kg); and 1 are = 100 square meters (*Note:* Are was defined by older forms of the metric system, but is now outside the modern International System, or SI).

Units of Measurement

The seven basic types of measurement most commonly made are illustrated in Table 4-6 below. Note that there is an SI base unit for each of these measurements.

The English System

Table 4-7 shows a listing of the relationships between measures of length in the English system of measurements.

Metric and English Systems Relationships

The procedure for converting measures of length from one system to another is based on the relationships in Table 4-8.

TABLE 4-6 SI Base Units

Measurement	Abbreviation	Metric Unit
Length	m	meter
Mass	kg	kilogram
Time	s	second
Electric current	A	ampere
Temperature	K	kelvin
Light intensity	cd	candela
Molecular substance	mol	mole

© Cengage Learning 2013

TABLE 4-7 English System Length Relationships

12 inches	equal	1 foot
3 feet	equal	1 yard
5 1/5 yards	equal	1 rod
320 rods	equal	1 mile
5280 feet	equal	1 mile
1760 yards	equal	1 mile

Note: Lengths of less than 1 inch are expressed as fractions or decimals.
© Cengage Learning 2013

TABLE 4-8	Metric System Length Relationships	
1 inch	equals	2.54 cm
1 foot	equals	0.3048 m
1 yard	equals	0.9144 m
1 meter	equals	39.37 in.

© Cengage Learning 2013

Conversion between any two standards is possible if the ratio between them is known.

LIFE SKILLS

Paraphrase what you just read about the metric system.

Self-Check 4

1. The metric system is based on units of what?

2. Rewrite the following using metric units (prefixes):

    a. 200,000 volts

    b. 64,000 watts

    c. 0.067 ampere

3. Convert the following to metric:

    a. 8 inches = _____ cm

    b. 2 feet = _____ m

    c. 5 yards = _____ m

    d. 1 meter = _____ inches

4-5 The Tape Measure

The tape measure allows electricians to measure walls, wire, lengths, and widths in order to do their work accurately. As an electrician, you have to make precise measurements when installing runs of conduit or marking locations for switches, wall boxes, and luminaire. Learn about this tool and *how* to use it. Remember, the tape measure is the device behind the "measure twice, cut once" adage. Mistakes can be costly. In other words, "measure twice, *buy* once!"

The Tool

The tape measure is one of the most useful tools in a tradesman's toolbox. It is an instrument consisting of a narrow strip (cloth or metal) marked in inches or centimeters and is used for measuring. **Figure 4-8** shows a picture of a tape measure.

The tape measure can be marked with U.S. or English measurements (feet and inches), metric measurements (meters and parts of meters: centimeters, millimeters, etc.), or both, shared on the same tape. To make tape measures easy to use, most have common elements. For example, most U.S. tape measures have both inch and foot marks. The 16-inch, 32-inch, 48-inch, 64-inch, and so on, markers are marked differently on various measures. For instance, some measures have small black arrows or pointers with a box around the number; other tape measures have these numbers highlighted in red. These numbers refer to the 16-inch centers for laying out studs and joists (*Note:* Most carpenters have the 16-inch centers memorized up to 8 feet, but appreciate the help on the tape measure).

Reading the Tape Measure

Knowledge of mathematics, especially fractions, is necessary when reading a tape measure. The fraction ½ is the same as ⁴⁄₈, ⁸⁄₁₆, ¹⁶⁄₃₂, and so on. The ½-inch mark could be referred to as the 16/32-inch mark but, generally, the simplest numbers are used. For example, since 16/32 is ½, when looking for 15/32 inches on the tape measure, it is the first 32nd mark before the ½-inch mark.

Figure 4-9 illustrates tape measure increments.

Graduations
linear measure markings on a tape measure.

Refer to Figure 4-9, and note that the size of the graduations on the tape measure change with the fraction increment. Graduations are linear measure markings on the tape measure. Other than the inch mark, the ½-inch graduation is the longest. The next longest graduation is for ¼ and ¾ inches, then eighths, and then sixteenths. The shortest graduations are the thirty-seconds. These divisions are on most tape measures. However, some tape measures include tenths and divide down to one-hundredths.

FIGURE 4-8 Tape measure.

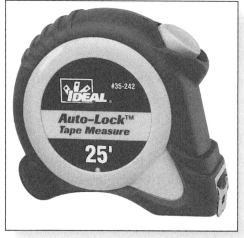

FIGURE 4-9   Tape measure increments.

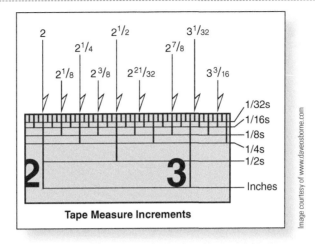

Tape Measure Increments

Image courtesy of www.daveosborne.com

Most tape measures used by electricians are the type where 1/16 of an inch is the smallest measurement on the tape. With this type of tape measure, the distance between every line on the tape is 1/16 of an inch.

Understand that 1/8 inch is twice as big as 1/16 of an inch. On the type of tape referenced above, 1/8 of an inch is every other mark. The ¼-inch mark is every four marks on this type of tape measure. The ½-inch mark is one-half of an inch; inches are the long lines on the tape measure that cross either half or all of the 1-inch width on the tape. The 1-inch marks are usually preceded or followed by numbers.

Making a Measurement

When making a measurement with a tape measure, use the following procedure:

- Hold the front of the tape at the point you wish to start the measurement from, and extend it to the point where you want to stop.
- Read the first large number before your stop point. This shows you the number of inches.
- Read the smaller lines of various sizes until your stop point (*Note:* This is why knowing and remembering what the various lines on a tape represent is important). *This reading is your fraction of an inch.*

Example
Measure a space on your wall.

- Place the front of the tape at the start point, and extend it until after you reach the stop point.
- Look at the tape and note the last number before the stop point. Let's say 13.
- After the number 13, count four lines, stopping on the third longest line.
- That makes the total measured space 13 and ¼ of an inch.

LIFE SKILLS

Find a partner. Explain the procedure for making a measurement using a tape measure to your partner.

Self-Check 5

1. Why is a tape measure important to an electrician?

2. On tape measures most often used by electricians, what is the smallest measurement (increment) on the tape?

3. Explain (briefly) how a measurement is made using a tape measure.

Summary

- The basic operations of mathematics are addition, subtraction, multiplication, and division.

- Arithmetic uses combinations of numbers and symbols when working with mathematical problems.

- Addition is the process of finding the sum of two or more numbers.

- Subtraction is finding the difference between two numbers.

- Multiplication is basically a method of addition that is used when like numbers are added.

- Division is the process of subtracting a smaller number from a larger number a specified number of times.

- A fraction is a number that can represent part of a whole.

- To convert a fraction to a decimal, divide the numerator by the denominator.

- A decimal fraction is a fraction written with a denominator of 10, or a multiple or power of 10.

- A percentage is a way of expressing a number or value as a fraction of 100.

- Algebra is a type of symbolic arithmetic that enables us to solve problems by simple operations that is difficult using arithmetic with only numbers.

- An algebraic expression uses signs and symbols to represent numbers or quantities.

- An equation is a mathematical statement that asserts the equality of two expressions.

- Area is the amount of surface of an object or the amount of material required to cover the surface.

- The metric system is an international decimalized system of weights and measures based on the meter, the kilogram, liter, and the second.
- The tape measure is an instrument consisting of a narrow strip (cloth or metal) marked in inches or centimeters and is used for measuring.
- Knowledge of mathematics, especially fractions, is necessary when reading a tape measure.

Review Questions

True/False

1. The Arabic numerals 0 through 9 are known as widgets.   (True, False)

2. A prime number in arithmetic is a number that has no factors except itself and 1.   (True, False)

3. In subtraction of whole numbers, it is sometimes necessary to borrow from the number in the adjacent column.   (True, False)

4. The divisor, or the number below the line in a fraction, is called the numerator.   (True, False)

5. The measurement of area when analyzing a rectangle is found by multiplying length and width.   (True, False)

Multiple Choice

6. The formula used to find the area of a triangle is
   A. $A = s^2$
   B. $A = ab/2$
   C. $A = lw$
   D. $A = \pi r^2$

7. When using algebra, what are the letters or literal symbols that can be assigned different values?
   A. Terms
   B. Binomials
   C. Constants
   D. Variables

8. Which of the following is not a sign of grouping?
   A. Brackets
   B. Division
   C. Parentheses
   D. Vinculum

9. When all the factors in a product are equal, the product of the factors is referred to as a _____.
   A. Term
   B. Coefficient
   C. Power
   D. Constant

10. On a tape measure, which of the following is the smallest increment?
    A. ½ inch
    B. ¼ inch
    C. ⅛ inch
    D. 1/16 inch

Fill in the Blank

11. A _____ is the result of dividing one number by another.

12. The sum of 12, 8, and 345 is _____.

13. The product of 267 and 34 is _____.

14. The decimal equivalent of 11/16 is _____.

15. The prefix *kilo* means _____.

16. The metric unit for measuring length is the _____.

17. One meter equals _____ inches.

18. _____ is the amount of surface of an object.

19. One hundred square decimeters equals _____ square meter(s).

20. Most tape measures used by electricians are the type where _____ of an inch is the smallest increment on the tape.

True/False

21. For an electrician, the easiest way to work with fractions is to change them into decimals. (True, False)

22. In algebra, the addition sign is not used. (True, False)

23. The most common base units in the metric system are the meter, gram, second, and liter. (True, False)

24. A tape measure can be marked with English or metric measurements. (True, False)

25. On a tape measure, 1/16 of an inch is twice as big as 1/8 of an inch. (True, False)

Fill in the Blank

26. A(n) _____ is a number that can represent part of a whole.

27. When writing a decimal fraction, 3/1000 is written as _____.

28. In an algebraic expression, the part of the expression not separated by a plus or minus sign is called a(n) _____.

29. A(n) _____ is a mathematical statement that asserts the equality of two expressions.

30. A(n) _____ is a rule expressed in letters, symbols, and constant terms.

31. A value or number per hundred is written as a _____.

ACTIVITY 4-1   Calculating Area Square Feet

Using **Figure 4-10**, determine the square footage of each of the following rooms:

Master Bedroom _____ sq. ft   Bathroom _____ sq. ft

Master Bath _____ sq. ft   Bedroom 1 _____ sq. ft

Garage _____ sq. ft   Bedroom 2 _____ sq. ft

Family Room _____ sq. ft   Laundry Room _____ sq. ft

Walk-in Closet _____ sq. ft   Dining Room _____ sq. ft

FIGURE 4-10   Floor plan.

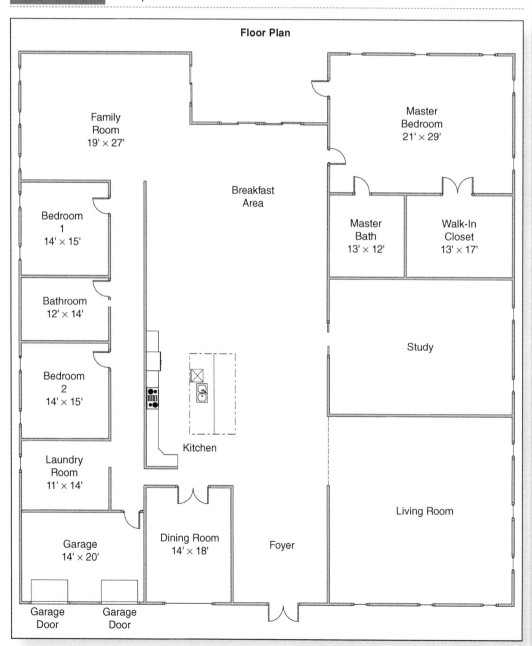

ACTIVITY 4-2   Calculating Averages

Background

Baseball Hall of Famer Ted Williams played his entire career for the Boston Red Sox. During his career, Ted had several notable achievements such as leading the League in batting six times, winning the Triple Crown twice, and being an American League MVP twice.

Williams holds the highest career batting average of anyone with 500 or more home runs. In fact, in the year 1941, Williams saw:

- 37 home runs
- 120 RBIs
- 143 games
- 185 hits
- 135 runs
- 456 at-bats
- 2 stolen bases
- 147 base on balls
- 27 strikeouts

To calculate a batting average for a player, divide the player's *hits* by the player's *at-bats,* and round the answer to the third decimal point.

Example: 85 hits divided by 400 at-bats: 85 ÷ 400 = 0.2125, or 213

ACTIVITY: Using the information above, calculate the batting average for Ted Williams' career year, 1941.

ACTIVITY 4-3   Calculating Your Paycheck

Finally, you have found what you think is the perfect job! The interview went well, and the interviewer said that you made the short list and that she would call and let you know, either way.

During the interview you asked several questions with regard to work days, work hours, rules, requirements, and so forth. One of the most important questions to ask during an interview is "What does the job pay?"

Although that seems like a simple question, really it isn't. What a job pays is not an accurate reflection of what you can expect to take home in the form of a pay-check at the end of the week. Remember, it's not just about you anymore! We all have to pay taxes on the money that we make. Any time there is an increase in your wealth, there are taxes to pay.

(Continues)

(CONTINUED)

So, how do you know how much you will take home at the end of the week? To answer that question you must determine the total amount of money that you earn for the week or pay period (gross income) and deduct the taxes that you owe. A part of your paycheck, commonly called a *paystub,* which you keep for your records, contains all of this information. Additionally, the information is usually broken down by category of tax, current pay period information, and YTD (year-to-date) information.

You want to know what you will take home based on the amount talked about in your interview. To perform this calculation is fairly simple: To determine the gross income for an hourly position, multiply the number of hours worked by the amount of pay per hour.

Example: $15.00 per hour $\times$ 40 hours (normal workweek) = $600.00

If you are paid $15.00 per hour and you work 40 hours in the week, you make $600 in gross income. Fantastic! But wait—it's not all yours!

Once you have determined your gross income you have to subtract the taxes that you owe, commonly referred to as *deductions,* from that gross income. To correctly determine your taxes, you need access to information published by the IRS. Because that information is not in this book, we are going to make some assumptions to make things easier.

FICA

Everyone pays a tax called FICA, Federal Insurance Contributions Act, more commonly known as *Social Security tax.* Two contributions are made on your behalf to FICA; one comes from you, and the other, from your employer. The total contribution on your behalf equals 15.3% of your gross income. This amount is split between you and your employer, 7.65% for each of you. This tax is levied to provide you with a modest income when you retire.

Federal Withholding

You are also required to pay a tax to the federal government, more commonly known as *income tax.* The amount of income tax you pay depends on the amount of money you make in a given year. Generally, you contribute between 12% and 15% of your income. If you make over $34,000, you creep into the next tax bracket, which

(Continues)

(CONTINUED)

is 25%, and yes, tax brackets do go even higher! So, let's assume for right now that you will contribute 12% in federal taxes.

Net Income

Now that we understand what the taxes are and we have an idea as to what our contributions will be, we can calculate our *net income,* the term for the amount of money that you actually get in the form of a paycheck or a deposit in your bank account, if you have direct deposit.

Now that you understand what taxes are and have an idea as to what your contributions will be, you can calculate your net income. *Net income* is the term for the amount of money you actually get, usually in the form of a paycheck or a direct deposit into your bank account, if your account is set up for direct deposit.

To calculate net income, enter your gross income and multiply this figure by 7.65% to calculate your Social Security tax; next, multiply your gross income by 12% to calculate your federal income tax. Once you know the amount of these two deductions, subtract them from your gross income to get your net income.

Example:

Gross income	$600.00		Gross income	$600.00
	$\times.0765$			$\times 0.12$
FICA	$45.90		Federal withholding	$72.00

Gross pay	$600.00
FICA	$-$45.90
Fed. w/h	$-$72.00
Net income	$482.10

ACTIVITY: Assume that you work a 40-hour week and are paid $18.75 per hour. Using the values of 7.65% for FICA and 12% for federal withholding, calculate your take-home pay after taxes.

Chapter 5
Basic Concepts of Electricity and Magnetism

Career Profile

Ray Shorkey "enjoys electrical work in and of itself." He believes that "being an electrician is a career that provides you with a variety of work, and is challenging both mentally and physically. There is a sense of satisfaction in building something [and] then seeing it completed and operating properly."

Shorkey is currently a journeyman electrician and lead-man at Merit Electric in Fort Collins, Colorado. Describing his responsibilities on a typical day, he says, "We start early, usually six or seven in the morning, and, depending on the job, work between eight and twelve hours. As a lead-man you must be aware of the job schedule, plan and prioritize tasks for your crew, and ensure that the needed materials are available to complete those tasks. In addition to your own tasks, you need to be available to assist your crew members with questions and troubleshooting."

Additionally, he notes, "Some jobs or projects don't always work out as planned. Sometimes you run into a situation that wasn't already taken into account, or even known about. You have to solve problems quickly, using the resources that are available. Time management is also a challenge. It requires planning to balance everything and to be able to prioritize. It takes practice to think clearly in stressful situations."

Because the challenges of managing projects and crews are so many and so diverse, "proper training and education are important," Shorkey states. "The more knowledge you have, the more value you hold in your field." Shorkey endorses IEC programs (like the one he completed at the Rocky Mountain chapter) because it "teaches theory and provides you with a deeper knowledge of the electrical trade than you learn in the field." He adds, "The education you receive at IEC is also instrumental in preparing for your journeyman license exam."

Chapter Outline

THE ATOM
ELECTRIC CHARGE
ELECTRICITY—ELECTRONS IN MOTION
ELECTRIC SOURCES
MAGNETISM

Key Terms

Air-core magnet
Ampere-turns
Atom
Atomic number
Compound
Conductor
Electrical current
Electromagnet
Electron
Element
Free electrons
Ions
Iron-core magnet
Law of charges
Magnetic field
Magnetism
Molecule
Neutron
Nucleus
Permanent magnet
Permeability
Polarity
Poles
Proton
Reluctance
Residual magnetism
Saturation
Static electricity
Temporary magnet
Transformer
Valence shell

Chapter Objectives

After completing this chapter, you will be able to:

1. Identify the three principle parts of an atom.
2. State the law of charges, and describe its importance to current flow.
3. Explain electron current flow, and contrast direct and alternating current.
4. Describe current flow through a conductor, and discuss why heat is produced.
5. Define magnetic terms, and discuss magnetic principles and concepts as they relate to electricity.

Life Skills Covered

 Managing Stress

 Critical Thinking

 Take Action

Life Skill Goals

The life skill goals covered in this chapter are Managing Stress, Critical Thinking, and Take Action. You will practice the reading strategies introduced in Chapter 2. You will also be asked to think about how you are fitting reading into your schedule. Finally, as in every chapter, you have opportunities to check your understanding throughout the chapter.

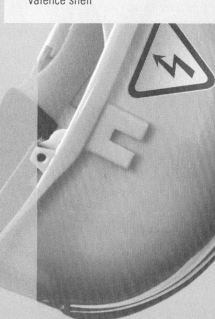

Where You Are Headed

Identifying the parts of an atom and understanding the laws of charges will help you realize that electricity is not magic! There are real and practical reasons *why* and *how* current flows through a conductor. Atomic theory, laws of charges, electron movement (current flow), and the basic principles of magnetism are all building block materials. And like the foundations of the houses and buildings you will someday be wiring, this information will serve as the foundation for more advanced electrical topics.

Introduction

It has been over 250 years since Ben Franklin discovered two kinds of electrical charges. It has been almost 200 years since Hans Oersted and Michael Faraday developed ideas of electric fields, currents, and magnetism. At about the same time, British electrician/engineer William Sturgeon invented the electromagnet. Because of these men and other pioneers in the fields of magnetism and electricity, we enjoy watching television, listening to our stereos, surfing the Internet on our computers, and heating our food in microwave ovens. This chapter covers the basic concepts of electricity and magnetism; learn about the forces that improve the quality of our lives.

We learn about atomic theory to help us understand electricity, but few of us realize that much of an atom's mass is in its dense nucleus, which stores enormous amounts of energy. This energy can be released by breaking up the nucleus. Why is this important? Our understanding of this basic atomic structure has been used as the basis of nuclear power plants and many modern medical applications, such as radiation therapy for cancer and medical imaging used for noninvasive medical testing. This knowledge has also enabled scientists to make weapons. The advent of atomic weapons influenced the way World War II ended, its aftermath, and the power plays between nations at work today.

LIFE SKILLS

How long do you think it will take you to read this chapter? When do you think you can fit that much time into your schedule?

LIFE SKILLS

Practice SQR3.

- Survey through Section 5-1, *The Atom*.
- Skim through the section. Take a look at the pictures. Read the Self-Check questions at the end of the section.

- Create the questions you are going to answer by reading this section.
- Read this section. Take some notes. Find the answers to your questions.
- Recite these answers.
- Review. Go back and highlight the main points in the section. Add to your notes.

5-1 ■ The Atom

Atom the smallest particle into which an element can be divided without losing its identity.

Element a group of identical atoms.

Molecule consists of two or more atoms.

An **atom** is the smallest particle into which an element can be divided without losing its identity. A group of identical atoms is called an **element**. All matter is composed of atoms. Whether we are discussing metal, wood, or glass, they are all made up of atoms. Before we go any further, let's take a step back and look at some useful terms. The smallest particle of a substance that still has all of its characteristics is called a **molecule**. A molecule consists of two or more atoms. If a molecule of metal is divided into smaller parts, it is no longer metal.

There are over a hundred different types of atoms and as many elements as there are atoms. A few of the more common elements are iron, nitrogen, and oxygen; important elements in terms of electrical conductivity are copper, silver, and gold. **Figure 5-1** lists the table of elements.

FIGURE 5-1 Table of elements.

ATOMIC NUMBER	NAME	VALENCE ELECTRONS	SYMBOL	ATOMIC NUMBER	NAME	VALENCE ELECTRONS	SYMBOL	ATOMIC NUMBER	NAME	VALENCE ELECTRONS	SYMBOL
1	Hydrogen	1	H	37	Rubidium	1	Rb	73	Tantalum	2	Ta
2	Helium	2	He	38	Strontium	2	Sr	74	Tungsten	2	W
3	Lithium	1	Li	39	Yttrium	2	Y	75	Rhenium	2	Re
4	Beryllum	2	Be	40	Zirconium	2	Zr	76	Osmium	2	Os
5	Boron	3	B	41	Niobium	1	Nb	77	Iridium	2	Ir
6	Carbon	4	C	42	Molybdenum	1	Mo	78	Platinum	1	Pt
7	Nitrogen	5	N	43	Technetium	2	Tc	79	Gold	1	Au
8	Oxygen	6	O	44	Ruthenium	1	Ru	80	Mercury	2	Hg
9	Fluorine	7	F	45	Rhodium	1	Rh	81	Thallium	3	Tl
10	Neon	8	Ne	46	Palladium	–	Pd	82	Lead	4	Pb
11	Sodium	1	Na	47	Silver	1	Ag	83	Bismuth	5	Bi
12	Magnesium	2	Ma	48	Cadmium	2	Cd	84	Polonium	6	Po
13	Aluminum	3	Al	49	Indium	3	In	85	Astatine	7	At
14	Silicon	4	Si	50	Tin	4	Sn	86	Radon	8	Rd
15	Phosphorus	5	P	51	Antimony	5	Sb	87	Francium	1	Fr
16	Sulfur	6	S	52	Tellurium	6	Te	88	Radium	2	Ra
17	Chlorine	7	Cl	53	Iodine	7	I	89	Actinium	2	Ac
18	Argon	8	A	54	Xenon	8	Xe	90	Thorium	2	Th
19	Potassium	1	K	55	Cesium	1	Cs	91	Protactinium	2	Pa
20	Calcium	2	Ca	56	Barium	2	Ba	92	Uranium	2	U
21	Scandium	2	Sc	57	Lanthanum	2	La				
22	Titanium	2	Ti	58	Cerium	2	Ce		Artificial Elements		
23	Vanadium	2	V	59	Praseodymium	2	Pr				
24	Chromium	1	Cr	60	Neodymium	2	Nd	93	Neptunium	2	Np
25	Manganese	2	Mn	61	Promethium	2	Pm	94	Plutonium	2	Pu
26	Iron	2	Fe	62	Samarium	2	Sm	95	Americium	2	Am
27	Cobalt	2	Co	63	Europium	2	Eu	96	Curium	2	Cm
28	Nickel	2	Ni	64	Gadolinium	2	Gd	97	Berkelium	2	Bk
29	Copper	1	Cu	65	Terbium	2	Tb	98	Californium	2	Cf
30	Zinc	2	Zn	66	Dysprosium	2	Dy	99	Einsteinium	2	E
31	Gallium	3	Ga	67	Holmium	2	Ho	100	Fermium	2	Fm
32	Germanium	4	Ge	68	Erbium	2	Er	101	Mendelevium	2	Mv
33	Arsenic	5	As	69	Thulium	2	Tm	102	Nobelium	2	No
34	Selenium	6	Se	70	Ytterbium	2	Yb	103	Lawrencium	2	Lw
35	Bromine	7	Br	71	Lutetium	2	Lu				
36	Krypton	8	Kr	72	Hafnium	2	Hf				

From HERMAN. *Delmar's Standard Textbook of Electricity, 5E.* © 2011 Cengage Learning

Compound exists when two kinds of atoms combine chemically.

Electron an elementary particle carrying one unit of negative electrical charge.

Proton found in the nucleus and has a positive electrical charge.

Neutron a small particle that possesses no electrical charge and is typically found within an atom's nucleus.

Nucleus the very dense region consisting of protons and neutrons at the center of an atom.

Because there are so many different materials in the world, most materials are made up of more than one element. When different kinds of atoms combine chemically, they form materials referred to as compounds. Two very familiar compounds are table salt and water. Table salt is made up of the elements sodium and chloride; water is made up of the elements hydrogen and oxygen.

Because we need to develop an understanding of what electricity is, let's take a closer look at the atom. There are three principle parts of the atom. They are the electron, proton, and neutron. An electron is an elementary particle carrying one unit of negative electrical charge. The proton is found in the nucleus and has a positive electrical charge equal to the negative charge of an electron. Neutrons and protons have about the same weight, but the neutron has no electrical charge. **Figure 5-2** illustrates a helium atom. The nucleus is the very dense region consisting of protons and neutrons at the center of an atom. The electrons revolve around the nucleus in an elliptical, or oval-shaped, path.

The orbiting electrons have a negative charge, the protons have a positive charge, and the neutrons display no charge. Because the neutron has no charge, the nucleus has a net positive charge. The element type is determined by how many protons are in the nucleus. Helium, as already noted, has 2 protons in its nucleus; copper has 29.

FIGURE 5-2 Helium atom illustrating electrons, protons, and neutrons.

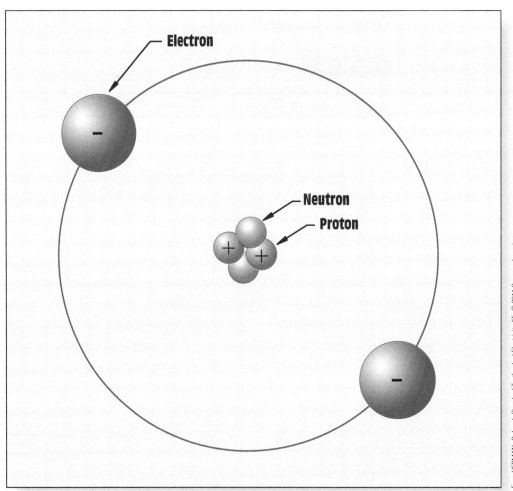

From HERMAN. *Delmar's Standard Textbook of Electricity*, 5E. © 2011 Cengage Learning

FIGURE 5-3 Positive charge of a proton with lines of force extending outward.

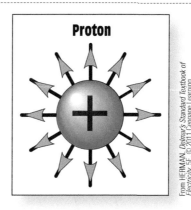

FIGURE 5-4 Electron lines of force.

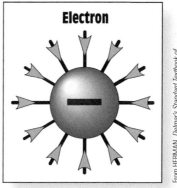

Atomic number
equal to the number
of protons in the
nucleus.

The **atomic number** of an element is equal to the number of protons in the nucleus. **Figure 5-3** shows that the positive charge of the proton produces lines of force that *extend outward* in all directions.

We saw in Figure 5-2 that the electrons orbit outside the nucleus. Although an electron is about three times larger than a proton, the proton weighs over 1800 times more. Because electrons are negatively charged, the lines of force produced by an electron *come in* from all directions. See **Figure 5-4.**

Electrons revolve around the nucleus of the atom similar to the way the Earth rotates around the sun. In atoms that contain more than one electron (all atoms except hydrogen), each electron has its own orbit. It is possible for two or more atoms to share common space; in many materials, closely spaced atoms share both electrons and space. **Figure 5-5** shows the electron orbits of an atom.

The two electrons closest to the nucleus are said to occupy the first shell, or orbit, of the atom. This first shell can only have two electrons. Atoms that have more than two electrons, like iron, with 26 electrons, must have a second, third, and fourth shell, or orbit.

Have you ever heard the expression "Opposites attract and likes repel?" Well, much like the magnetic poles of a magnet repel if like poles are brought together and attract if unlike poles are brought together, the law of charges states that opposite charges attract and like charges repel. This basic law will help you understand the importance of attraction and repulsion of electrons and protons in the atom.

FIGURE 5-5 Electron orbits.

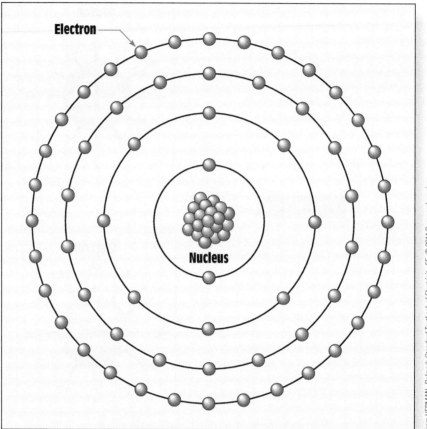

Electron

Nucleus

From HERMAN. *Delmar's Standard Textbook of Electricity*, 5E. © 2011 Cengage Learning

LIFE SKILLS

There are many key terms in this chapter. It would be a good idea to create some flashcards. Put the term on one side and the meaning on the other side. Find someone in class or at home to quiz you on your words.

Self-Check 1

1. List three common elements.
2. What are the three principal parts of the atom?

Polarity the type (positive or negative) of charge.

Law of charges opposite charges attract and like charges repel.

5-2 Electric Charge

Electrons and protons exhibit electric charges of opposite polarity. Polarity refers to the type (positive or negative) of charge. The proton possesses a positive charge, and the electron is negatively charged. As illustrated in Figure 5-3, these charges produce lines of force. A basic law of physics, known as the law of charges, states that opposite charges attract and like charges repel. **Figures 5-6** and **5-7** illustrate this law.

FIGURE 5-6 Unlike charges attract each other.

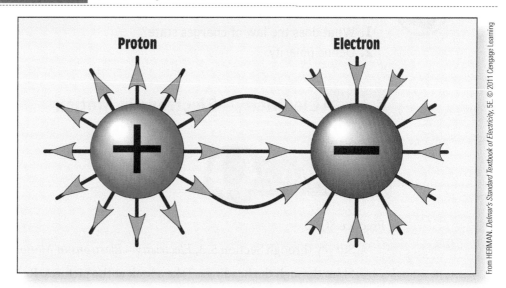

From HERMAN. *Delmar's Standard Textbook of Electricity*, 5E. © 2011 Cengage Learning

FIGURE 5-7 Like charges repel each other.

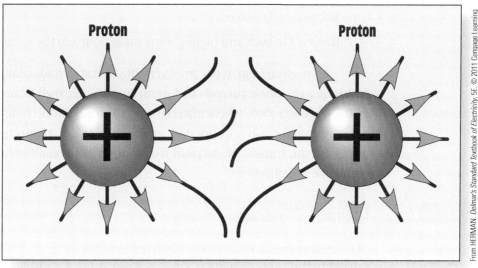

From HERMAN. *Delmar's Standard Textbook of Electricity*, 5E. © 2011 Cengage Learning

When considering the electric charge of an atom, the neutron can be ignored because it has no electric charge. An atom such as helium, illustrated in Figure 5-2 in its natural state, is electrically neutral. Helium has two protons in the nucleus and two electrons orbiting the nucleus. Even though the electrons and protons are electrically charged, the *net* electric charge is zero.

Electricity is not magic. Current is simply electrons in motion. Next, we learn how those electrons are put into motion.

What was the most important thing that you learned about electric charge in this section?

Self-Check 2

1. What does the law of charges state?
2. Define polarity.

5-3 ⬛ Electricity—Electrons in Motion

Practice SQR3.

- Survey through Section 5-3, *Electricity—Electrons in Motion*.
- Skim through the headings. Take a look at the pictures. Read the Self-Check questions at the end of the section.
- Create the questions you are going to answer by reading this section.
- Read this section. Take some notes. Find the answers to your questions.
- Recite these answers.
- Review. Go back and highlight the main points in the section. Add to your notes.

As an electrician, it becomes very important to understand the chemical reactions and electric currents that occur in the outer shell of an atom. This section will define this outer shell, where materials such as conductors come together, free electrons can escape, and positive or negative ions are created. Static electricity will also be discussed; this transfer of electrons from one object to another has both practical and problematic applications.

Valence Electrons

Valence shell the outermost shell, or ring, of an atom.

Conductor a material that readily conducts electricity.

The outermost shell of an atom is referred to as the valence shell, or ring. Electrons located in this valence shell are called valance electrons. These electrons are atomic particles involved in chemical reactions and electric currents. The valence shell cannot hold more than eight electrons. A conductor, for example, is comprised of a material that contains between one and three valence electrons. A conductor is a material that readily conducts electricity.

One of the forces that helps to keep electrons in orbit is the force of attraction between unlike charges. The closer together two particles of opposite electric charges are, the greater the electrical attraction between them. So, the attraction between the proton in the nucleus and the orbiting electron decreases as the electron gets farther from the nucleus. Because of this, valence electrons are more loosely held than the electrons in the inner shells. Here are a few characteristics of valence electrons:

- They can be more easily removed from the atom than electrons in the inner shell.
- Valence electrons possess more energy than electrons in the inner shells because the farther an electron is from the nucleus, the more energy it possesses.

Self-Check 3

1. What is a conductor?
2. When does the attraction between the proton in the nucleus and the orbiting electron decrease?

Free Electrons and Electron Flow

Free electrons
valence electrons that have been temporarily separated from an atom.

Valence electrons that have been temporarily separated from an atom are called free electrons. They are not attached to any particular atom. Only the valence electrons in the outermost orbit can become free electrons. Recall that the nucleus and the attraction of the protons keep electrons in the inner shells held tightly. A valence electron can be freed from an atom when energy is added to the atom. This energy permits the loosely held valence electron to escape the force of attraction between the electron and the nucleus. A free electron possesses more energy than it did as a valence electron.

Electrical current the flow of electrons.

Electrical current is the flow of electrons. Heating an atom or subjecting an atom to an electric field provides the necessary energy to free the valence electron. After traveling a short distance, the free electron enters the valence orbit of a different atom. When it returns to orbit, some or all of the gained energy is released in the form of heat. This is why conductors become warm when current flows through them. If current through a conductor becomes excessive, a fire may result.

Self-Check 4

1. What are free electrons?
2. Define electrical current.

Ions

Ions atoms that have more or less than their normal complement of electrons.

The departure of a valence electron leaves a previously neutral atom with an excess of positive charge, or more protons than electrons. Atoms that have more or less than their normal complement of electrons are referred to as ions. When an atom has an excess of electrons, it becomes a negative ion; when an atom loses electrons, it becomes a positive ion. The amount of energy necessary to cause a valence electron to become free is different with each element.

The number of an atom's valence electrons determines how much energy is required to create a free electron. Typically, if there are fewer electrons in the valence orbit, less energy is required to free that valence electron.

Self-Check 5

1. How does an ion become positive?
2. How does an ion become negative?

Static Electricity

Static electricity
the accumulation of an electric charge on an insulated body.

Static electricity is a common phenomenon that all of us have observed at one time or another. Static electricity is the accumulation of an electric charge on an insulated body. An example of static electricity is the shock we feel when walking across a high pile carpet and reaching for a metal doorknob. Another example is hair clinging to a comb. The best example of natural static electricity is lightning, but it is potentially harmful. Lightning occurs when a static charge builds up in clouds that contain large amounts of moisture. Most lightning takes place within a cloud, but discharges from cloud to cloud and cloud to ground are common. All of these discharges have one thing in common; they all involve the transfer of electrons from one object to another. A lightning discharge, or bolt, has an average voltage of 15,000,000 volts.

Practical Applications of Static Electricity

Most of the practical applications of static electricity do not operate by taking advantage of static discharges. Useful applications of static electricity generally make use of the forces caused by the repulsion of like charges and the attraction of unlike charges. These forces are used to move charged particles to desired locations. An example of this application would be removing dust particles from the air. **Figure 5-8** illustrates how air is forced between *negatively* charged rods and *positively* charged plates. The device in Figure 5-8 is referred to as an electrostatic dust precipitator. Dust particles are given a negative charge; they are attracted to positively charged plates removing the dust.

Spray painting is another application that works on the principle of electric charges attracting and repelling. A high-voltage grid is placed in front of a spray gun.

FIGURE 5-8 An electronic air cleaner.

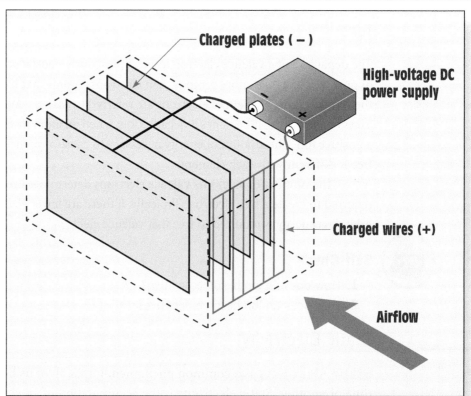

Charged plates (–)

High-voltage DC power supply

Charged wires (+)

Airflow

From HERMAN. *Delmar's Standard Textbook of Electricity*, 5E. © 2011 Cengage Learning

FIGURE 5-9 Spray painting using static charges.

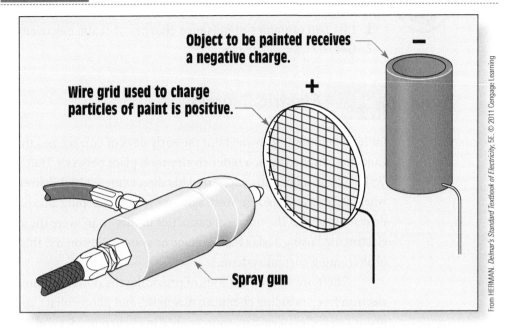

A positive charge is applied to this grid. The object to be painted has a negative charge. **Figure 5-9** shows how static charges are used in spray painting.

As the paint droplets pass through the grid, the positive charge causes electrons to be removed from the paint droplets. The positively charged droplets are attracted to the negatively charged object. Using the static charge paint operation, less paint is wasted, and the object has a more uniform finish.

Nuisance Charges of Static Electricity

Static electricity is the static charge that accumulates on a person's body as he or she walks across a carpet. Although static charges have some useful applications, they are often a nuisance. Here are some examples of nuisance static charges:

- When reaching for a metal doorknob or other metal object, the discharge in the form of an electric shock can be painful. Static buildup while walking across a carpeted floor happens because carpets are made of nylon or other types of insulators, and heating systems generally remove moisture from the air. The installation of a humidifier usually prevents the buildup of static charges.

- A static charge causes clothes moving through the dry air in a clothes dryer to cling together. Synthetic fabrics, which are the best insulators, retain electrons more than natural fabrics do. So, synthetic fabrics build up the greatest static charge.

You may already know that batteries, fuel cells, and solar cells produce direct current. This current always flows in the same direction. You may have also heard the term *alternating current*. This type of current reverses, or alternates, 60 times per second (in the United States). The next section examines both types of electric sources and their defining characteristics.

Self-Check 6

1. List two examples of nuisance charges of static electricity.
2. Define static electricity.

5-4 Electric Sources

Little is commonly known about the early days of current and the so-called "War of Currents." In the 1890s, a bitter rivalry took place between Thomas Edison and Nikola Tesla. Edison adamantly supported his direct current (dc) delivery of electricity, whereas Tesla and George Westinghouse were developing alternating current (ac) delivery systems. The only real casualties in this "war" were the animals Edison publicly electrocuted using Tesla's high-voltage ac system. Edison did this to prove the dangers of alternating current systems.

There are several methods of providing the power required to set a valence electron free, including chemical, magnetic, and photovoltaic, to list a few. The two major categories of electric sources are direct current (dc) and alternating current (ac).

Direct Current

A direct current (dc) power source provides a voltage whose polarity and output voltage never change direction. Dc power sources provide two terminal devices: one terminal negative and the other positive, with respect to each other. Direct current sources of energy include batteries, dc generators, and electronic power supplies.

Self-Check 7

1. What never changes in direct current?

Alternating Current

Alternating current (ac) power sources produce a voltage that periodically reverses polarity and magnitude. An example of an ac power source is an ac generator. Alternating current has one major advantage over direct current; the voltage can be increased or decreased by using a transformer. A transformer is an electrical device by which alternating current of one voltage is changed to another voltage.

When electric current runs through a conductor, a magnetic field is produced. Magnetism makes possible the operation of motors, transformers, and even your doorbell. The next section discusses magnetism, electromagnets, and some of the devices that work because of this force.

Transformer an electrical device by which alternating current of one voltage is changed to another voltage.

Self-Check 8

1. What is one major advantage alternating current has over direct current?

LIFE SKILLS

Practice SQR3.

- Survey through Section 5-5, *Magnetism*.
- Skim through the headings. Take a look at the pictures. Read the Self-Check questions at the end of the section.
- Create the questions you are going to answer by reading this section.
- Read this section. Take some notes. Find the answers to your questions.
- Recite these answers.
- Review. Go back and highlight the main points in the section. Add to your notes.

5-5 ▪ Magnetism

Magnetism a force field that acts on some materials, but not on other materials.

Magnetism is a force field that acts on some materials, but not others. Physical devices that possess this force are referred to as magnets. Lodestone is a natural magnet. However, magnets used today are all manufactured. They are made from alloys containing elements such as nickel, copper, aluminum, and iron. Many of today's instruments operate because of magnetism. A compass is an example of such a device.

A compass works on the principle that a small, lightweight magnet balanced on a nearly frictionless pivot point turns toward the north. Why? It is because the Earth itself contains magnetic poles. **Figure 5-10** illustrates the positions of the true North

FIGURE 5-10 The Earth's poles.

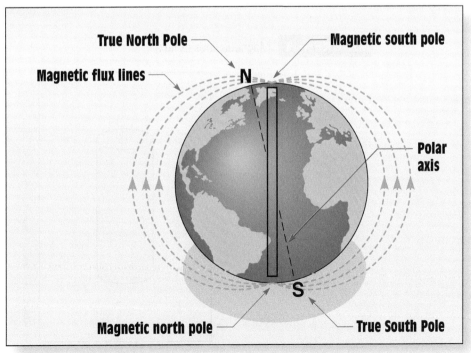

From HERMAN. *Delmar's Standard Textbook of Electricity*, 5E. © 2011 Cengage Learning

and South Poles, referred to as the axis of the earth. This figure also shows the positions of the magnetic poles.

Note that magnetic north is not located at the true North Pole of the earth. This is the reason why navigators must distinguish between true north and magnetic north. The angular difference between the two is known as the angle of declination.

Magnets

The two general types of magnets are known as permanent and temporary. **Permanent magnets** are magnets that do not require any power or force to maintain their field. A **temporary magnet** acts as a magnet only as long as it is in the magnetic field produced by a permanent magnet or an electric current. Magnetic materials from which permanent magnets are made are called hard magnetic materials; materials from which temporary magnets are made are known as soft magnetic materials.

Self-Check 9

1. What is the difference between permanent and temporary magnets?

Fields, Flux, and Poles

To better understand magnetism and devices that operate because of magnetic fields, such as transformers, motors, and loudspeakers, we must develop a basic understanding of magnetism.

A **magnetic field** is the result of the force of magnetism. The magnetic field extends out from the magnet in all directions. The invisible lines of force that make up this magnetic field are known as magnetic flux. **Figure 5-11** illustrates how lines of flux surround the entire magnet.

FIGURE 5-11 Magnetic lines of flux.

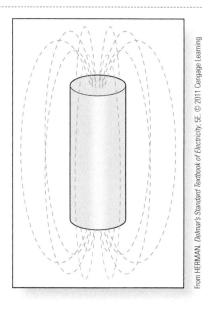

From HERMAN, *Delmar's Standard Textbook of Electricity*, 5E. © 2011 Cengage Learning

A magnetic field is stronger where lines of flux are dense. Where lines of flux are sparse, the field is weak. The lines of flux are most dense at the ends of a magnet. So, the magnetic field is strongest at the ends of the magnet. The ends of a magnet are referred to as the magnetic **poles** and are designated north (N) and south (S). Magnetic poles are the two areas (ends) of a magnet where the magnetic force is the greatest. Lines of flux are always assumed to leave the north pole and enter the south pole of a magnet.

Bringing the north (or south) poles of two magnets together will create a repelling force. If you bring a north pole of a magnet close to the south pole of another magnet, a force of attraction is created. So, unlike poles attract each other.

Poles the two areas (ends) of a magnet where the magnetic force is the greatest.

Self-Check 10

1. A magnetic field is the result of what?
2. Where is magnetic force the greatest on a magnet?

Electromagnetism

Electromagnet an object that acts like a magnet, but its magnetic force is created and controlled by electricity.

Electricity and magnetism cannot be separated. Whenever an electric current runs through a wire, a magnetic field is created. An **electromagnet** is simply an object that acts like a magnet, but its magnetic force is created and controlled by electricity. Electromagnets operate because current flowing through a conductor produces a magnetic field around the conductor, as illustrated in **Figure 5-12.**

FIGURE 5-12   Magnetic field around a current-carrying conductor.

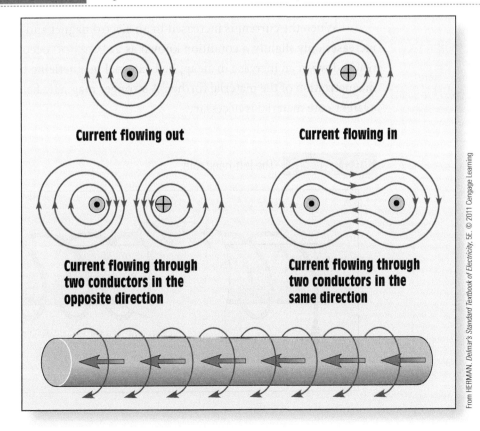

Current flowing out

Current flowing in

Current flowing through two conductors in the opposite direction

Current flowing through two conductors in the same direction

From HERMAN. *Delmar's Standard Textbook of Electricity,* 5E. © 2011 Cengage Learning

Ampere-turns determined by multiplying the number of turns of wire by the current flow.

Iron-core magnet an electromagnet made by wrapping a coil of wire around an iron core.

Air-core magnet an electromagnet comprised of a coil wound around a nonmagnetic material.

Permeability a term used to describe the measure of a material's ability to become magnetized.

Reluctance resistance to magnetism.

Saturation the state reached when an increase in applied external magnetizing field cannot increase the magnetization of the material further.

If a conductor is wound into a coil, the magnetic field is concentrated and stronger because of the added effect of each turn of the wire.

Electromagnetic field strength is also affected by the amount of current flowing through the wire. If current is increased, the magnetic field strength is stronger. The strength of an electromagnet is proportional to its ampere-turns. Ampere-turns are determined by multiplying the number of turns of wire by the current flow.

Wrapping a wire around an iron core greatly increases the magnetic field. This kind of coil is referred to as an iron-core magnet. Coils can also be wound around a nonmagnetic material such as wood or plastic. These coils are called air-core magnets. Air-core magnets are electromagnets that have a coil wound around a nonmagnetic material. Permeability is a term used to describe the measure of a material's ability to become magnetized. A magnetic core material, such as iron, provides an easy path for the flow of magnetic lines in the same way a conductor provides an easy path for the flow of electrons.

The polarity of an electromagnet can be determined by using what has become known as the left-hand rule. When fingers on your left hand are placed around the windings in the direction of current flow, your thumb points to the north magnetic pole, as can be seen in **Figure 5-13**. If the direction of current flow is reversed, the polarity of the magnetic field also reverses.

Another measurement used when working with magnetic devices is reluctance, which is simply resistance to magnetism. Materials like soft iron or nickel have high permeability and low reluctance because they can be easily magnetized. However, graphite, for example, has low permeability and high reluctance.

When the current is increased in an electromagnet and the magnetic strength increases only slightly, a condition known as saturation occurs. Saturation is the state reached when an increase in an applied external magnetizing field cannot increase the magnetization of the material further. A stronger magnetic field can be produced, but a larger core material is necessary.

FIGURE 5-13 The left-hand rule.

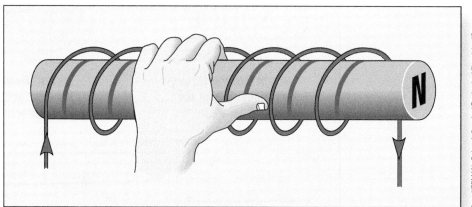

From HERMAN, *Delmar's Standard Textbook of Electricity*, 5E.
© 2011 Cengage Learning

When the current flow through the coil of an electromagnet is interrupted, some magnetism may remain in the core material. The amount of magnetism left in a material after the magnetizing force has been removed is called residual magnetism.

Self-Check 11

1. What is an electromagnet?

2. How can you determine the polarity of an electromagnet?

Magnetic Devices

Magnetism is responsible for the operation of many devices we use every day. Some of the more common devices include measuring instruments, transformers, motors, and electromagnets. One useful characteristic of an electromagnet is that magnetic force can be varied by changing the amount and direction of current going through the coils around its core. Tape recorders, speakers, and earphones operate because of varying current that flows through the winding of an electromagnet. Some other applications of electromagnetic devices are listed here:

- Scrap-yard crane with a huge electromagnet to pick up and move junked autos

- Electromagnetic door lock

- Doorbell ringer

- Electric motor

- Maglev trains

Self-Check 12

1. What devices operate because of varying current?

2. List three applications of electromagnetic devices.

LIFE SKILLS

How long did it take you to read this chapter? In what part of the day did you read it? Was it a time when you were able to concentrate, or should you find another time of day to do your reading?

Summary

- All matter is comprised of atoms, and an atom is the smallest particle into which an element can be divided without losing its identity.

- Some common elements include iron, nitrogen, oxygen, copper, silver, and gold.

- The three principal parts of an atom are the electron, proton, and neutron.
- The center of an atom, called the nucleus, contains positively charged protons as well as neutrons, which have no electrical charge.
- Negatively charged electrons orbit the nucleus.
- The law of charges states that opposite electrical charges attract and like electrical charges repel.
- The outermost shell of an atom is called the valence shell, or ring.
- A conductor is a material that contains between one and three valence electrons and readily conducts electricity.
- Valence electrons temporarily separated from an atom are free electrons.
- Electrical current is the movement, or flow, of free electrons.
- Atoms that have more or less than their normal complement of electrons are ions.
- Static electricity is the accumulation of an electric charge on an insulated body.
- Applications of static electricity include dust precipitators and spray painting.
- The two major categories of electric sources are direct current and alternating current.
- Magnetism is a force field that acts on some materials, but not on other materials.
- An electromagnet is an object that acts like a magnet, but its magnetic field is created and controlled by electricity.
- Devices that operate because of magnetism include measuring instruments, transformers, motors, and electromagnets.
- Applications of electromagnets include door lock, doorbell ringer, electric motor, and maglev train.

Review Questions

True/False

1. A molecule is the smallest particle into which an element can be divided without losing its identity. (True, False)

2. An electron is an elementary particle carrying one unit of negative electrical charge. (True, False)

3. Protons and neutrons are found in the nucleus of an atom. (True, False)

4. The neutron found in the nucleus has a net positive charge. (True, False)

5. Protons revolve around the nucleus of the atom. (True, False)

Multiple Choice

6. What is the law of charges?
   A. Opposite charges repel.
   B. Like charges repel.
   C. Like charges are always negative.
   D. Like charges are always positive.

7. Neutrons _____.
   A. Have a positive charge
   B. Have a negative charge
   C. Revolve around the nucleus
   D. Have no electrical charge

8. The outermost shell of an atom is referred to as the _____.
   A. Variance shell
   B. Ion shell
   C. Valence shell
   D. Free electron shell

9. A material that readily conducts electricity is a(n) _____.
   A. Element
   B. Molecule
   C. Compound
   D. Conductor

10. What is the term for valence electrons that have been temporarily separated from an atom?
   A. Free electrons
   B. Ions
   C. Neutrons
   D. Free protons

Fill in the Blanks

11. Electrical _____ is the flow of free electrons.

12. Atoms that have more or less than their normal complement of electrons are called _____.

13. _____ electricity is an accumulation of an electric charge on an insulated body.

14. A(n) _____ current power source provides a voltage whose polarity and output voltage never change direction.

15. A(n) _____ is an electrical device by which alternating current of one voltage is changed to another voltage.

16. _____ is a force field that acts on some materials, but not on other materials.

17. _____ magnets are magnets that do not require any power or force to maintain their field.

18. The ends of a magnet are called magnetic _____.

19. A(n) _____ is an object that acts like a magnet, but its magnetic force is created and controlled by electricity.

20. _____ is resistance to magnetism.

▪ ACTIVITY 5-1 ▪ Electrical Theory—Law of Charges

FIGURE 5-14 Bar magnets.

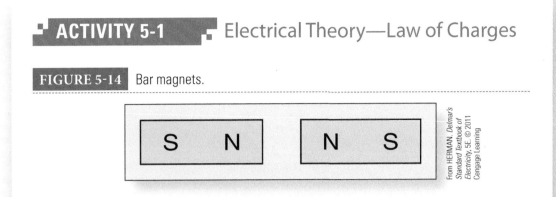

From HERMAN. *Delmar's Standard Textbook of Electricity*, 5E. © 2011 Cengage Learning

Materials

2 bar magnets with north and south poles identified (See Figure 5-14.)

Whiteboard or chalkboard

Part A

Your instructor will give you directions on how to proceed.

Questions

1. Explain the law of charges.

2. Explain what action will take place once the magnets are slowly brought together.

Part B

Your instructor will give you directions on how to proceed.

Question

1. Explain what action will take place once the magnets are slowly brought together.

Follow-up Questions

1. Identify some objects that use magnets to function.

2. What happens to the magnetic field when a conductor is wound into a coil?

Chapter 6

Electrical Theory

Master electrician, home improvement contractor, and business owner, **George Hockaday-Bey, G-11 Enterprises, Inc.** knows that "staying busy in a slow economy" is one of the greatest challenges of working in the electrical trade. But George also feels he knows the key to dealing with this challenge: "Education is the key to remaining relevant in [the] market. My education has been the reason for my success. Knowledge is critical." George began his 4-year IEC Chesapeake apprenticeship in 1992: "I was looking to better myself. I wanted to learn a skill that would keep me employed, and I didn't want to go to college for years and not know I would find a job," he explains. George notes two important advantages to entering the trade through an apprenticeship: "I chose the apprenticeship path [because] it allowed me to go to school at night and work during the day to support myself and my family. I [also] got the chance to find out that I really liked what I was going to school for and that it was the right path for me." He reminds current apprentice electricians that "the time spent learning today could lead to a lifetime of achievement and success," and advises, "Be the best at what you are doing and always look forward. Stay focused!"

Chapter Outline

ELECTRICAL QUANTITIES AND UNITS

CIRCUIT ESSENTIALS

OHM'S LAW

SERIES CIRCUITS

PARALLEL CIRCUITS

SERIES-PARALLEL CIRCUITS

Key Terms

Ampere

Chassis ground

Combination circuit

Coulomb

Current

Directly proportional

Earth ground

Electric power

Electrical blueprint

Inversely proportional

Kirchoff's Current Law (KCL)

Kirchoff's Voltage Law (KVL)

Multiple-load circuit

Node

Ohms

Resistance

Schematic diagram

Schematic symbol

Series circuit

Subscript

Volt

Voltage

Voltage drop

Watt

Wiring diagram

Chapter Objectives

After completing this chapter, you will be able to:

1. Define the following electrical quantities: ampere, volt, ohm, and watt.
2. List the four components necessary to have a complete circuit, and describe the functions of each.
3. Explain Ohm's law and calculate electrical values, using formulas and an Ohm's law formula chart.
4. Describe how current flows in a series circuit, and solve for circuit values.
5. Describe how current flows in a parallel circuit, and solve for circuit values.
6. Describe how current flows in a series-parallel circuit, and solve for circuit values.

Life Skills Covered

 Communication Skills

 Cooperation and Teamwork

Life Skills Goals

The life skill goals for this chapter center around Communication Skills, with a focus on Cooperation and Teamwork. Start by reading and considering the chapter's content on your own, but don't stop there; once you have formed your own opinions on the content, share them with your classmates. Then listen to what they have to say. How do their opinions of the chapter compare with your own? Agree with them or disagree—it's up to you. Just make sure to do it in a respectful, professional manner. Once you're on a jobsite, being able to effectively discuss differing opinions on an issue is critical to getting the job done.

After you have discussed the content, move on to the Review Questions, but don't tackle these alone. Choose a partner and answer the questions together. When you are done, ask yourself: Is it easier to tackle the harder questions when you have someone to bounce ideas off of? Are two heads really better than one? How does having a teammate make your job easier?

Where You Are Headed

An electrician must be able to read and understand schematic and wiring diagrams as well as electrical blueprints. This chapter introduces circuit symbols and basic electrical circuits. Ohm's law is presented, and you learn to analyze circuits and calculate circuit values. The information presented in this chapter and the skills you are developing will be invaluable as you begin your career in the electrical field.

Introduction

Expressing quantities and units is important when describing just about anything. If you were giving directions to a location, the distance (a quantity) in blocks or miles (units) would be important to mention. In describing electric circuits, you also use quantities and units. If you are going to work in the electrical field, you must know and understand electrical quantities and basic units.

6-1 Electrical Quantities and Units

Here are several important electrical quantities:

- Current
- Voltage
- Resistance
- Power

Current

Coulomb the practical unit of electrical charge. One coulomb of charge is the total charge on 6.25 × 10^{18} electrons.

Before we discuss current, let's recall that an electric charge is the electrical property possessed by electrons and protons. Because the negative charge on an electron is so small, scientists combine many of these charges so that measurements are possible. The practical unit of electrical charge is the coulomb. One coulomb of charge is the total charge on 6.25×10^{18} electrons. This important unit is used to describe other electrical units.

Current the rate of flow of electrons through a conductor.

Current is the rate of flow of electrons through a conductor. The letter I (referring to intensity) is the symbol used to represent current. Current is measured in **amperes (A),** often abbreviated as amps. The term *ampere* refers to the number of electrons passing a given point in 1 second. *One ampere* is equal to 1 coulomb of charge moving past a given point in 1 second. The flow of electric current can be compared to the flow of water through a pipe.

Ampere the number of electrons passing a given point in one second. *One ampere* is equal to 1 coulomb of charge moving past a given point in 1 second.

When electrons begin to flow, the effect is felt instantly all along the conductor. Although individual electrons travel only a few inches per second, current effectively travels through the conductor at near the speed of light (186,000 miles/second). **Figure 6-1** illustrates the instantaneous effect of electric impulses.

Figure 6-1 shows a pipe filled with table-tennis balls. If a ball is forced into the end of the pipe, the ball at the other end is forced out. Each time a ball enters one end of the pipe, another ball is forced out the other end. This same principle can be applied to electrons in a wire. There are billions of electrons in a wire. If an electron enters one end of a wire, another electron is forced out the other end. An instrument called an ammeter is used to measure current flow in a circuit.

Voltage

Voltage the electrical pressure, or the potential force, or the difference in electrical charge between two points.

The electrical pressure, or the potential force, or the difference in electrical charge between two points, is referred to as **voltage**. Voltage is the "electrical" pressure that pushes current through a wire much like water pressure pushes water through a pipe. Unlike current, voltage cannot flow. **Figure 6-2** shows voltage provided by a battery being compared to a pump that provides the water pressure that pushes water through a pipe.

The voltage level, or value, is proportional to the difference in the electrical potential energy between two points. Voltage is often thought of as the potential to do something. Combining the previous statements, we introduce another term: potential difference. Voltage must be present *before* current can flow. A voltage of 120 volts is present at a common wall outlet, but there is no current flow until some device is connected and a complete circuit exists. The letter E (electromotive force, or EMF) or the letter V (volt) represents voltage in formulas and equations. Voltage is measured in **volts (V),** the base unit for voltage. The device used to measure potential difference,

Volt the base unit for voltage.

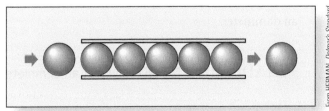

FIGURE 6-1 Pipe filled with table-tennis balls.

FIGURE 6-2 Voltage in an electric circuit compared to pressure in a water system.

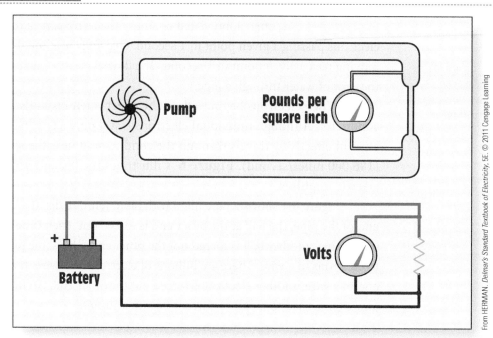

From HERMAN. *Delmar's Standard Textbook of Electricity*, 5E. © 2011 Cengage Learning

or voltage, is called a *voltmeter*. The meter shown in the electric circuit of Figure 6-2 is a voltmeter. Note that the meter measures a potential difference between two points (the probes).

Resistance

The opposition to the flow of electrons (current) is called resistance. It may be thought of as friction that slows the flow or lowers the level of electrical current. Resistance is measured in ohms. The Greek symbol Ω (omega) is used to represent ohms. The letter R (resistance) is used to represent ohms in formulas and equations. An ohm is the amount of resistance that allows 1 ampere of current to flow when the applied voltage is 1 volt.

Electric loads such as heating elements, lamps, motors, and so on, have resistance and are measured in ohms. **Figure 6-3** shows how resistance in an electric circuit can be compared to reducer in a water system.

Every electrical component has resistance, and this resistance changes electric energy into another form of energy such as heat, light, or motion. Even good conductors of electricity have resistance. This explains why some wires become warm when current flows through them. Resistance is measured with an instrument called an ohmmeter.

TRADE TIP   NEVER connect an ohmmeter to a live circuit!

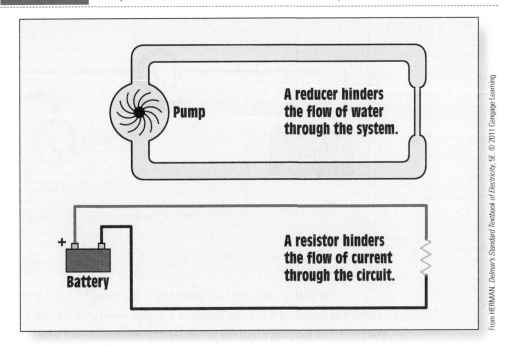

FIGURE 6-3 Comparison of a circuit resistance to a water system reducer.

Pump

A reducer hinders the flow of water through the system.

+

Battery

A resistor hinders the flow of current through the circuit.

From HERMAN. *Delmar's Standard Textbook of Electricity, 5E.* © 2011 Cengage Learning

Power

The amount of electric energy converted to another form of energy in a given length of time is **electric power (P)**. To better understand this definition, let's talk about energy. Energy is the ability to do work. Power is concerned with how fast work is done. Therefore, power is the rate of using energy or doing work. The symbol for power is P, and the base unit is the **watt**. Power in an electric circuit is equal to

$$\text{Power} = \text{Voltage} \times \text{Current}$$
$$\text{Watts} = \text{Volts} \times \text{Amperes}$$
$$P = EI$$

Where:

P is the power in watts

E is the voltage in volts

I is the current in amperes

The power of a circuit can be determined by measuring current and voltage and then using the formula: $P = IV$. **Figure 6-4** illustrates such a circuit.

As an example, let's set the battery voltage in Figure 6-4 at 120 volts. The resistance in this figure could be an electric heating element. Let's also say that the amperes being read by the ammeter are 2 amperes. With 120 volts across the heating element and 2 amperes flowing through it, the power produced is:

$$P = IE$$
$$P = 2\,A \times 120\,V$$
$$P = 240 \text{ watts or } 240\,W$$

Electric power
the amount of electric energy converted to another form of energy in a given length of time.

Watt the base unit for power.

FIGURE 6-4 Circuit illustrating current through and voltage across a resistance.

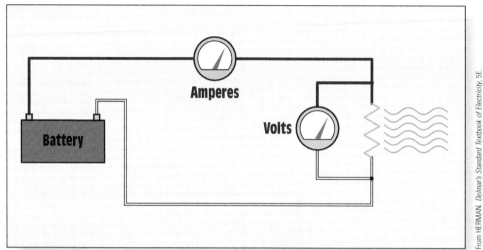

If the battery voltage in Figure 6-4 is increased to 240 volts, and the heating element is replaced with one that has a *higher resistance* so that current remains constant, the heating element provides 480 watts of heat (240 V × 2 A = 480 W). If the voltage remains at 120 volts, but the current is increased to 4 amperes by using a heating element with a *lower resistance*, the heating element again produces 480 watts (120 V × 4 A = 480 W). The amount of power used by the heating element is therefore determined by the amount of current flow and the voltage source.

There are many devices that are rated in watts. Examples include:

- Blow dryers
- Refrigerators
- Microwave ovens
- Fans
- Clothes washers.

Ratings are not always specified in watts; some ratings are specified in terms of voltage and current, and some devices list only voltage and watts.

True power, as measured in watts, can only exist when some type of energy change or conversion takes place. It is important to understand that true power described here refers to purely resistive dc circuits where there are no losses in the conversion of energy. When you study ac circuits, you learn about other oppositions to current that affects the true power. Two common examples of this energy conversion are electric energy to light energy (lamp) and electric energy to heat energy (toaster).

LIFE SKILLS

In this section, you learn some basic industry terminology. On the jobsite, why is it important for you to be able to use and understand these terms in conversation?

Self-Check 1

1. Define: current, voltage, resistance, and power.

All devices and machines powered by electricity use electric circuits. A closed electric circuit can be defined as a complete path from one side of a voltage source to the other.

When wiring circuits and working with components, you need to know how to read and understand schematic and wiring diagrams as well as electrical blueprints. To an apprentice electrician, reading complex diagrams may seem like a foreign language. A person learning how to read schematics can be compared to a child just beginning to read. The child looks at a newspaper and is confused with columns and rows, captions and pictures. That child learns to read, one article at a time, and to connect that article with its picture or caption. You too will learn to see and work with just one circuit at a time. You will learn how this circuit is wired to the next circuit or device, and so on. In the next section, you learn about circuits, components, and their symbols.

6-2 Circuit Essentials

Now that you are familiar with voltage, current, resistance, and power, let's take a closer look at these quantities in a basic electrical circuit.

Basic Electrical Circuit

Most complete circuits contain six parts:

- *Power source*—potential difference or voltage necessary to force current (electrons) through the circuit.
- *Conductor*—a low resistance path through which current flows from source to load.
- *Insulator*—a material that confines current to desired paths.
- *Load*—a component that uses or changes electric energy and limits current flow.
- *Control device*—a component that varies or turns on and off the flow of current.
- *Protection device*—a component that opens the current path if current exceeds a specified level.

If a complete circuit is defined as an uninterrupted current path, why do we need control devices? These components allow us to be able to turn off power entirely to a circuit we intend to work on. Remember; *before you work on a circuit **always** turn off (**disconnect**) the power!*

A protection device is not necessary for current to flow through a circuit, but it is absolutely essential to prevent damage to conductors and load devices. Think about those times when, in your home, a circuit breaker tripped (opened). That breaker opened a circuit that had too much current flowing through it. It protected devices on that circuit and may have prevented a fire.

Circuit Symbols and Diagrams

Schematic diagram
an electrical drawing that uses symbols to represent physical components.

Wiring diagram
a schematic drawing that shows the physical relationship of all components and the information needed to hardwire the circuit.

Electrical blueprint
a document containing all the necessary information needed to create an electric circuit.

Schematic symbol an electrical drawing used to represent a physical component.

Working with electricity requires you to understand symbols and be able to read and interpret schematic and wiring diagrams, as well as electrical blue prints. A schematic diagram is an electrical drawing that uses symbols to represent physical components. A wiring diagram is a schematic drawing that shows the physical relationship of all components and the information needed to hardwire the circuit. An electrical blueprint is a document containing all the necessary information needed to create an electric circuit. In the chapters on wiring, you will learn more about and use these diagrams.

You will find that it is more convenient to use schematic symbols to represent electric components than to draw pictures of the components. A schematic symbol is used to represent a physical component. **Figure 6-5** shows schematic symbols used to represent both fixed and variable resistors.

Other common electric components and their symbols are shown in **Figure 6-6**.

Note that insulated and uninsulated conductors are drawn the same. Insulation is assumed to be present wherever it is needed to keep components and conductors from making contact.

There are two methods used to list electrical values of components used in a circuit as shown on a schematic diagram. One method prints the component values next to the schematic symbols. The other method identifies the schematic symbol by printing a letter next to the schematic symbol, with the values of the components provided in an accompanying parts list.

LIFE SKILLS

For a circuit to function efficiently, all six of its parts must work in concert. The same is true on a jobsite. What is required of *you* to keep the jobsite "circuit" running?

FIGURE 6-5 Schematic symbols used to represent resistors.

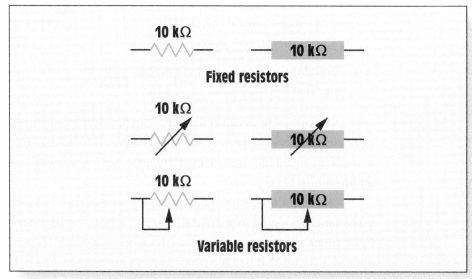

10 kΩ

10 kΩ

Fixed resistors

10 kΩ

10 kΩ

10 kΩ

10 kΩ

Variable resistors

FIGURE 6-6 Electric components and their symbols.

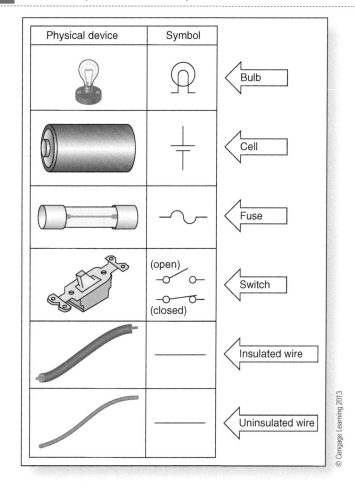

Self-Check 2

1. List the six parts of a circuit.

When you hand a store clerk a $20 bill for merchandise that costs $16.25, he or she does not say, "I'm going to use subtraction to determine your change." The clerk just applies the law of subtraction and gives you your change. So it is with electricians. When an electrician is asked to determine whether a 15-amp circuit can carry the load for a specific number of devices with a given circuit voltage, the electrician calculates the current and takes into account that a continuous-use circuit can be loaded to only 80% of its rating. The electrician doesn't say, "I am going to use Ohm's law and some basic math skills now." He or she just makes the calculations. For electricians, Ohm's law, like basic math skills, becomes second nature. Read on and learn about this law!

6-3 Ohm's Law

The amount of current (electron) flow in a circuit is determined by the voltage and resistance. This relationship is named Ohm's law because it was discovered in 1827 by Georg Ohm, a German scientist.

Recall that voltage is the force that causes current to flow. So, if voltage is increased, current is a higher value. And if voltage is decreased, there is a decrease in current. This relationship assumes that circuit resistance remains constant. The voltage is not affected by resistance or current. Therefore, if voltage remains unchanged, the current increases if resistance decreases; and current decreases if the resistance increases.

Ohm's law is often stated as follows: The current flowing in a circuit is directly proportional to the applied voltage and inversely proportional to the resistance.

Directly proportional means that changing one factor results in a direct change to another factor in the same direction and by the same magnitude. For example, in a circuit where the resistance is unchanged, if the voltage level is increased by 50%, the current increases 50%. Likewise, if the voltage decreases by 50%, the current decreases 50%.

Inversely proportional means that increasing one factor results in a decrease in another factor by the same magnitude, or alternately, decreasing one factor results in an increase of the same magnitude in another factor. For example, as long as voltage remains constant, if resistance increases by 50%, the current decreases by 50%. If the resistance decreases by 50%, the current increases by 50%.

Ohm's law formulas are:

$$E = I \times R$$
$$I = E/R$$
$$R = E/I$$

Where:

$$E = \text{EMF, or voltage}$$
$$I = \text{intensity of current, or amperage}$$
$$R = \text{resistance}$$

Example 1

What is the voltage in a circuit that has a resistance of 100 ohms and a current flow of 1 ampere?

$$E = I \times R$$
$$E = 1\,A \times 100\,\Omega$$
$$E = 100\,V$$

Example 2

What is the current in a circuit that has a 120-volt source and a resistance of 40 ohms?

$$I = E/R$$
$$I = 120\,V/40\,\Omega$$
$$I = 3\,A$$

Example 3

What is the circuit resistance if the applied voltage is 240 volts and circuit current is 6 amperes?

$$R = E/I$$
$$R = 240\,V/6\,A$$
$$R = 40\,\Omega$$

Directly proportional means that changing one factor results in a direct change to another factor in the same direction and by the same magnitude.

Inversely proportional means that increasing one factor results in a decrease in another factor by the same magnitude, or a decrease in one factor results in an increase of the same magnitude in another factor.

FIGURE 6-7   Ohm's law divided circle chart.

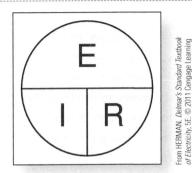

An aid to help you remember the Ohm's law relationships is shown in the Ohm's law chart with a divided circle, illustrated in **Figure 6-7**. Using this chart is simple. Cover the quantity you want to find, and perform the multiplication or division indicated. For example, if you need to determine voltage, cover the E. The remainder of the circle indicates I multiplied by R. Therefore, voltage (V) equals current (I) multiplied by resistance (R). The same method works for finding resistance (R) or current (I).

LIFE SKILLS

Imagine that you are an electrician and your apprentice asks you to explain Ohm's law and why it matters. How would you explain it?

Self-Check 3

1. If a circuit's voltage remains constant, but the resistance increases, how is the current affected?

2. Write the Ohm's law formula for voltage (E).

While trimming your tree during the holiday season and checking strings of lights, did you or perhaps your mom or dad say, "This string of lights won't light up. Throw it out!" Chances are, only one bulb was burned out. Because the bulbs were connected in series, when one bulb burned out, no other bulb in the string would light. The good news is that newer series-connected light strings have shorting bars. When one bulb burns out, the bulb leads are connected together by the shorting bar, and the rest of the string stays lit. Read on and learn about series circuits!

6-4 Series Circuits

The basic electric circuit described in Section 6-2 is a series circuit. Most circuits however, are designed to operate with more than one load. Let's take a closer look at series circuits with more than one load.

Series Multiple-Load Circuits and Subscripts

A series circuit is a circuit that has two or more loads but only one path from which the current flows, from the voltage source, through circuit loads, and back to the source. In a series circuit, the current is the *same* through all circuit components. Circuits that contain two or more loads are sometimes referred to as multiple-load circuits. However, multiple-load circuits also include parallel and series-parallel circuits, described later in this chapter, along with series circuits.

When working with multiple-load circuits, there has to be a way to distinguish one load from another. Subscripts are used to identify the different loads, components, devices, and so on, in a circuit. A subscript is a small letter or number written or printed to the right of and slightly below a letter symbol. Subscripts may be used to identify different variables that are similar enough to each other to have the same main letter symbol. **Figure 6-8** illustrates a series circuit.

In Figure 6-8, notice that there is only one path from which current can flow. Also note that this is a multiple-load circuit (three resistors). The use of subscripts identifies each resistor as R_1, R_2, and R_3. Subscripts are also used in this circuit to designate total current (I_T) and the voltage source, or total voltage (V_T).

Voltage Drops

Voltage is that pressure or electromotive force that causes electrons (current) to move through a load (resistance). The voltage required is determined by the level of current and the amount of resistance. The voltage or potential energy difference across a resistor is called a voltage drop. In other words, a voltage is developed across the resistor. Keep in mind that only part of the potential energy difference of the source appears across each resistor (in a multiple load circuit). It must be understood that there is a distinction between source voltage and voltage across the loads. The *source* voltage provides the electric energy, and the *load* voltage converts the electric energy into another form.

In a series circuit, the *sum* of all the voltage drops across all circuit resistors must *equal* the voltage source. This statement is generally referred to as

FIGURE 6-8 Series circuit.

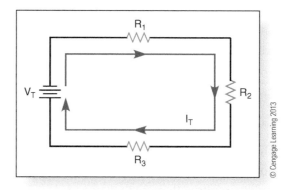

FIGURE 6-9 Series circuit illustrating voltage drops.

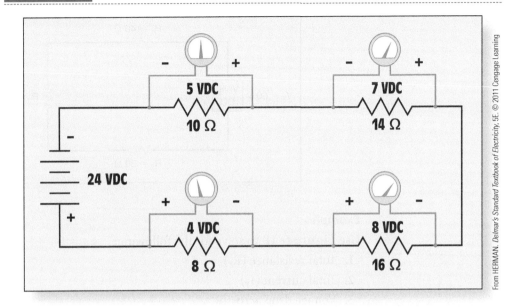

Kirchoff's Voltage Law (KVL) states that the sum of all the voltage drops across all circuit resistors must equal the voltage source.

Kirchoff's voltage law (KVL). The amount of voltage drop across each resistor is proportional to its resistance and the circuit current. In other words, the voltage supplied by the source is distributed among the circuit resistors according to what is called the resistance law of proportion. This law of proportion states that the source voltage is *divided* among all resistors, according to the proportion of resistance *each* resistor has relative to the total resistance. **Figure 6-9** shows a series circuit configured with resistors of different values.

Voltmeters connected across resistors show that the voltage drop across each resistor is proportional to its resistance. If you add measured voltages across each resistor, the total (measured) voltage equals the source voltage (24 Vdc).

Resistance

In a series circuit, there is only one path for current to flow. Current must flow through each resistor in the circuit, and each resistor opposes current flow. The total amount of opposition (total circuit resistance) is equal to the sum of all series-connected resistors.

Calculating Series Circuit Values

When calculating *series* circuit values, Ohm's law is very important. However, there are other rules that help to determine values of voltage, current, resistance, and power. They are as follows:

- At any point in a circuit, *the current is the same.*
- Total circuit resistance is equal to the *sum* of all series resistances.
- *Source* voltage is equal to the sum of the voltage drops across all resistors.

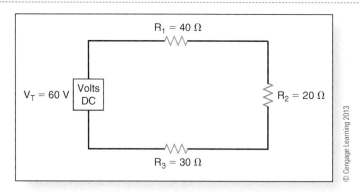

FIGURE 6-10 Series circuit.

Example

Use **Figure 6-10** to solve for the following:

1. Total resistance (R_T)
2. Total current (I_T)
3. Voltage drop across R_2
4. Power dissipated by R_2

1. Total resistance (R_T) = $R_1 + R_2 + R_3$
$$R_T = 40\ \Omega + 20\ \Omega + 30\ \Omega$$
$$R_T = 90\ \Omega$$

2. Total current (I_T) = E/R_T
$$I_T = 60\ V/90\ \Omega$$
$$I_T = 0.67\ A$$

3. Voltage across R_2 (V_{R2}) = $I_2 \times R_2$
(*Note:* Because this is a series circuit, $I_T = I_1 = I_2 = I_3 = 0.67\ A$) Therefore:
$$V_{R2} = 0.67\ A \times 20\ \Omega$$
$$V_{R2} = 13.4\ V$$

4. Power dissipated by R_2 (P_2) = $V_{R2} \times I_2$
$$P_2 = 13.4\ V \times 0.67\ A$$
$$P_2 = 9\ W$$

Ground as a Reference

Earth ground
a ground point made by physically driving a pipe or rod into the ground.

Chassis ground
a point used as a common connection from other parts of a circuit.

We have seen some schematic symbols (refer to Figure 6-6) for some commonly used components. There are many other symbols as well. There are two schematic symbols used to represent ground, illustrated in **Figure 6-11**.

The symbol shown in Figure 6-11 (A) is an earth ground. This symbol denotes a ground point made by physically driving a pipe or rod into the ground. This symbol is also used to represent common ground connections that are connected to the main (physically connected) electrical ground point.

The ground symbol illustrated in Figure 6-11 (B) is known as chassis ground. Chassis ground is a point used as a common connection from other parts of a circuit. It is not connected to a rod driven into the ground.

FIGURE 6-11 Ground symbols.

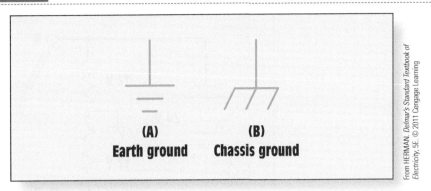

Figure 6-12 shows that a 120-volt, single-phase, 2-wire electrical system must have one conductor grounded. The (earth) ground symbol is used.

In circuits that divide voltage, ground is typically used to provide a common reference point to produce voltages that are above and below ground. Figure 6-13 illustrates how a common (chassis) ground is used to produce above- and belowground voltage.

An aboveground voltage is a voltage that is positive with respect to ground. A belowground voltage is negative with respect to ground. In Figure 6-13, one terminal of a zero-center voltmeter is connected to ground. If the probe is connected to A, the needle on the meter gives a negative reading of voltage. If the probe is connected to B, the needle indicates a positive voltage.

FIGURE 6-12 Earth ground symbol signifying where to "ground" one conductor.

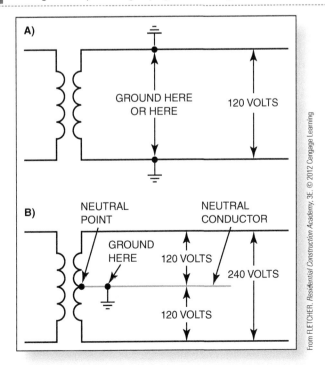

146

FIGURE 6-13 Using a common (chassis) ground when making a voltage measurement.

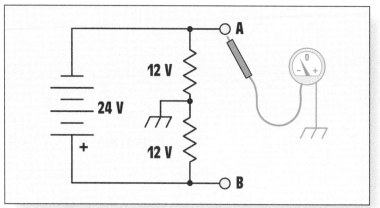

From HERMAN, *Delmar's Standard Textbook of Electricity*, 5E. © 2011 Cengage Learning

LIFE SKILLS

When explaining a process, sometimes writing it down is the most effective method. Write a short note explaining how to calculate the total resistance of a series circuit.

Self-Check 4

1. What is a multiple-load circuit?
2. Explain Kirchoff's voltage law.
3. What can be said about the current in a series circuit?

Most people may not be familiar with the term *parallel-connected*, but when we turn a light off in our home, we know that other lights stay on. And when we plug or unplug a device into an outlet, that action has no effect on devices plugged into other outlets. What do you think would happen if the lights in our homes were wired in series and a bulb burned out? You guessed it … all lights would be out. Parallel wiring is used so lights and receptacles work independently of each other. As a matter of fact, if you ever jump-started a car, using your car's battery, you connected your battery in *parallel* with the other battery. Read on and learn about parallel circuits.

6-5 Parallel Circuits

Parallel circuits are multiload circuits that have more than one path for current. Individual current paths are called branches. **Figure 6-14** illustrates a parallel circuit.

Voltage

In Figure 6-14, note that the parallel resistors are connected like rungs on a ladder, and the voltage source (battery) is across each load. Because the terminals of the battery are effectively connected across each resistor, the voltage drop across any branch of a

From HERMAN. *Delmar's Standard Textbook of Electricity*, 5E. © 2011 Cengage Learning

FIGURE 6-14 Parallel circuit.

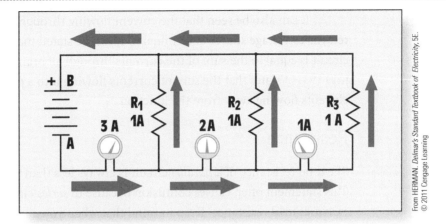

parallel circuit is equal to the applied voltage. The relationship of source voltage to load is expressed as

$$\text{Total source voltage } (V_T) = V_{R1} = V_{R2} = V_{R3}$$

Because of the voltage-to-load relationship, parallel circuits are more frequently used in wiring homes and buildings. **Figure 6-15** shows an example of how lights and receptacles are connected in parallel.

Note that in Figure 6-15, a voltage, for example, 120 V, is supplied across receptacles and lamps. You can see that the receptacles always have this 120 V across them. However, until a switch is closed and *current flows,* a lamp is not lit. The 120 V is across the open switch. We learn more about voltage drops across open switches (and open fuses) later in the text.

Current

It can also be seen in Figure 6-14 that if it is assumed that current leaves battery terminal A and returns to battery terminal B, there are three separate paths that current flow can take. By reading ammeters connected in this circuit, you can see the following:

- 3 amperes of current leave terminal A.
- The second ammeter reads 2 amperes.

FIGURE 6-15 Lights and receptacles connected in parallel.

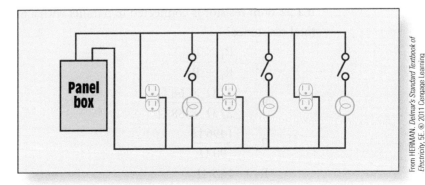

From HERMAN. *Delmar's Standard Textbook of Electricity*, 5E. © 2011 Cengage Learning

Therefore, 1 ampere must be flowing through R_1. The last ammeter reads 1 ampere, which flows through R_3. This means that 1 ampere must be flowing through R_2.

It can also be seen that the current flowing through each resistor recombines and returns to voltage source at terminal B. It can be stated that the total current flow in the circuit is equal to the sum of the currents through all branches. **Kirchoff's current law (KCL)** states that the sum of currents flowing into a junction equals the sum of currents flowing away from the junction.

Kirchoff's Current Law (KCL) states that the sum of currents flowing into a junction equals the sum of currents flowing away from the junction.

Resistance

The total resistance of a *parallel* circuit is always less than the lowest branch resistance. This statement often causes confusion because in *series* circuits, adding resistors *increases* total resistance. Keep in mind that when you connect resistors in parallel, you have *more* paths through which current can flow. This results in *less* total opposition to the total current flow and a *decrease* in total resistance.

There are three formulas that are used to determine the total resistance of a parallel circuit. They are as follows:

- Equal value resistors
- Product over sum
- Reciprocal

If resistors of *equal value* are connected in parallel, the total resistance is equal to the value of one resistor divided by the number of resistors. This formula can be expressed as:

$$R_T = R/n$$

Where: R is the value of one resistor and n is number of parallel resistors.

Example 1

If three resistors are connected in parallel, each having a resistance value of 33 ohms, what is the total circuit resistance?

$$R_T = 33 \ \Omega/3$$
$$R_T = 11 \ \Omega$$

If two resistors are connected in parallel, a formula called the *product over sum* can be used to calculate total resistance. This formula can be expressed as:

$$R_T = \frac{R_1 \times R_2}{R_1 + R_2}$$

Example 2

If a 22-ohm resistor is connected in parallel with a 68-ohm resistor, what is the total resistance?

$$R_T = \frac{R_1 \times R_2}{R_1 + R_2}$$
$$R_T = \frac{22 \ \Omega \times 68 \ \Omega}{22 \ \Omega + 68 \ \Omega}$$
$$R_T = \frac{1496 \ \Omega}{90 \ \Omega}$$
$$R_T = 16.7 \ \Omega$$

If there are *more* than two resistors connected in parallel, a formula referred to as the *reciprocal formula* can be used. This formula can be expressed as:

$$R_T = \cfrac{1}{\dfrac{1}{R_1} + \dfrac{1}{R_2} + \dfrac{1}{R_3} + \dfrac{1}{R_n}}$$

(*Note:* The value R_n is used to signify the number of resistors in a circuit. If a circuit has 12 resistors connected in parallel, the last resistor in the formula would be 12.)

Example 3

Three resistors having values of 47 ohms, 68 ohms, and 100 ohms are connected in parallel. What is their total resistance?

$$R_T = \cfrac{1}{\dfrac{1}{R_1} + \dfrac{1}{R_2} + \dfrac{1}{R_3}}$$

$$R_T = \cfrac{1}{\dfrac{1}{47\ \Omega} + \dfrac{1}{68\ \Omega} + \dfrac{1}{100\ \Omega}}$$

$$R_T = \cfrac{1}{0.0212\ S + 0.0147\ S + 0.01\ S}$$

$$R_T = \cfrac{1}{0.046\ S}$$

$$R_T = 21.7\ \Omega$$

It is also possible to use the product over sum method in the previous example. However, you must first use product over sum for R_1 and R_2. Find the total (equivalent) resistance for that combination and then use product over sum for the equivalent resistance of R_1 and R_2 *and* the resistance, R_3.

Calculating Parallel Circuit Values

Use **Figure 6-16** (see page 150) and solve for the following:

1. R_T

2. I_{R2}

3. I_T

4. P_T

1. Total resistance

$$R_T = \cfrac{1}{\dfrac{1}{33\ \Omega} + \dfrac{1}{47\ \Omega} + \dfrac{1}{100\ \Omega}}$$

$$R_T = \cfrac{1}{0.0303\ S + 0.0212\ S + 0.01\ S}$$

$$R_T = \cfrac{1}{0.062\ S}$$

$$R_T = 16.12\ \Omega$$

FIGURE 6-16 Parallel circuit.

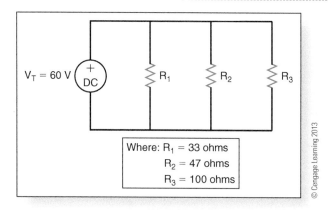

Where: R_1 = 33 ohms
R_2 = 47 ohms
R_3 = 100 ohms

© Cengage Learning 2013

2. Current through R_2

$I_{R2} = V_{R2}/R_2$ (*Note:* In a parallel circuit, source voltage V_T is across all loads.)
Therefore, $V_{R2} = V_T$.

I_{R2} = 60 V/47 Ω

I_{R2} = 1.28 A

3. Total current

$I_T = V_T/R_T$

I_T = 60 V/16.12 Ω

I_T = 3.72 A

4. Total power

$P_T = V_T \times I_T$

P_T = 60 V $\times$ 3.72 A

P_T = 223.2 W

Use **Figure 6-17** and solve for the following:

1. R_T

2. I_{R3}

3. I_T

4. P_T

FIGURE 6-17 Parallel circuit.

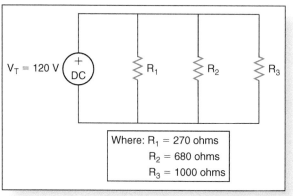

Where: R_1 = 270 ohms
R_2 = 680 ohms
R_3 = 1000 ohms

© Cengage Learning 2013

1. Total resistance

   (Using product over sum for R_1 and R_2)

   $$R_{1,2} = \frac{R_1 \times R_2}{R_1 + R_2}$$

   $$R_{1,2} = \frac{270 \times 680}{270 + 680}$$

   $$R_{1,2} = \frac{183,600}{950}$$

   $$R_{1,2} = 193\ \Omega$$

   The equivalent resistance for the parallel combination of R_1 and R_2 equals 193 Ω. (Using product over sum for $R_{1,2}$ and R_3)

   $$R_{1,2,3} = \frac{R_{1,2} \times R_3}{R_{1,2} + R_3}$$

   $$R_{1,2,3} = \frac{193 \times 1000}{193 + 1000}$$

   $$R_{1,2,3} = \frac{193,000}{1,193}$$

   $$R_{1,2,3} = 162\ \Omega$$

   The equivalent or total resistance (R_T) for the parallel resistors R_1, R_2, and R_3 equals 162 Ω.

2. Current through R_3

   $$I_{R3} = V_T/R_3$$
   $$I_{R3} = 120\ V/1000\ \Omega$$
   $$I_{R3} = 0.12\ A$$

3. Total current

   $$I_T = V_T/R_T$$
   $$I_T = 120\ V/162\ \Omega$$
   $$I_T = 0.74\ A$$

4. Total power

   $$P_T = V_T \times I_T$$
   $$P_T = V_T \times I_T$$
   $$P_T = 120\ V \times 0.74\ A$$
   $$P_T = 89\ W$$

LIFE SKILLS

Choose a partner, and discuss some of the primary differences between series and parallel circuits. Make extra note of the differences in calculating their values.

Self-Check 5

1. What can be said about voltage in a parallel circuit?
2. Explain Kirchoff's current law.
3. Explain how adding a resistance in parallel decreases the total resistance.

Combination circuit a circuit configured (wired) with both series and parallel connections.

Circuits are sometimes connected in series and sometimes connected in parallel. However, there are instances when circuits are configured (wired) with both series and parallel connections. These circuits are sometimes referred to as combination circuits. Information already learned about series circuits still applies when working with series-parallel circuits. Relationships learned about parallel circuits also apply to the parallel-connected components of series-parallel circuits. Read on and learn about combination circuits.

6-6 ◢ Series-Parallel Circuits

It is important to learn how to distinguish between resistors connected in series and resistors connected in parallel. When working with series-parallel circuits, remember that current flow determines whether the resistor is connected in series or parallel. Always begin at the negative terminal of the battery, and apply the following rules:

- If the total current has only one path to follow through a component, then that component is connected in series.

- If the total current has two or more paths to follow through two or more components, then those components are connected in parallel.

A series-parallel circuit is illustrated in **Figure 6-18**. It is assumed that current flows from the battery's negative terminal (Point A) through components and back to the source voltage (Point B).

FIGURE 6-18 A series-parallel circuit.

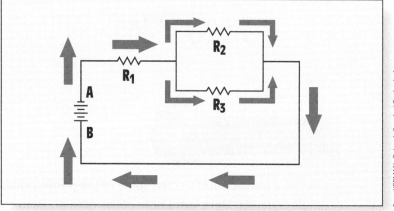

From HERMAN. *Delmar's Standard Textbook of Electricity*, 5E. © 2011 Cengage Learning

To determine the series and parallel elements, follow the current path. All current leaving Point A must flow through R_1. R_1 is therefore a series-connected component and is in *series* with the rest of the circuit. When the current flows through R_1, it arrives at the junction of R_2 and R_3. Note, however, the current splits. A junction point such as this is referred to as a node. A node refers to any point on a circuit where two or more circuit elements meet. Because the current now has two paths to follow, some current flow through R_2 and some through R_3, these resistors are in parallel.

Node refers to any point on a circuit where two or more circuit elements meet.

Before we solve a series-parallel circuit, let's review circuit rules and laws for series-connected components and for parallel-connected components. They are as follows:

Series Circuits

- The current is the same at any point in the circuit.
- The total resistance is the sum of the individual resistances.
- The applied voltage is equal to the sum of the voltage drops across all resistors.

Parallel Circuits

- The voltage drop across any branch of a parallel circuit is the same as the voltage source.
- The total current flow is equal to the sum of the currents through all circuit branches.
- The total resistance is equal to the reciprocal of the sum of the reciprocals of branch resistances.

Use **Figure 6-19** to solve for the following:

1. R_T

2. V_T

3. V_{R4}

FIGURE 6-19 A series-parallel circuit.

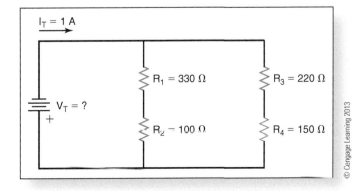

1. Total resistance (R_T)

   (*Note:* Because R_1 and R_2 are connected in series, their values can be added to form an equivalent resistance $R_{1,2}$. Resistors R_3 and R_4 are also in series and are added to form $R_{3,4}$. Because this circuit has been reduced to a simple parallel circuit, total resistance can be calculated using product over sum.)

$$R_T = \frac{430\ \Omega \times 370\ \Omega}{430\ \Omega + 370\ \Omega}$$

$$R_T = \frac{159{,}100\ \Omega}{800\ \Omega}$$

$$R_T = \frac{159{,}100\ \Omega}{800\ \Omega}$$

$$R_T = 200\ \Omega$$

2. Total voltage (V_T)

$$V_T = I_T \times R_T$$
$$V_T = 1\ A \times 200\ \Omega$$
$$V_T = 200\ V$$

3. Voltage drop across R_4 (V_{R4})

$$V_{R4} = I_{R4} \times R_4$$

   (*Note:* Because R_3 and R_4 are in series and V_T is across their combined resistance, current through R_4 can be calculated as follows:)

$$R_{3,4} = R_3 + R_4$$
$$R_{3,4} = 220\ \Omega + 150\ \Omega$$
$$R_{3,4} = 370\ \Omega$$
$$I_{3,4} = V_T/R_{3,4}$$
$$I_{3,4} = 200\ V/370\ \Omega$$
$$I_{3,4} = 0.541\ A$$

   (*Note:* Because R_3 and R_4 are in series, $I_{3,4} = I_3 = I_4 = 0.541$ A.)

$$V_{R4} = I_{R4} \times R_4$$
$$V_{R4} = 0.541\ A \times 150\ \Omega$$
$$V_{R4} = 81.2\ V$$

Use **Figure 6-20** to solve for the following:

1. R_T
2. I_T
3. I_{R1}
4. V_{R2}
5. I_{R4}
6. V_{R3}

FIGURE 6-20 A series-parallel circuit.

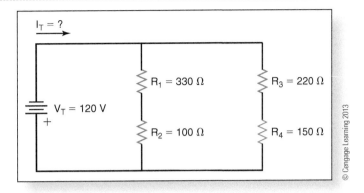

© Cengage Learning 2013

1. Total resistance

   (*Note:* Because R_1 and R_2 are connected in series, their values can be added to form an equivalent resistance $R_{1,2}$. Resistors R_3 and R_4 are also in series and are added to form $R_{3,4}$. Because this circuit has been reduced to a simple parallel circuit, total resistance can be calculated using product over sum.)

   $$R_T = \frac{430\ \Omega \times 370\ \Omega}{430\ \Omega + 370\ \Omega}$$

   $$R_T = \frac{159{,}100\ \Omega}{800\ \Omega}$$

   $$R_T = \frac{159{,}100\ \Omega}{800\ \Omega}$$

   $$R_T = 200\ \Omega$$

2. Total current (I_T)

   $$I_T = V_T/R_T$$
   $$I_T = 120\ V/200\ \Omega$$
   $$I_T = 0.6\ A$$

3. Current flow through R_1 (I_{R1})

   (*Note:* Because V_T (120 V) is across both R_1 and R_2, the current flow through these series-connected resistors can be calculated by first adding R_1 and R_2, then dividing the total voltage (V_T) by their combined resistance. (R_1 and R_2 are added because they are in series.))

   $$R_{1,2} = 330\ \Omega + 100\ \Omega$$
   $$R_{1,2} = 430\ \Omega$$

   (Calculate the current flow through the *series-connected* resistors.)

   $$I_{R1,R2} = V_T/R_{1,2}$$
   $$I_{R1,R2} = 120\ V/430\ \Omega$$
   $$I_{R1,R2} = 0.28\ A - I_{R1} = I_{R2}\ (\text{because } R_1 \text{ and } R_2 \text{ are connected in series})$$

   Remember, the current flow through *series-connected* components is the *same!*

4. Voltage across R_2 (V_{R2})

$$V_{R2} = I_{R2} \times R_2$$
$$V_{R2} = 0.28 \text{ A} \times 100 \text{ } \Omega$$
$$V_{R2} = 28 \text{ V}$$

5. Current through R_4 (I_{R4})

Total current (I_T) was calculated and equals 0.6 A.

Using Kirchhoff's current law (KCL), which states that the sum of currents flowing into a junction equals the sum of currents flowing away from the junction, current flow through I_{R1} was calculated and equals I_{R2} and is 0.28 A. Current flow through R_3 and R_4 can be calculated by:

$$I_{R3} = I_{R4} = I_T - I_{R1}$$
$$I_{R3} = I_{R4} = 0.6 \text{ A} - 0.28 \text{ A}$$
$$I_{R3} = I_{R4} = 0.32 \text{ A}$$

6. Voltage across R_3 (V_{R3})

$$V_{R3} = I_{R3} \times R_3$$
$$V_{R3} = 0.32 \text{ A} \times 220 \text{ } \Omega$$
$$V_{R3} = 70.4 \text{ V}$$

LIFE SKILLS

It would be difficult to understand series-parallel circuits without first understanding series and parallel circuits. How do the basics you learn from your journeyman prepare you for more complicated issues that you may face alone after the apprenticeship ends?

Self-Check 6

1. What is a combination circuit?

2. What is a node?

3. How can total resistance be determined if a series-parallel circuit is reduced to a simple parallel circuit?

Summary

- Current, the rate of flow of electrons through a conductor, is measured in amperes and is symbolized with the letter I.

- Voltage, the electrical pressure or difference in electrical charge between two points, is measured in volts and is symbolized by the letter V.

- Resistance, the opposition to the flow of current, is measured in ohms and is symbolized by the Greek symbol Ω (omega).
- Power, the electric energy converted from one form of energy to another form of energy in a given length of time, is measured in watts and is symbolized by the letter W.
- Most complete circuits contain six parts: power source, conductors, insulators, load, control device, and protection device.
- Schematic symbols are used to represent physical components.
- Ohm's law states that the current flowing in a circuit is directly proportional to the applied voltage and inversely proportional to the resistance.
- Ohm's law formulas are $E = I \times R$, $I = E/R$, and $R = E/I$.
- A series circuit has two or more loads but only one path for current to flow.
- Multiple-load circuits contain two or more loads.
- Subscripts are used to identify different loads, components, devices, and so on, in a circuit.
- Kirchoff's voltage law states that in a series circuit, the sum of all the voltage drops across all circuit resistors must equal the voltage source.
- In a series circuit, the total amount of resistance is equal to the sum of all series-connected resistors.
- Parallel circuits are multiload circuits that have more than one path for current.
- The voltage drop across any branch of a parallel circuit is equal to the applied voltage.
- Kirchoff's current law states that the sum of currents flowing into a junction equals the sum of currents flowing away from the junction.
- The total resistance of a parallel circuit is always less than the lowest branch resistance.
- Parallel total resistance formulas are: equal value resistors, product over sum, and reciprocal.
- A series-parallel circuit, also called a combination circuit is a circuit with both series-connected components and parallel-connected components.

Review Questions

True/False

1. The letter A is the symbol used to represent current.   (True, False)

2. The difference in electrical charge between two points is referred to as wattage.   (True, False)

3. The base unit for resistance is represented by Θ (theta).   (True, False)

4. Power is the rate of doing work.   (True, False)

5. Power is equal to voltage multiplied by current.   (True, False)

Multiple Choice

6. In an electrical circuit, the device that uses or changes electric energy and limits current flow is the _____.
   A. Power source
   B. Conductor
   C. Load
   D. Protection device

7. A circuit diagram that uses symbols to represent physical components is called a _____.
   A. Wiring diagram
   B. Schematic diagram
   C. Blueprint
   D. Pictorial diagram

8. What is Ohm's law?
   A. Current is directly proportional to voltage.
   B. Current is inversely proportional to voltage.
   C. Voltage is inversely proportional to current.
   D. Resistance is directly proportional to current.

9. Which of the following is *not* an Ohm's law formula?
   A. $E = I \times R$
   B. $I = E/R$
   C. $R = I \times E$
   D. $R = E/I$

10. A circuit that has two or more loads but only one path for current to flow is a _____.
    A. Series circuit
    B. Parallel circuit
    C. Series-parallel circuit
    D. Combination circuit

Fill in the Blank

11. The voltage across a resistor is called a voltage _____.

12. In a series circuit, the _____ of all the voltage drops across all circuit resistors must equal the voltage source.

13. In a series circuit, the total resistance is equal to the _____ of all series-connected resistors.

14. A(n) _____ ground symbol denotes a ground point made by physically driving a pipe or rod into the ground.

15. _____ circuits are multiload circuits that have more than one path for current.

True/False

16. The voltage drop across any branch of a parallel circuit
is equal to the applied voltage. (True/False)

17. Kirchoff's current law states that the sum of currents flowing
into a junction equals the sum of currents flowing away from
the junction. (True/False)

18. The total resistance of a parallel circuit is always greater than
the lowest branch resistance. (True, False)

19. If the total current has two or more paths, then components
are connected in series. (True/False)

20. A node refers to any point on a circuit where two or more
circuit elements meet. (True/False)

Chapter 7

Introduction to the
National Electrical Code (NEC)

Career Profile

Jimmie Slemp is an Instructor and Lab Manager in the Western Electrical Contractors Association's (WECA) apprenticeship program. After serving in the United States Army, where he became a power generation mechanic and worked on portable generators, Slemp completed an electrical apprenticeship and spent 15 years working primarily in new construction. He then "started working more in service, which entailed a wide variety of smaller electrical installations and repairs." Jimmie says that he enjoyed the variety and challenges of "troubleshooting" in the service department and that working in this area led him to seek teaching opportunities.

Having been a teacher for 7 years, Slemp believes that it is important for students to obtain both classroom instruction and hands-on experience while they are training to become electricians. Teaching and learning in each of these situations can be difficult, however. For example, Jimmie notes, teaching electrical theory is challenging: "Electricity is not always seen in the conventional mechanical way; it is not always easy to explain how electricity is 'working' without seeing it." But just as challenging is conducting labs that provide students with practical experiences: "The projects have to be relevant to what is being taught in the classroom, and lab safety is always a concern. Quite simply, electricity can be dangerous."

In terms of subject matter that it is important for students to learn, Slemp highlights knowledge of the *National Electric Code (NEC)*. This knowledge is important because "you always want to make sure that your installation [or service work] is *Code* compliant. . . . When [one is] faced with troubleshooting, understanding electrical theory helps take some of the guesswork out of it."

When he advises students about the electrical trade, Slemp asserts, "Being a good electrician is as much about being a good mechanic as it is about understanding electricity." He continues, "Take it *all* in, not just what you happen to be working on at the moment. Look at the big picture, learn how things work together, and ask questions. [Also remember to] keep on learning—either in a classroom, on the computer, or on the job. There are so many different aspects of electricity and electrical work, you *never know where you are going to end up.*"

Chapter Outline

THE HISTORY OF THE *NATIONAL ELECTRICAL CODE (NEC)*

CHANGING AND WRITING *CODE* AND CODE-MAKING PANELS

THE PURPOSE AND IMPORTANCE OF THE *NATIONAL ELECTRICAL CODE*

CODE BOOK CHAPTERS—ARRANGEMENT AND SUMMARY

NEC STANDARDS AND LOCAL AUTHORITIES

USING THE *CODE* BOOK

Key Terms

Authority Having Jurisdiction (AHJ)

Bonded (bonding)

Code-Making Panel (CMP)

Listed

Luminaires

National Fire Protection Association (NFPA)

Nationally Recognized Testing Laboratories (NRTL)

Underwriters Laboratories (UL)

Chapter Objectives

After completing this chapter, you will be able to:

1. Describe the history of the *NEC*.
2. Explain how codes are formed by code-making panels.
3. Highlight the importance and the intent of the *Code*.
4. Identify, list, and summarize chapters of the *Code book*.
5. Explain the process for changing codes.
6. Explain why a local authority may raise standards.
7. Locate information in the *Code book*.

Life Skills Covered

 Self-Advocacy

 Cooperation and Teamwork

 Managing Stress

Life Skills Goals

With the possible exclusion of the text you are working from right now, there will never be a written resource that is of more value to you as an electrician than the *National Electric Code*. The life skills for this chapter show you what having a firm grasp of the text and its organization can do for you!

The first skill we cover is Self-Advocacy. Understanding the *Code* and how to use it as a source of information not only helps you in your apprenticeship, but being able to cite its standards makes you a more attractive candidate for full-time employment. This chapter also covers Cooperation and Teamwork. Creating the *NEC* is a group effort, and learning it should be too. Once you have finished reading this chapter, write four or five "test questions" on the Chapter 7 material. Then, trade questions with a partner and see how you do with answering

each other's questions. As is the case in the field, two people looking at the same situation can often see two different things. It's not until views are combined that you see the complete picture. The final life skill that we cover is Managing Stress. So relax, take a deep breath, and get to work!

Where You Are Headed

The National Electrical Code is the reference book most often used by electricians on the job. In this chapter, you learn not only what this *Code* is, but more importantly, *how to use* the *NEC*. You will soon realize that the *National Electrical Code* provides countless rules that guide you in each and every part of an electrical installation.

Introduction

National Fire Protection Association (NFPA) an international standards-making organization dedicated to the protection of people from the ravages of fire and electric shock.

The *National Electrical Code (NEC)*, or NFPA 70, is a United States standard for the safe installation of electrical wiring and equipment. It is part of the National Fire Codes series published by the **National Fire Protection Association (NFPA)**. The NFPA is an international standards-making organization dedicated to the protection of people from the ravages of fire and electric shock. The *NEC* is not itself a U.S. law; however, use of the *NEC* is commonly mandated by state or local law, as well as by many jurisdictions outside of the United States.

Simply put, the *NEC* codifies the requirements for safe electrical installations into a single, standardized source. The *National Electrical Code*, or *NEC*, serves as a guide for safe wiring practices in the electrical field. The *NEC* is not meant to be an instructional manual or a "how-to" guide. The *NEC* is updated and published every 3 years, and updated versions reflect the latest changes and trends in the electrical industry. This *Code* book contains specific rules to help safeguard people and property from hazards arising from the use of electricity. (See *NEC 90.1.*) Throughout the *NEC*, you see the terms *shall* and *shall not*. By definition, *shall* is used to express a command. So it is here; *shall* means "must" and *shall not* means "must not." Those individuals who are involved in the design, installation, or maintenance of electrical wiring should become very familiar with the *NEC*.

Many electricians refer to the *National Electrical Code* as "the electrician's bible." This demonstrated regard for knowing and applying the *NEC* emphasizes the importance of this *Code*!

7-1 The History of the *National Electrical Code (NEC)*

The *National Electrical Code* was published for the first time in 1897. The driving force behind developing such a code book was the many electrical fires caused by electric lighting systems installed during the late 1800s. By 1881, 65 textile mills in the New England area had been destroyed or badly damaged from fires that were started by faulty electric lighting systems. Updates and revisions occurred on a regular basis during the early years of the *NEC*. This was done because the electrical construction field and the uses of electricity were still in the developing stages. In 1975 and thereafter, up to this day, the *NEC* uses a 3-year cycle to publish new editions containing *Code* additions, revisions, and/or deletions. The *NEC* is available as a bound book containing approximately 1000 pages. It has been available in electronic form since 1993. The *National Electrical Code* is considered to be among the finest building code standards in use today.

TRADE TIP The *NEC* online version is available at the **NFPA** website (http://www.nfpa.org). You must install **JAVA** on your computer, and you must register (free of charge) at the **NFPA** website to use it. However, you are not permitted to copy or print text.

The *National Electric Code* is the bible of the electrical trade. How does understanding the *Code* make you a more effective job candidate?

Self-Check 1

1. What was the main reason for developing the *NEC*?

7-2 ⬛ Changing and Writing *Code* and Code-Making Panels

In an effort to keep the *Code* up to date, it is revised every 3 years. Let's take a closer look at how changes are made and who writes the *Code*.

Changing and Writing *Code*

The procedure of updating and revising the *National Electrical Code* is never-ending. Technology advances and trends in the electrical industry move forward, but the ever-present danger of fire or shock hazard remain. To reflect these advancements and changes and remain vigilant toward hazards of all kinds, the National Fire Protection Association (NFPA) solicits proposals from anyone interested in electrical safety for each *NEC* cycle (every 3 years). Anyone may submit a proposal to change the *NEC*, using the Proposal Form found in the back of the *NEC*. **Figure 7-1** shows a copy of the Form for proposal for 2014 *NEC*.

Proposals received are submitted to a specific **code-making panel (CMP)** to accept, reject, accept in part, accept in principle, or accept in principle in part. The code-making panel (CMP) is one of 19 groups responsible for producing sections of *NEC*. CMPs consist of elected volunteers from a variety of electrical backgrounds. These actions are published in the Report on Proposals (ROP). As a general rule, proposals are published in April following the edition year for which the proposals were

Code-Making Panel (CMP) one of 19 groups responsible for producing sections of the *NEC*. CMPs consist of elected volunteers from a variety of electrical backgrounds.

FIGURE 7-1 Form for Proposal for 2014 *National Electrical Code.*

FORM FOR PROPOSAL FOR 2014 NATIONAL ELECTRICAL CODE®

INSTRUCTIONS — PLEASE READ CAREFULLY

Type or print **legibly** in **black** ink. Use a separate copy for each proposal. Limit each proposal to a **SINGLE** section. All proposals **must be received by NFPA by 5 p.m., EST, Friday, November 4, 2011**, to be considered for the 2014 National Electrical Code. Proposals received after 5:00 p.m., EST, Friday, November 4, 2011, will be returned to the submitter. If supplementary material (photographs, diagrams, reports, etc.) is included, you may be required to submit sufficient copies for all members and alternates of the technical committee.

For technical assistance, please call NFPA at 1-800-344-3555.

FOR OFFICE USE ONLY

Log #:

Date Rec'd:

Please indicate in which format you wish to receive your ROP/ROC ☐ electronic ☐ paper ☐ download
(Note: If choosing the download option, you must view the ROP/ROC from our website; no copy will be sent to you.)

Date _____ Name _____ Tel. No. _____

Company _____ Email _____

Street Address _____ City _____ State _____ Zip _____

If you wish to receive a hard copy, a street address MUST be provided. Deliveries cannot be made to PO boxes.

Please indicate organization represented (if any)

1. **Section/Paragraph**

2. **Proposal Recommends (check one):** ☐ new text   ☐ revised text   ☐ deleted text

3. **Proposal (include proposed new or revised wording, or identification of wording to be deleted):** [Note: Proposed text should be in legislative format; i.e., use underscore to denote wording to be inserted (<u>inserted wording</u>) and strike-through to denote wording to be deleted (~~deleted wording~~).]

4. **Statement of Problem and Substantiation for Proposal:** (Note: State the problem that would be resolved by your recommendation; give the specific reason for your Proposal, including copies of tests, research papers, fire experience, etc. If more than 200 words, it may be abstracted for publication.)

5. **Copyright Assignment**

   (a) ☐ I am the author of the text or other material (such as illustrations, graphs) proposed in the Proposal.

   (b) ☐ Some or all of the text or other material proposed in this Proposal was not authored by me. Its source is as **follows:** (please identify which material and provide complete information on its source)

I hereby grant and assign to the NFPA all and full rights in copyright in this Proposal and understand that I acquire no rights in any publication of NFPA in which this Proposal in this or another similar or analogous form is used. Except to the extent that I do not have authority to make an assignment in materials that I have identified in (b) above, I hereby warrant that I am the author of this Proposal and that I have full power and authority to enter into this assignment.

Signature (Required)

PLEASE USE SEPARATE FORM FOR EACH PROPOSAL

Mail to: Secretary, Standards Council · National Fire Protection Association
1 Batterymarch Park · Quincy, MA 02169-7471 OR
Fax to: (617) 770-3500 OR Email to: proposals_comments@nfpa.org

8/5/2010-C

applicable. There were 3688 proposals for changes to the 2008 edition of the *National Electrical Code*. Individuals may send in their comments on these actions using the Comment Form found in the ROP. The comment period remains open until October. There were 2349 comments processed for the 2011 *NEC* by the National Fire Protection Association staff at NFPA headquarters in Quincy, Massachusetts. The CMPs meet again to review and take action on the comments received. These actions are published in the Report on Comments (ROC). Voting, which is the final action on proposals and comments, takes place at the NFPA annual meeting.

Before the *NEC* is published, if there is disagreement on any specific *Code* requirement adopted by way of the above listed process, the NFPA considers an appeal that is reviewed and acted upon by the NFPA Standards Council approximately 6 weeks after the annual meeting. After an appeal is acted upon by the Standards Council, if there is still a controversy, a step not frequently used in the *Code* adoption process is taken: a petition, which is reviewed and acted upon by the NFPA Board of Directors.

Code-Making Panels

Code-making panels (CMPs) are comprised of individuals who are electrical inspectors, electrical contractors, and electrical engineers, individuals from utilities, purchasers of products, manufacturers, members of the Institute of Electrical and Electronic Engineers (IEEE), Underwriters Laboratories, insurance companies, and other similar organizations. CMP members are appointed by the NFPA. The CMPs have 10 to 20 principal members and a similar number of alternate members.

LIFE SKILLS

Writing and changing the *NEC* is a total team effort that relies on electricians submitting ideas and a team of panelists determining their merits. Why do you think that it is set up this way?

Self-Check 2

1. How are proposed *NEC* changes submitted? Who can submit them?
2. Who is responsible for reviewing, accepting, or rejecting proposed changes to the *NEC*?

7-3 The Purpose and Importance of the *National Electrical Code*

The *National Electrical Code* is the electrical industry standard and is recognized and used by everyone in the electrical field. The introduction to the *NEC*, found in *Article 90.1(A)*, clearly states the purpose of this *Code* as the practical safeguarding of persons and property from hazards arising from the use of electricity. *Article 90.1(B)* goes on to state that the *NEC* contains provisions that are considered necessary for safety and that compliance therewith and proper maintenance results in an installation that is free from hazard, but not necessarily efficient, convenient, or adequate for good service or future expansion of electrical use.

To be clear, the intent of the *NEC* is to serve as a *guide* for *safe* wiring. It is not meant to be an instructional manual or a "how-to" guide. Although the *NEC* is not itself a U.S. law, *NEC* use is often mandated and at times adopted by official action of the legislative body of a city, county, or state. In an effort to prevent a fire or shock hazard, electricians must use listed materials and devices and must perform all work with recognized standards. Being *listed* indicates that a device has met the testing and other requirements set forth by a listing agency. To be listed, a material or device has to meet the testing and other requirements set by a Nationally Recognized Testing Laboratory (NRTL). The Nationally Recognized Testing Laboratory (NRTL) is a listing organization that has passed the OSHA recognition process. NRTLs certify the safety of devices through listing and labeling. Underwriters Laboratories (UL) is perhaps the most well-known NRTL. UL wrote many of the standards for safety devices before OSHA formalized the process.

LIFE SKILLS

Choose a classmate or two, and have a discussion regarding the *NEC*. Why was it created? Is it still necessary today? Why or Why not?

Self-Check 3

1. What is the main purpose of the *NEC*?

The first four chapters of the *NEC* contain the basic electrical installation requirements for all electrical installations. This information applies generally to all electrical installations from a single-family residence to a manufacturing or health care facility. Learning about all chapters of the *NEC* makes *your* searches easier so you spend less time looking for information!

7-4 ◾ *Code* Book Chapters—Arrangement and Summary

New electricians usually say that using the *National Electrical Code* for the first few times is difficult. They are right! It is not an easy book to read. However, using the *Code* along with a text that explains it, is the best approach. Understanding how this book is organized is the best way to get started.

Chapters

To understand how to *use* the *NEC*, you need to know how it is structured or arranged. **Figure 7-2** shows the layout of the *National Electrical Code*, and the following sections describe how the *NEC* is arranged and written.

FIGURE 7-2 Layout of the *National Electrical Code.*

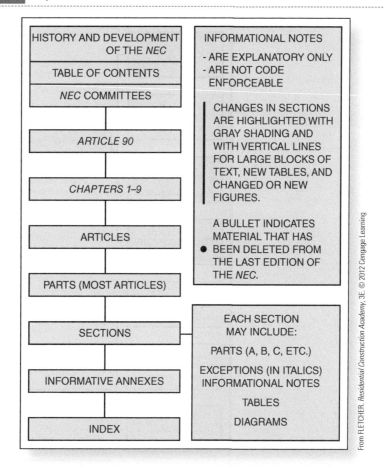

The *NEC* is comprehensive and at a glance may seem difficult to use. However, after much practice, you will develop the navigational skills required to find applicable information in the *NEC.* As an electrician, you will, for the most part, be using *Chapters 1* through *4*, which consist of rules that apply *generally* to all electrical installations. This chapter (in this text) describes all blocks (sections) shown in Figure 7-2.

The *NEC* is divided into an introduction and nine chapters. **Figure 7-3** illustrates the organization of the *National Electrical Code.*

The following list highlights each chapter as shown in Figure 7-3:

- *Chapter 1, General*, contains definitions and rules covering the basic requirements for electrical wiring installations.

- *Chapter 2, Wiring and Protection*, contains rules that apply to the installation of branch circuits, feeders, services, fuses, circuit breakers, grounding, and so on.

- *Chapter 3, Wiring Methods and Materials*, contains rules that apply to various wiring methods such as nonmetallic sheathed cable, metal-clad cable, rigid metal conduit, and electrical metallic tubing.

- *Chapter 4, Equipment for General Use*, contains rules that apply to the installation of equipment and materials such as receptacles, panelboards, luminaires, electric motors, and transformers.

Luminaires lighting fixtures

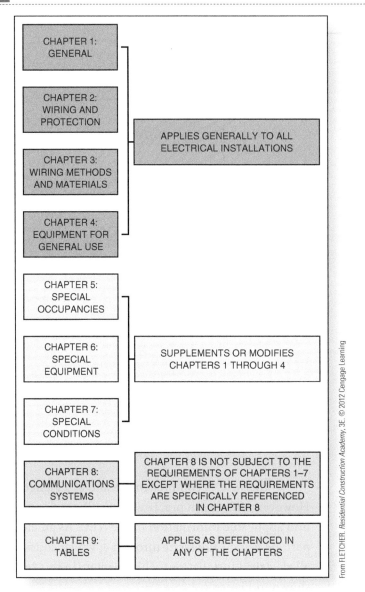

FIGURE 7-3 Chapter organization of the *National Electrical Code.*

- *Chapter 5, Special Occupancies*, contains rules that apply to wiring in special areas such as gas stations, aircraft hangers, hospitals, movie theaters, agricultural buildings, and mobile homes.
- *Chapter 6, Special Equipment*, contains rules that apply to the installation of special equipment such as electric signs, cranes, elevators, electric welders, swimming pools, and solar photovoltaic systems.
- *Chapter 7, Special Conditions*, contains rules that apply to special wiring installations such as emergency power systems, fire alarm systems, and optical fiber cables.

TRADE TIP *Chapters 5, 6,* and *7* often amend the rules found in *Chapters 1* through *4*, and whenever *Chapters 5, 6,* and *7* differ from *Chapters 1* through *4*, the requirements in *Chapters 5, 6,* and *7* apply.

- *Chapter 8, Communication Systems*, contains rules for the installation of wiring for communication systems such as cable television and telephones. It is a stand-alone chapter. As an independent chapter, it is not subject to the requirements of *Chapters 1* through *7* unless material in those chapters is specifically referenced in *Chapter 8*.

- *Chapter 9, Tables,* is comprised only of tables that contain information such as conduit fill and properties of conductors.

Articles

Chapters 1 through *8* are broken down into a series of articles. Each chapter heading describes the general subject area covered in the chapter. Articles cover specific subjects that fall under the general subject area of a chapter. For example, *Chapter 3, Wiring Methods and Materials*, is a general subject area. The articles contained in this chapter cover topics such as conductors for general wiring; outlet, device, pull, and junction boxes; conduit bodies; fittings; and handhole enclosures. Articles are very specific in their coverage. The *first* number of the article number denotes the chapter in which the article is found. For example, *Article 320, Armored Cable: Type AC*, is found in *Chapter 3*; and *Article 210, Branch Circuits*, is found in *Chapter 2*.

TRADE TIP   *Article 90, Introduction,* is the only article that is not actually located in a chapter. It stands alone at the beginning of the *NEC* and contains information on such things as the purpose of the *NEC*, what kinds of electrical installation the *NEC* covers and doesn't cover, and how the rules listed in the *NEC* are to be enforced.

Parts

Most articles are divided into parts. The parts of an article are indicated by roman numerals. *Part I, General*, describes general rules that apply to all the following parts in that article. The remaining parts are independent and apply to very specific topics. For example, *Part III of Article 314, Outlet, Device, Pull, and Junction Boxes; Conduit Bodies; Fittings; and Handhole Enclosures*, covers construction specifications. The only rules covered in *Part III* are those describing construction specifications. You do not find information covering pull and junction boxes, conduit bodies, and handhole enclosures. These topics, with their descriptions and rules, are located in *Part IV, Pull and Junction Boxes, Conduit Bodies, and Handhole Enclosures for Use on Systems over 600 Volts, Nominal*.

It is important to understand which part of the article you may be looking at in the *NEC*. Rules that apply under one specific part may not apply under another part.

Sections

Each rule found in the *National Electrical Code* is called a section. Articles contain the *NEC* rules covering a specific electrical area. As a result, articles (except *Article 100, Definitions*) consist of many sections. Sections are identified by a number that consists of the article number, a dot, and then the section number. For example, *Section 250.4* is located in *Chapter 2, Article 250*, and is *Section 4*. This section is *General Requirements for Grounding and Bonding*. This section, as well as some others, is further divided into parts identified by *letters* in parentheses such as (A), (B), (C), and so on. As in our example, *Section 250.4* is broken down into *250.4(A)*, which is *Grounded Systems*, and *250.4(B)*, which is *Ungrounded Systems*. Also note that in this section, *Grounded Systems, 250.4(A)* is further divided into *numbers* with parentheses. *Section 250.4(A)(2)*, for example, describes *Grounding of Electrical Equipment*. Some sections are divided even further with the use of lowercase letters such as (a), (b), (c), and so on. An example of this would be *Section 250.32(B)(2)(a)*. This is a rule describing grounded systems supplied by a separately derived system with overcurrent protection. It is found in *Chapter 2, Article 250, Section 32, Part (B), Subpart (2), Subpart (a)*.

Often electricians in the field make the mistake of using the term *article* when they should be using the term *section*. For example, it should be stated that grounding electrodes are found in *Section 250.32(A), not* in *Article 250.32(A)*.

Tables

There are numerous tables found in the *NEC*, and a lot of important information is contained in them. Many tables include numerical data that can be applied to a particular wiring installation. Make sure you read the title of each table carefully. You must be certain of what the table contains, where its information can be applied, and what limitations apply, if any. Be sure to read and understand all notes and footnotes, because this information must also be applied.

Figures

The *NEC* would not be considered "illustrative." As a matter of fact, there are very few figures that help us to apply the *Code*. However, there are a few illustrations included for areas that may be somewhat confusing. A good example is *Section 230.1*, which describes services. A figure is used to show a listing of topics that include *General, Overhead Service Conductors, Underground Service Conductors*, and so on. The figure also shows in which parts of the article these topics can be found. This figure also illustrates the interconnection of the service from source to branch-circuit feeders. A clear picture of how components are connected together is provided by this figure, and it also lists parts and articles where specific information on each component can be located.

Exceptions

When an exception to a *Code* rule is given, it is printed in italics. Exceptions always follow the rule that they amend, and they only apply to the *Code* rule that they follow. They are in the *NEC* to provide an alternate method to a specific requirement. There are two types of exceptions found in the *NEC*, and they are as follows:

- A mandatory exception is identified by the use of the terms *shall* or *shall not* and means that you must apply the rule in a specific way.
- A permissive exception is identified by a phrase like *is permitted*. This means that you could apply the rule as written, but you may decide not to.

On occasion, there is more than one exception to a rule. In those cases, they are listed as *Exception No. 1*, *Exception No. 2*, and so on.

Informational Notes

Informational notes provide explanatory material and are found following the section or table to which they apply. They often refer the reader to other *NEC* sections that give additional information on the same subject. Informational notes may also refer the reader to other NFPA documents for more information. Informational notes are information only and, as such, are not enforceable as requirements of the *NEC*. If more than one informational note is listed in a section, they are shown as *Informational Note No. 1*, *Informational Note No. 2*, and so forth.

Extractions

Many of the requirements contained in the *National Electrical Code* have been taken or extracted from other existing NFPA codes and standards. Brackets that contain section references to other NFPA documents are used to indicate where the extracted material comes from. The bracketed references immediately follow the extracted text in the *NEC*.

Table of Contents

The table of contents (TOC) lists each article in numerical order, starting with *Article 90*. Except for *Article 90*, each article is listed in the TOC under the chapter in which it resides. The title of each article is given, and a corresponding page number is listed. Under each article in the TOC, the various parts of the article are listed with corresponding page numbers. The parts are numbered with a roman numeral and have specific titles.

TRADE TIP   The *NEC* is published by the NFPA and has "70" as its publication number. Therefore, all page numbers begin with the number 70, a dash, and then the actual page number.

Index

The index is located in the back of the *National Electrical Code*, immediately after the annexes. The index contains the major topic areas that are covered in the *NEC* and conveniently lists them in alphabetical order. Several subtopics are listed under most of the major topic areas. The subtopics are also listed in alphabetical order. The major topics are printed in bold print to help make locating them easier, whereas subtopics are printed in regular print. Many of the major topic areas are followed by a list of *NEC* articles, parts, or sections, to let you know where specific information on the major topic can be found. All of the subtopics have article, part, section, and/or table listings that contain information about the subtopic.

TRADE TIP There are no page numbers given in the *NEC* index. So, knowing how the *NEC* is organized is very important when looking up information on a specific topic area.

The index also has dictionary-style headers with helpful identifiers (keywords) at the top of every index page. However, some major topic areas and subtopics only have references to other topic areas, where you find article, part, section, and/or table listings that contain information about your specific subject.

Annexes

The annexes are located in the back of the *NEC* after *Chapter 9*. They contain additional explanatory material that is designed to help users of the *NEC*.

Terms and Definitions

Article 100 contains definitions of terms essential to the proper use of the *NEC*. The definitions contained in *Article 100* are generally those terms that are used in two or more *NEC* articles. The article does not include commonly defined general terms or common technical terms from related codes or standards. The terms listed in *Article 100* are in alphabetical order and are displayed in bold print for easier locating.

Terms that are unique to an article are located and defined within that article. Generally, these definitions are at the beginning of the article. *Article 326, Integrated Gas Spacer Cable: Type IGS*, is an example of a term and definition in the article that is not in *Article 100*. *Section 326.2* lists a term specific to Type IGS cable. It is not defined in *Article 100* because this term is not used throughout the *NEC*.

Scope

The scope of an article, outlined in the first section, describes what the article covers and is very important to anyone using the *NEC*. Reading the scope of an article can help you determine whether the article likely contains the information you are looking for.

Boldface Type

The writing style used in the *NEC* is designed to allow for information to be found quickly and easily. One of the writing styles used is to include bold print lettering in certain locations. Boldface type lettering is used for the chapter titles, article titles, part titles, each section number and title, and each subpart. Table titles and row and column titles in the tables also use boldface lettering.

Gray Highlighting

Changes to the current *NEC* from the previous edition (other than editorial) are indicated by gray highlighting within sections and with vertical ruling for large blocks of changed or new text and for new tables and changed or new figures. This is designed to make the *Code* more user-friendly.

Bullets

A bullet or black dot (•) is used to denote material that has been deleted from the previous edition of the *NEC*. The dot (•) is found on the page where the material from the last *NEC* was located.

One of the reasons that the *NEC* is regarded as the finest building code standard in use today is because it is so well organized, making it very user-friendly. When going through this chapter, try developing a simple sensible, method for organizing your notes. Use the *NEC* as your inspiration! This makes the notes much easier to use when you're studying!

 Self-Check 4

1. Which *NEC* chapters do electricians use more than others?
2. What is the difference between an article and a section?
3. What is meant by an informational note?
4. How are changes in the 2011 *NEC* recognized?

7-5 *NEC* Standards and Local Authorities

As an electrician, you will find out there are other authorities in addition to the *NEC* and other requirements that must be adhered to. Let's look a little closer at those authorities.

Standards

Article 90 clearly points out that it is the authority having jurisdiction (AHJ) that inspects the electrical system installation to make sure that all electrical materials and wiring techniques used are acceptable. The AHJ is an organization, office, or individual responsible for enforcing the requirements of a code or standard and for enforcing the use of listed materials that have been approved by an NRTL for the specific installation. Usually, the local or state electrical inspector inspects the electrical work and is considered to be the AHJ.

However, there may be situations where you, as an electrician, are required to follow standards and codes other than *NEC*. In some cases, these codes are more specific and may be at a higher standard than those listed in the *NEC*.

As with any uniform code, a few jurisdictions regularly omit or modify some *NEC* sections or add their own requirements. These modifications or amendments to existing *Code*, sometimes referred to as specialty codes, come about because local, county, or state authorities may require more stringent procedures. In some cases, local, county, or state authorities add their own requirements based on earlier versions of the *NEC* or on locally accepted practices.

TRADE TIP Remember, no court has faulted anyone for using the latest version of the *NEC*, even when the local code was modified or not updated.

Examples

Before we list a couple of examples of when an electrician would be required to provide service at a higher standard than the *NEC*, let's look at where one might find information about codes that override the *NEC*. Resources are as follows:

- Check city, county, and state amendments.
- When you pull your permit, ask about "gray" areas and/or local amendments.
- Call or visit the local jurisdiction building office. Ask about any "specialty" codes.
- Refer to Statewide Building Code and then the International Building Code (IBC) and/or the International Residential Code (IRC).
- Ask your local municipality whether they have any special requirements other than what is in the United States Building Code (USBC), IBC, IRC, and *NEC*.

TRADE TIP If and when a disagreement exists regarding "special" requirements, the municipality should put you in contact with someone in their office who can assist you with the interpretation needed to complete your project.

In addition to the above listed resources, websites that can be helpful include these:

International Association of Electrical Inspectors: http://www.iaei.org/regulations
Municode Library: http://www.municode.com/library/library.aspx

Example 1

Some geographical areas have very strict fire and electrical codes. Some cities *require 100% conduit* in all residential wiring. As always, secure the proper permits and "know and understand" codes that need to be followed.

> TRADE TIP When you pull your permits, *always* make sure that you know not only the *NEC* but any and all special requirements that exist where the work is being done!

Example 2

Many questions are being raised about the requirements for bonding metal gas piping systems. Bonded (bonding) means *connected to establish electrical continuity and conductivity.** More specifically, when corrugated stainless steel tubing (CSST) is installed for the gas piping in a building, bonding methods and bonding jumper sizes that are more restrictive than required by the *National Electrical Code* are causing confusion and concern.

> **Bonded (bonding)** connected to establish electrical continuity and conductivity.

Electrical bonding is addressed in the *NEC* in *Part V* of *Article 250*. The *NEC* section dealing with bonding metal gas piping is *250.104(B)*. We will learn all about bonding and grounding in the next chapter (Chapter 8). All we need to understand now is in the rule cited above, the *NEC* clearly indicates the minimum size for the bonding jumper. However, the very best approach would be to *verify with the local inspection authority* how it approaches sizing bonding jumpers for metal gas piping systems. Do this because recent changes in other applicable standards and in the installation instructions provided by manufacturers of certain types of gas piping systems result in the *minimum* bonding jumper size *being larger than what is required by the NEC.*

LIFE SKILLS

One way to keep stress levels down on a jobsite is to be prepared. Take some time to research on your own, and see whether the local government in your area has any special codes that modify the *NEC* Standards.

Self-Check 5

1. Who is the authority having jurisdiction (AHJ), and why is the AHJ important?

Don't wait until you are asked to find out "what the *Code* says" about a certain topic to learn how to use the *NEC*! You have heard that "time is money" and you will

---

*Reprinted with permission from NFPA 70-2011.

hear that more than once. So *don't waste time* fumbling through such an important book as the *National Electrical Code*. Learn how to use it now!

7-6 Using the *Code* Book

The only way to learn how to find information quickly and accurately in the *National Electrical Code* is by *using* this *Code* book! The good news is that you become better able to use the *NEC* after a few searches. In this, the age of the Internet, we all want information instantly. Don't overlook an item or make careless mistakes because you are in a hurry! Remember, the information you need is in the *NEC*. Know *what* you are looking for, and take the time to read what is listed. You will be successful in your search *only* if you stay focused and read the information carefully during your search.

The following steps help you to find information in the *National Electrical Code*.

1. Determine the main topic area for the information you need.
   Without knowing the main topic area, you will not be able to find the information you need.

2. Locate the main topic area in the index.
   Main topic areas are listed in the index in boldface type, in alphabetical order. If you cannot locate the main topic area in the index, try using other, similar terms.

3. Determine the appropriate subtopic.
   Typically, the index lists several subtopics, in alphabetical order, under the main topic area heading. If the main topic area you selected does not have any subtopics listed, go to step 4 below.

4. Determine which article, part, section, or table is referenced for the topic area.
   If two or more references are listed, select the one that seems closest to your application. If the reference you selected does not contain the information you are looking for, try the next reference; keep looking until you find the proper information.

5. Go to the article's page, and look through the information until you locate the referenced part, section, or table.

6. Carefully read over all material in the referenced area until you locate the information that applies to your specific application.

Just like any multistep procedure, once you become familiar with the procedure—and in this case, with the *NEC*—the steps become second nature. You will be following the procedure, but you won't be thinking in terms of steps.

Let's put this procedure to the test. Using the *NEC*, find out about any rules pertaining to testing ground-fault protection in hospitals.

Using the six-step procedure outlined above, your search should go as follows:

1. Determine that *ground-fault protection* is the main topic area.

2. Using the alphabetically listed index, locate *ground-fault protection* listed in boldface print.

3. Because *hospital* is not listed, select *Health care facilities*.

4. The index lists *517.17*.

5. Refer to Article 517 and read through until you locate Wiring and Protection, Part II, and read further to Ground-Fault Protection, 517.17. Article and Section identification is covered in Section 7-5 on page 170. In this example of the six-step procedure, "517" is the Article and ".17" is the Section. Remember when working with the *NEC*, the number to the right of the dot is the section. The word "Section" is dropped. In this example, more information is required and is found in Part (D) of Section 17.

6. Read over *517.17* until you come to *517.17(D)*, which is *Testing*. This is the information needed.

Odds are that when you finish your apprenticeship, you're still not going to know all of the codes in the *NEC*. How will what you learned in this lesson help you find the answers that you need?

Self-Check 6

1. What is the first step in searching for information in the *NEC* and why is that step so important?

Summary

- The *National Electrical Code (NEC)*, or NFPA 70, is a United States standard for the safe installation of electrical wiring and equipment.
- The *National Electrical Code* was published for the first time in 1897.
- The *NEC* uses a 3-year cycle to publish new editions containing *Code* additions, revisions, and/or deletions.
- The *NEC* online version is available at the NFPA website (http://www.nfpa.org).
- Code-making panels review proposed changes to the *NEC* and accept or reject proposals.
- In the *NEC* introduction, found in *Article 90.1(A)*, the *NEC* clearly states the purpose of this *Code* as *the practical safeguarding of persons and property from hazards arising from the use of electricity.**
- The *NEC* is divided into an introduction and nine chapters.
- *NEC Chapters 1* through *4* consist of rules that apply generally to all electrical installations.
- *NEC Chapters 5* through *7* apply to special occupancies and other special wiring conditions.

*Reprinted with permission from NFPA 70-2011.

- *NEC Chapter 8* contains rules for the installation of wiring for communication systems, and *Chapter 9* has tables referenced in previous chapters.
- When securing a permit for a project, ask about specialty codes and/or other local amendments and requirements.
- You learn to find information in the *National Electrical Code* by being able to determine the main topic area. Use the index to locate main topic area and subtopics; determine which article, part, or section is referenced; and read the information carefully.

Review Questions

True/False

1. The *National Electrical Code* is a U.S. standard for safe installation of electrical wiring and equipment. (True, False)

2. The *National Electrical Code* is in itself a United States law. (True, False)

3. The *NEC* is updated and published every 2 years. (True, False)

4. The *NEC* is available online at the NFPA website http://www.nfpa.org. (True, False)

5. Only journeyman electricians may submit proposals to the NFPA in an effort to change the *NEC*. (True, False)

Multiple Choice

6. Proposed changes to a current *NEC* are submitted to what group?
    A. *NEC*
    B. OSHA
    C. NFPA
    D. ANSI

7. Which of the following statements is true of code-making panels?
    A. They consist of paid consultants from a variety of electrical backgrounds.
    B. They are appointed by the NFPA.
    C. They have 25 principal members.
    D. They are comprised only of electrical inspectors and journeymen electricians.

8. Which *NEC* article states the purpose of the *Code*?
    A. *Article 90.1(A)*
    B. *Article 100*
    C. *Article 110.1*
    D. *Article 392.10(A)*

9. When a device has met the testing and other requirements set forth by a Nationally Recognized Testing Laboratory, it is referred to as _____.
    A. Listed
    B. Certified
    C. Registered
    D. Warranted

10. Which *NEC* chapters consist of rules that apply generally to all electrical installations?
    A. *Chapters 1* through 5
    B. *Chapters 1* through 4
    C. *Chapters 4, 5*, and 6
    D. *Chapters 8* and 9

Fill in the Blank

11. *Chapter* _____ contains definitions and rules covering the basic requirements for electrical wiring installations.

12. *Chapter* _____ contains definitions and rules that apply to the installation of branch circuits, feeders, services, grounding, and so on.

13. The AHJ is usually a local or state electrical _____.

14. _____ are identified by a number that consists of the article number, a dot, and then another number.

15. The first step when searching for information in the *NEC* is to determine the _____ topic area for the information you need.

ACTIVITY 7-1   Practice the Six-Step Process

Using the *NEC* 2011 and the six-step process identified in this chapter, locate the answer to the following electrical question:

Question: In general, switches shall be wired so that all switching is done in which conductor?

Six-step procedure:

1.

2.

3.

4.

5.

6.

ACTIVITY 7-2   Practice the Six-Step Process

Using the *NEC* 2011 and the six-step process identified in this chapter, locate the answer to the following electrical question:

Question: Which, if any, receptacles may be connected to the small appliance branch circuit?

Six-step procedure:

1.

2.

(Continues)

(CONTINUED)

3.

4.

5.

6.

ACTIVITY 7-3   Practice the Six-Step Process

Use the six-step process identified in Chapter 7 and locate the answer to the electrical question listed below by using the *NEC®* 2011.

Question: Luminaires with GFCI protection located over a spa shall be permitted provided the mounting height is not less than how many feet/inches?

Six-step procedure:

1.

2.

3.

4.

5.

6.

Chapter 8

Grounding

Career Profile

Sarah E. High, Coordinator of the Construction Pre-Apprenticeship Training at the Community College of Baltimore County, says, "Anyone interested in pursuing a career as an electrician has made an excellent career choice.... As a journeyman, you will provide rewarding work to your customers," she continues, "and make a respectful living to sustain yourself and your family. You could own your business or partner with someone with future partnership opportunities. The field is so broad that you can work low voltage as a linesman, do control work, or become the foreman or supervisor on the job." High knows just how versatile the field can be because she has worked in a variety of positions at American University, the University of Maryland College Park, the Johns Hopkins Applied Physics Laboratory, and the National Security Agency. High identifies as a challenge in her current job "containing the excitement from students who are eager to enter the electrical field." She explains that "often students fail to recognize the effort put forth by those who are skilled craftspersons. The title of journeyman comes with a price." But Sarah strongly endorses apprenticeships programs because they provide exactly "the academic and on-the-job training hours needed to be a journeyperson." She also advises, "If you want to own your business or advance in this industry you [should] take courses in business management, financial management, and team building."

Chapter Outline

GROUNDING AND BONDING

SERVICE GROUNDING

INCORRECT GROUNDING OR LACK OF GROUNDING

NEC REQUIREMENTS—BONDING OF WIRING DEVICES
TO OUTLET BOXES

GROUND-FAULT CIRCUIT INTERRUPTER (GFCI)

Key Terms

Bonding

Bonding conductor or jumper

Bonding jumper, equipment

Bonding jumper, main

Bonding jumper, supply-side

Bonding jumper, system

Effective ground-fault current
path

Electrical continuity

Ground

Ground fault

Ground-Fault Circuit Interrupter
(GFCI)

Ground-fault current path

Grounded conductor

Grounded, solidly

Grounding

Grounding electrode

Grounding electrode conductor

Service

Service entrance

Service equipment

Supplemental grounding
electrode

Chapter Objectives

After completing this chapter, you will be able to:

1. Define key terms used when discussing grounding and bonding as explained in *Article 100, Definitions*.
2. Describe service grounding for a single-family dwelling.
3. Describe the consequences of incorrect grounding or lack of grounding.
4. Reference, interpret, and apply the *NEC* requirements concerning the bonding of wiring devices to outlet boxes.
5. List GFCI requirements, and explain their application for a single-family dwelling unit.

Life Skills Covered

 Communication Skills

 Self-Advocacy

 Goal Setting

Life Skills Goals

Safety is always the first priority in the electrical field, and grounding is a cornerstone of safe electrical installations. An electrician who can't grasp grounding won't be an electrician for very long. That's why we've chosen Self-Advocacy as one of this chapter's life skills. A mastery of the finer points of grounding will prove invaluable as you begin your apprenticeship. Because this is such an important topic, we have chosen Communication Skills as the second life skill associated with the chapter. Discuss what you are learning with your classmates. If you have questions, ask! When possible, it's much better to clear up your misunderstandings in the classroom than on a jobsite. The final life skill that we cover is Goal Setting. As you go through the chapter, make it your goal to know the material so well that you could answer any question on its content without consulting your notes.

Where You Are Headed

In this chapter, you learn key terms used to describe grounding and bonding as defined in the *Article 100* definitions. This chapter also describes service grounding for a single-family dwelling. Not only is the importance of grounding explained, but you also learn to appreciate the consequences of improper or a complete loss of grounding. As an electrician, you need to understand and apply the *NEC* requirements concerning bonding of wiring devices to outlet boxes. You also need to understand GFCI requirements and GFCI application in single-family dwellings.

Introduction

Article 250 of the *NEC* requires that alternating current (ac) systems of 50 to 1000 volts that supply premises' wiring systems where the system can be grounded so that the maximum voltage to ground on the ungrounded conductors does not exceed 150 volts shall be grounded. Therefore, all residential electrical systems rated at 120/240 volts must be grounded. Within a system, some devices are grounded and some are bonded.

Grounding is the act of connecting an electrical device to the ground (earth), generally by means of a conductor such as a wire or rod, so that the device has zero electrical potential. *Grounding* works on the principle that the earth is a very good conductor of electricity. The *grounding* of electrical devices and appliances *protects* those who handle and use them from electrical shocks!

Bonding is the connection of two or more conductive objects to establish electrical continuity and conductivity. The terms *grounding* and *bonding* are often used interchangeably by many in the electrical field, but this should *never* be done! *Grounding and bonding are not the same!* Each serves a different purpose. These purposes will be explored and explained throughout this chapter. The consequences of incorrect or missing ground will also be described, and you will learn about the shock-preventing and often life-saving ground-fault circuit interrupter (GFCI).

Grounding is the act of connecting an electrical device to the ground (earth), generally by means of a conductor such as a wire or rod, so that the device has zero electrical potential.

Bonding the connection of two or more conductive objects to establish electrical continuity and conductivity.

8-1 Grounding and Bonding

Any intentional or accidental connection between an electrical circuit or instrument and the earth is known as grounding. Grounding ensures that all metal parts of an electrical circuit that an individual might contact are connected to the earth, which will prevent a difference of potential (voltage). The main reason for grounding is safety! When all metal parts in electrical equipment are grounded, there will be no voltage present on the equipment case. So, if the insulation inside the equipment fails, there will be *no shock hazard*!

Bonding is generally done as protection from electrical shocks. Bonding ensures that connected electrical conductors are at the same electrical potential. In other words, you won't get electricity building up in one and not the other. There can be *no* current flow between two bonded conductors because they have the same potential. Bonding of electrical outlet boxes is generally accomplished by connecting a wire between them. If no other connections were made to these boxes *and* a "hot" (black) wire came in contact with one box, *both* boxes would have 120 volts of electrical potential, and *either* box could give you a shock!

Bonding by itself, doesn't protect anything. However, if one of those outlet boxes is grounded, there can be no electrical energy buildup. If the grounded box is bonded to the other box, then the other box is also at zero electrical potential, and there will be no shock hazard! The proper grounding of objects (conductors) in any electrical system will generally incorporate both *bonds between objects* and a *specific bond to the earth (ground)*.

Article 100 has very clear definitions for *grounded* and *bonded*, and they are as follows:

- **Grounded** *(grounding)—connected (connecting) to ground or to a conductive body that extends the ground connection.** *
- **Bonded** *(bonding)—connected to establish electrical continuity and conductivity.** *

Grounding

In order to become more familiar with *grounding*, the following are key terms as defined in *Article 100, 250.2*, and *250.4(A)*:

- *Ground—The earth.** * (*Note:* The earth has become a worldwide standard as the common reference point for all electrical systems.)
- *Ground Fault—*An unintentional, electrically conducting connection between an ungrounded conductor of an electrical circuit and the normally non-current-carrying conductors, metallic enclosures, metallic raceways, metallic equipment, or earth.
- *Ground-Fault Current Path—*An electrically conductive path from the point of a ground fault on a wiring system through normally non-current-carrying conductors, equipment, or the earth to the electrical supply source.

---

*Reprinted with permission from NFPA 70-2011.

Ground the earth.

Ground fault an unintentional, electrically conducting connection between an ungrounded conductor of an electrical circuit and the normally non-current-carrying conductors, metallic enclosures, metallic raceways, metallic equipment, or earth.

Ground-fault current path an electrically conductive path from the point of a ground fault on a wiring system through normally non-current-carrying conductors, equipment, or the earth to the electrical supply source.

Effective ground-fault current path an intentionally constructed, low-impedance, electrically conductive path designed and intended to carry current under ground-fault conditions from the point of a ground fault on a wiring system to the electrical supply source and that facilitates the operation of the overcurrent protective device or ground-fault detectors on high-impedance grounded systems.

Grounded, solidly connected to ground without inserting any resistor or impedance device.

Grounded conductor a system or circuit conductor that is intentionally grounded.

Grounding electrode a conducting object through which a direct connection to earth is established.

Grounding electrode conductor a system or circuit conductor that is intentionally grounded.

● *Effective Ground-Fault Current Path*—An intentionally constructed, low-impedance electrically conductive path designed and intended to carry current under ground-fault conditions from the point of a ground fault on a wiring system to the electrical supply source and that facilitates the operation of the overcurrent protective device or ground-fault detectors on high-impedance grounded systems.

● *Grounded, Solidly*—*Connected to ground without inserting any resistor or impedance device.**

● *Grounded Conductor*—*A system or circuit conductor that is intentionally grounded.** (*Note:* When working in residential wiring, this is the white (or neutral) conductor.)

● *Grounding Electrode*—*A conducting object through which a direct connection to earth is established.** Grounding electrodes used in residential wiring include: underground metal water pipe supply, a concrete-encased bare copper conductor (UFER ground), and ground rods. *NEC 250.52(A)(1) through (7) list acceptable grounding electrodes.*

● *Grounding Electrode Conductor*—*A conductor used to connect the system grounded conductor or the equipment to a grounding electrode or to a point on the grounding electrode system.** (*Note:* Generally, when working in residential wiring, the grounding electrode conductor is run between the main service panel neutral/ground terminal bar and the grounding electrode.)

Bonding

As previously defined, elements are *bonded* when connected to establish electrical continuity and conductivity.

Figure 8-1 shows two metal boxes bonded together using the metal raceway installed between the two boxes. The equipment bonding jumper could have been used.

TRADE TIP If a nonmetallic raceway had been used, a separate equipment grounding conductor would be required. An equipment bonding jumper must be sized large enough to handle any fault current that it may need to carry. Equipment bonding jumpers are sized in accordance with *NEC Table 250.122*. Additional *NEC* references are found in *250.96, 250.102(D)*, and *250.90*.

FIGURE 8-1 Methods for bonding metal boxes.

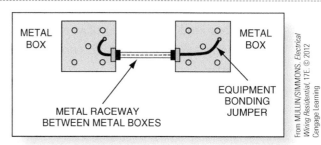

METAL BOX · METAL RACEWAY BETWEEN METAL BOXES · EQUIPMENT BONDING JUMPER · METAL BOX

From MULLIN/SIMMONS. *Electrical Wiring Residential*, 17E. © 2012 Cengage Learning

*Reprinted with permission from NFPA 70-2011.

Bonding conductor or jumper a reliable conductor to ensure the required electrical conductivity between metal parts required to be electrically connected.

Bonding jumper, equipment the connection between two or more portions of the equipment grounding conductor.

Bonding jumper, main the connection between the grounded circuit conductor and the equipment grounding conductor at the service.

Bonding jumper, supply-side a conductor installed on the supply side of a service or within a service equipment enclosure, or for a separately derived system, that encloses the required electrical conductivity between metal parts required to be electrically connected.

Bonding jumper, system the connection between the grounded circuit conductor and the supply-side bonding jumper, or the equipment grounding conductor, or both, at a separately derived system.

To become more familiar with *bonding,* the following are key terms as defined in *Article 100:**

- *Bonding Conductor or Jumper*—*A reliable conductor to ensure the required electrical conductivity between metal parts required to be electrically connected.*
- *Bonding Jumper, Equipment*—*The connection between two or more portions of the equipment grounding conductor.* (*Note:* Bonding alone does not ground equipment.)
- *Bonding Jumper, Main*—*The connection between the grounded circuit conductor and the equipment grounding conductor at the service.* (*Note:* **Figure 8-2** shows how the main bonding jumper connects the neutral service conductor to the service-entrance main disconnect enclosure and the equipment grounding conductors.)

TRADE TIP NEC 250.28 lists some installation requirements for the main bonding jumper.

- *Bonding Jumper, Supply-Side*—*A conductor installed on the supply side of a service or within a service equipment enclosure(s), or for a separately derived system, that enclosures the required electrical conductivity between metal parts required to be electrically connected.*
- *Bonding Jumper, System*—*The connection between the grounded circuit conductor and the supply-side bonding jumper, or the equipment grounding conductor, or both, at a separately derived system.*

FIGURE 8-2 Main bonding jumper.

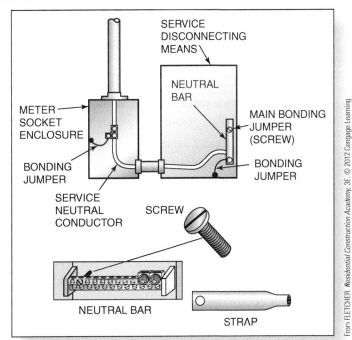

*Reprinted with permission from NFPA 70-2011.

LIFE SKILLS

At times, you will find that it is equally effective, if not more so, to write down an explanation or instructions instead of giving them verbally. Practice your written communication skills by writing a short explanation describing the major aspects of both grounding and bonding.

Self-Check 1

1. What is the difference between grounding and bonding?
2. What is a grounding electrode conductor?
3. What is the purpose of the main bonding jumper?

8-2 Service Grounding

Service entrance *that section of the wiring system where electrical power is supplied to a dwelling or building from the electric utility.*

Service *the conductors and equipment for delivering electric energy from the serving electric utility to the wiring system of the premises served.*

Before we describe service grounding, let's take a brief look at where this grounding takes place, at the service entrance. The service entrance is that section of the wiring system where electrical power enters a dwelling or building from the electric utility. The *NEC* defines a service as the conductors and equipment for delivering electric energy from the serving electric utility to the wiring system of the premises served.

Electrical energy is delivered to residential wiring systems in one of the two following types of service entrance:

- Overhead service
- Underground service

Electricians must recognize the differences between the two service types, as well as the specific installation techniques required for each. **Figure 8-3** shows a typical overhead service entrance, and **Figure 8-4** illustrates a typical underground service entrance.

Our focus in this chapter is to describe grounding and bonding. So, why talk about the service entrance? Well to answer that, take a look at **Figure 8-5**. For now, note that the terms *ground* and *grounding* are listed several times. As we discuss grounding in greater detail, we will be referring back to this figure.

Service equipment *the necessary equipment connected to the load end of the service conductors supplying a building and intended to be the main control and cutoff site.*

In later chapters, we will be describing in great detail service entrances and *Article 230*, which covers many of the requirements for their installation. At this time, refer to Figure 8-5, and note that both overhead and underground service entrances have a component called service equipment. This is the necessary equipment connected to the load end of the service conductors supplying a building and is the main control and cutoff mechanism. The service equipment can consist of a fusible disconnect switch or a main breaker panel that also accommodates branch-circuit overcurrent protection devices (fuses or circuit breakers). You are already somewhat familiar with this panel if you have ever reset a circuit breaker or changed a fuse.

FIGURE 8-3 A typical overhead service entrance.

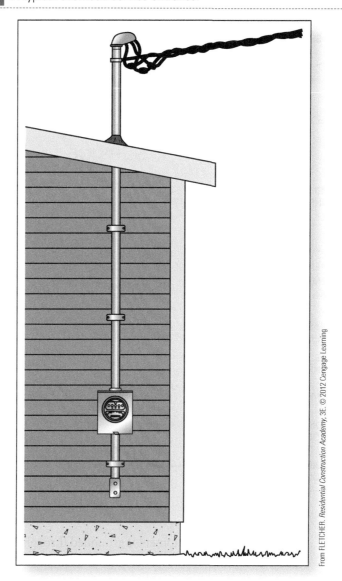

From FLETCHER. *Residential Construction Academy*, 3E. © 2012 Cengage Learning

FIGURE 8-4 A typical underground service entrance.

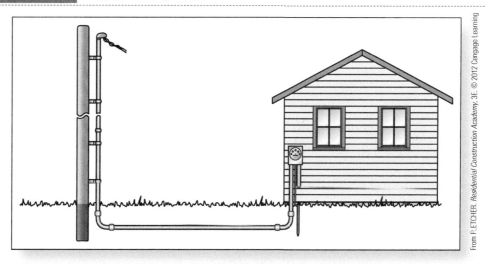

From FLETCHER. *Residential Construction Academy*, 3E. © 2012 Cengage Learning

FIGURE 8-5 Overhead and underground service and key terms.

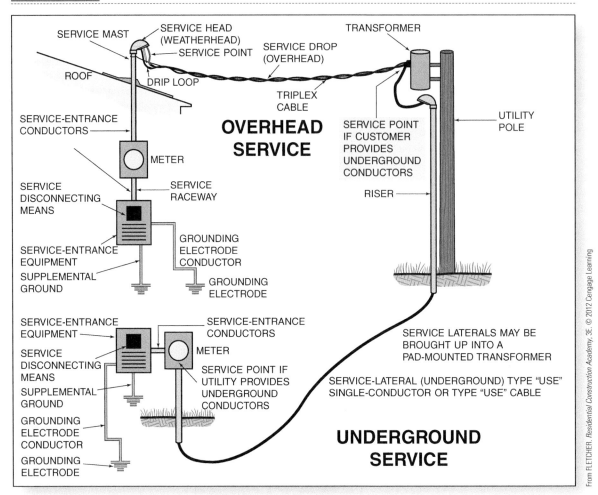

The service equipment is grounded through a grounding electrode conductor. Recall that this is a conductor used to connect the system grounded conductor or the equipment to a grounding electrode or to a point on the grounding electrode system.

TRADE TIP Generally, when working in residential wiring, the grounding electrode conductor is run between the main service panel neutral or ground terminal bar and the grounding electrode.

Figure 8-5 shows the grounding electrode conductor connected to the grounding electrode. Also recall that the grounding electrode is a conducting object through which a *direct* connection to earth is established. Residential electrodes include metal water pipes, concrete-encased bare copper conductors, and ground rods.

Another connection made to the service equipment, as illustrated in Figure 8-5, is the supplemental ground. The supplemental grounding electrode is a grounding electrode used to back up a metal water pipe grounding electrode. If a metal water pipe electrode breaks and is replaced with a length of *plastic* plumbing pipe, grounding continuity will be lost. Because this situation occurs on a regular basis, the

Supplemental grounding electrode a grounding electrode used to back up a metal water pipe grounding electrode.

NEC requires that metal pipe electrodes be supplemented by another electrode (see *Article 250.53(A)(2)).*

TRADE TIP Supplemental electrodes are generally 8-foot metal rods driven into the ground (earth).

Grounding Requirements for a Residence Service Installation

A residential wiring system supplied by a grounded ac service must have a grounding electrode conductor connected to the grounded service conductor. (See *250.24(A).*) The grounded service conductor is normally referred to as the Neutral conductor.

TRADE TIP The connection of the grounding electrode conductor to the grounded service conductor can be made at any accessible point from the load end of the service drop or service lateral to the terminal strip to which the grounded service conductor is connected at the service disconnecting means.

Figure 8-6 shows the connection points generally used to connect the grounded conductor of the service to the grounding electrode conductor and grounding electrode when wiring a residence.

Although Figure 8-6 illustrates three possible connection points where the grounded service conductor could be connected to the grounding electrode conductor

FIGURE 8-6 A residential service entrance supplied by an overhead distribution system.

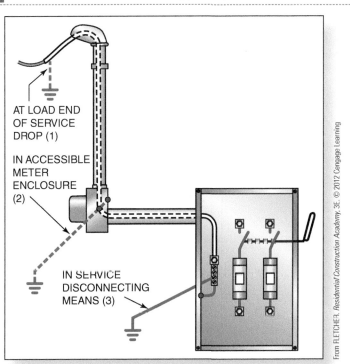

as per *250.24(A)(1)*, location 2 (in accessible meter enclosure) and location 3 (in service disconnecting means) are the most common locations to make the grounding electrode conductor connection.

Bonding Requirements for a Residence Service Installation

NEC 250.24(B) describes the requirements for the main bonding jumper. For a grounded system, an unspliced main bonding jumper must be used to connect the equipment grounding conductor and the service-disconnect enclosure to the grounded conductor of the system within the enclosure for each service disconnect. Bonding at service-entrance equipment is very important because service-entrance conductors do not have overcurrent protection at their line side, other than the electric utility company's primary transformer fuses. *Overload protection* for service conductors is at the electric company's load end.

The *NEC* lists some installation requirements for the main bonding jumper in *250.28*. **Figure 8-7** illustrates a main bonding jumper.

A main bonding jumper provides a connection from the Neutral service conductor to the service-entrance main disconnect enclosure and the equipment grounding conductors.

TRADE TIP A screw is typically used as a main bonding jumper in most residential service equipment. The screw must have a green finish that has to be visible with the screw installed. The green screw makes it possible to easily distinguish the main jumper screw for inspection.

FIGURE 8-7 A main bonding jumper connection.

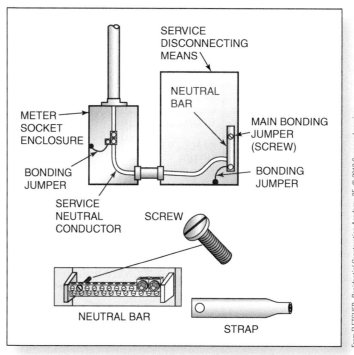

From FLETCHER, *Residential Construction Academy, 3E.* © 2012 Cengage Learning

Grounding Electrode System

Article 250, Part III, covers the requirements for establishing a grounding electrode system. *NEC 250.50* requires that metal underground water piping, the metal frame of a building, a concrete-encased electrode, a ground ring, and rod-pipe-plate electrodes be *bonded* together if any or all are present in a new installation.

NEC 250.52(B)(1) states that metal gas piping shall *never* be used as a grounding electrode. It has been recognized that galvanic action in gas pipes has caused deterioration and resulted in serious incidents. Metal gas piping is, however, required to be bonded. Generally, the equipment grounding conductor for the branch circuit supplying a gas furnace serves as the required bond. *NEC 250.104(B)* covers the rules for bonding the metal gas piping.

Should any one of the components listed above become disconnected, the integrity of the grounding system is maintained through the other paths. The following are key points as stated by the *NEC:*

- *250.90* states that bonding shall be provided where necessary to ensure electrical continuity and the capacity to conduct safely any fault current likely to be imposed.
- *250.92(A)* explains what parts of a service must be bonded together.
- *250.94* explains what is acceptable as a bonding means.
- *250.96(A)* states in part that bonding of metal raceways, cable armor, enclosures, frames, fittings, and so on, that serve as the grounding path shall be effectively bonded where necessary to ensure electrical continuity and the capacity to conduct safely any fault current likely to be imposed on them.

The potential differences between non-current-carrying metal parts is eliminated by *bonding* all metal parts of a grounding electrode system. The grounding electrode system:

- Serves as a means to bleed off lightning
- Stabilizes the system voltage
- Ensures that the overcurrent protective devices operate

Grounding Electrode Conductor Connection

The grounding electrode conductor shall be connected to the grounded (Neutral) service conductor. (See *250.24(A)*.) The connection is permitted to be made at any accessible point from the load end of the service drop or service lateral to, and including, the terminal or bus to which the grounded (Neutral) service conductor is connected at the service disconnecting means. (See *250.24(A)(1)*.)

All residential panelboards have the following:

- A neutral bus for the white grounded circuit conductors
- An equipment grounding bus for the bare equipment grounding conductors when nonmetallic sheathed cable is used as the wiring method

The connection of the grounding electrode conductor is made to the Neutral bus for most residential services. *NEC 250.24(A)(4)* permits the grounding electrode to be connected to the equipment grounding bus *if* the main bonding jumper is a wire or busbar.

TRADE TIP   Typically, residential panelboards have a green hexagon-shaped No. 10-32 screw that becomes the main bonding jumper between the Neutral bus and the enclosure when properly installed.

Figure 8-8 illustrates the grounding electrode conductor connected to the Neutral bus in the main service panelboard.

FIGURE 8-8   A typical residential service installation.

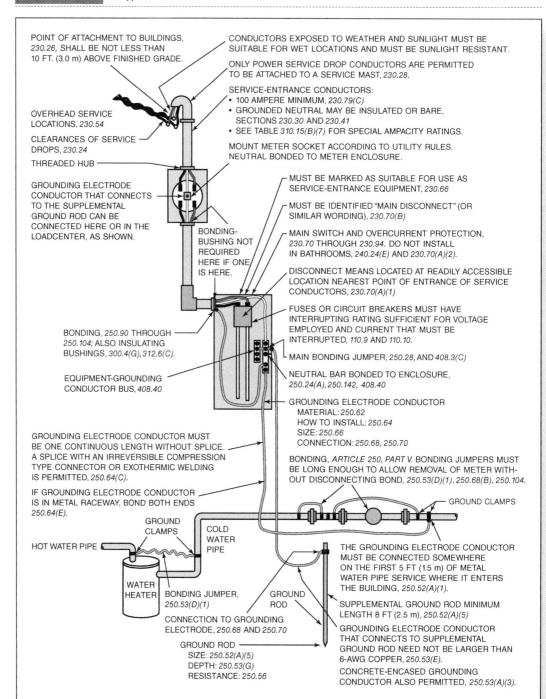

POINT OF ATTACHMENT TO BUILDINGS, *230.26*, SHALL BE NOT LESS THAN 10 FT. (3.0 m) ABOVE FINISHED GRADE.

CONDUCTORS EXPOSED TO WEATHER AND SUNLIGHT MUST BE SUITABLE FOR WET LOCATIONS AND MUST BE SUNLIGHT RESISTANT.

ONLY POWER SERVICE DROP CONDUCTORS ARE PERMITTED TO BE ATTACHED TO A SERVICE MAST, *230.28*.

SERVICE-ENTRANCE CONDUCTORS:
- 100 AMPERE MINIMUM, *230.79(C)*
- GROUNDED NEUTRAL MAY BE INSULATED OR BARE, SECTIONS *230.30* AND *230.41*
- SEE TABLE *310.15(B)(7)* FOR SPECIAL AMPACITY RATINGS.

OVERHEAD SERVICE LOCATIONS, *230.54*

CLEARANCES OF SERVICE DROPS, *230.24*

THREADED HUB

MOUNT METER SOCKET ACCORDING TO UTILITY RULES. NEUTRAL BONDED TO METER ENCLOSURE.

GROUNDING ELECTRODE CONDUCTOR THAT CONNECTS TO THE SUPPLEMENTAL GROUND ROD CAN BE CONNECTED HERE OR IN THE LOADCENTER, AS SHOWN.

MUST BE MARKED AS SUITABLE FOR USE AS SERVICE-ENTRANCE EQUIPMENT, *230.66*

MUST BE IDENTIFIED "MAIN DISCONNECT" (OR SIMILAR WORDING), *230.70(B)*

BONDING-BUSHING NOT REQUIRED HERE IF ONE IS HERE.

MAIN SWITCH AND OVERCURRENT PROTECTION, *230.70* THROUGH *230.94*. DO NOT INSTALL IN BATHROOMS, *240.24(E)* AND *230.70(A)(2)*.

DISCONNECT MEANS LOCATED AT READILY ACCESSIBLE LOCATION NEAREST POINT OF ENTRANCE OF SERVICE CONDUCTORS, *230.70(A)(1)*

FUSES OR CIRCUIT BREAKERS MUST HAVE INTERRUPTING RATING SUFFICIENT FOR VOLTAGE EMPLOYED AND CURRENT THAT MUST BE INTERRUPTED, *110.9* AND *110.10*.

BONDING, *250.90* THROUGH *250.104*; ALSO INSULATING BUSHINGS, *300.4(G), 312.6(C)*.

MAIN BONDING JUMPER, *250.28*, AND *408.3(C)*

NEUTRAL BAR BONDED TO ENCLOSURE, *250.24(A), 250.142, 408.40*

EQUIPMENT-GROUNDING CONDUCTOR BUS, *408.40*

GROUNDING ELECTRODE CONDUCTOR
MATERIAL: *250.62*
HOW TO INSTALL: *250.64*
SIZE: *250.66*
CONNECTION: *250.68, 250.70*

GROUNDING ELECTRODE CONDUCTOR MUST BE ONE CONTINUOUS LENGTH WITHOUT SPLICE. A SPLICE WITH AN IRREVERSIBLE COMPRESSION TYPE CONNECTOR OR EXOTHERMIC WELDING IS PERMITTED, *250.64(C)*.

IF GROUNDING ELECTRODE CONDUCTOR IS IN METAL RACEWAY, BOND BOTH ENDS *250.64(E)*.

BONDING, *ARTICLE 250, PART V*. BONDING JUMPERS MUST BE LONG ENOUGH TO ALLOW REMOVAL OF METER WITHOUT DISCONNECTING BOND, *250.53(D)(1), 250.68(B), 250.104*.

GROUND CLAMPS

GROUND CLAMPS

COLD WATER PIPE

HOT WATER PIPE

THE GROUNDING ELECTRODE CONDUCTOR MUST BE CONNECTED SOMEWHERE ON THE FIRST 5 FT (1.5 m) OF METAL WATER PIPE SERVICE WHERE IT ENTERS THE BUILDING, *250.52(A)(1)*.

WATER HEATER

BONDING JUMPER, *250.53(D)(1)*

GROUND ROD

CONNECTION TO GROUNDING ELECTRODE, *250.68* AND *250.70*

SUPPLEMENTAL GROUND ROD MINIMUM LENGTH 8 FT (2.5 m), *250.52(A)(5)*

GROUNDING ELECTRODE CONDUCTOR THAT CONNECTS TO SUPPLEMENTAL GROUND ROD NEED NOT BE LARGER THAN 6-AWG COPPER, *250.53(E)*.

GROUND ROD
SIZE: *250.52(A)(5)*
DEPTH: *250.53(G)*
RESISTANCE: *250.56*

CONCRETE-ENCASED GROUNDING CONDUCTOR ALSO PERMITTED, *250.53(A)(3)*.

From FLETCHER. *Residential Construction Academy*, 3E. © 2012 Cengage Learning

Grounding and Bonding in a Single-Family Dwelling

Figure 8-9 illustrates the service entrance, main panel, subpanel, and grounding for the service in a single-family dwelling.

A metal underground water piping system 10 ft (3.0 m) or longer, in direct contact with the earth, is acceptable as the primary grounding electrode. (See *250.52(A)(1)*.) The connection of the grounding electrode conductor must be made within the first 5 ft (1.52 m) of where the metal underground water piping enters the building. (See *250.52(A)(1)* and *250.54*).

FIGURE 8-9 The service entrance, main panel, subpanel, and grounding for the service of a single-family dwelling.

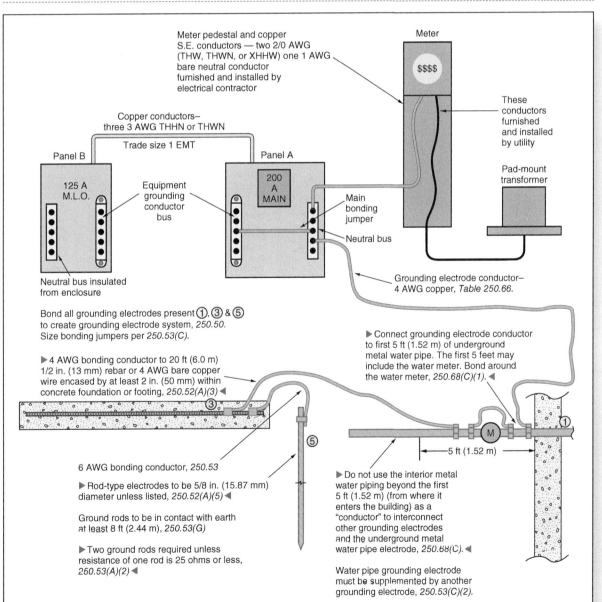

Meter pedestal and copper S.E. conductors — two 2/0 AWG (THW, THWN, or XHHW) one 1 AWG bare neutral conductor furnished and installed by electrical contractor

Meter

These conductors furnished and installed by utility

Copper conductors— three 3 AWG THHN or THWN

Trade size 1 EMT

Pad-mount transformer

Panel B

Panel A

125 A M.L.O.

Equipment grounding conductor bus

200 A MAIN

Main bonding jumper

Neutral bus

Neutral bus insulated from enclosure

Grounding electrode conductor— 4 AWG copper, *Table 250.66.*

Bond all grounding electrodes present ①, ③ & ⑤ to create grounding electrode system, *250.50.* Size bonding jumpers per *250.53(C).*

▶ 4 AWG bonding conductor to 20 ft (6.0 m) 1/2 in. (13 mm) rebar or 4 AWG bare copper wire encased by at least 2 in. (50 mm) within concrete foundation or footing, *250.52(A)(3)* ◀

▶ Connect grounding electrode conductor to first 5 ft (1.52 m) of underground metal water pipe. The first 5 feet may include the water meter. Bond around the water meter, *250.68(C)(1).* ◀

6 AWG bonding conductor, *250.53*

▶ Rod-type electrodes to be 5/8 in. (15.87 mm) diameter unless listed, *250.52(A)(5)* ◀

Ground rods to be in contact with earth at least 8 ft (2.44 m), *250.53(G)*

▶ Two ground rods required unless resistance of one rod is 25 ohms or less, *250.53(A)(2)* ◀

—5 ft (1.52 m)—

▶ Do not use the interior metal water piping beyond the first 5 ft (1.52 m) (from where it enters the building) as a "conductor" to interconnect other grounding electrodes and the underground metal water pipe electrode, *250.68(C).* ◀

Water pipe grounding electrode must be supplemented by another grounding electrode, *250.53(C)(2).*

When metal underground water piping is used as the primary grounding electrode, it must be supplemented by at least one additional grounding electrode. (See *250.53(D)(2)*.) For the service illustrated in Figure 8-9, a driven rod is the supplemental grounding electrode, as permitted by *250.52(A)(5)* and *250.54*. The supplemental grounding electrode shown in **Figure 8-10** can also be used.

The supplemental grounding electrode illustrated in Figure 8-10 is a concrete-encased 4 AWG bare copper conductor at least 20 ft (6.0 m) long, laid horizontally near the bottom of the concrete footing. This grounding arrangement is often called a UFER ground, named after Herbert G. Ufer who worked for Underwriters Laboratories. *NEC 250.50* requires that when connecting the grounding electrode conductor to a concrete-encased electrode, a *minimum* of 4 AWG bare copper conductor or reinforcing bars (rebars) must be used.

FIGURE 8-10 An electric service grounded to an underground metal water pipe and a concrete-encased electrode used as a supplemental grounding electrode.

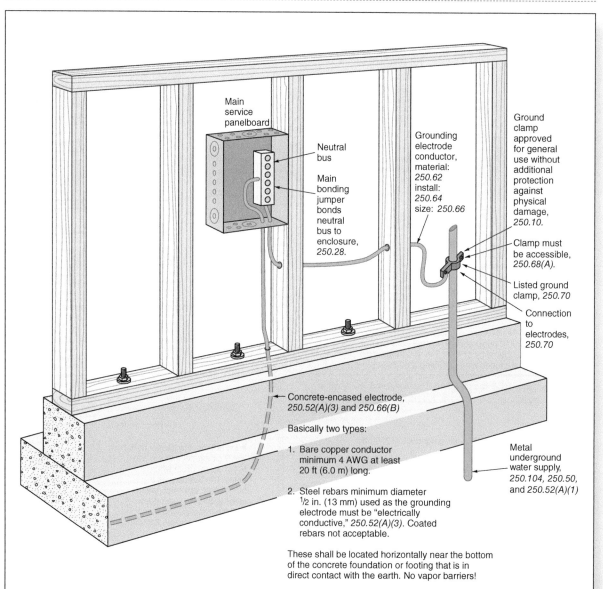

From MULLIN/SIMMONS. *Electrical Wiring Residential*, 17E. © 2012 Cengage Learning

TRADE TIP Unlike underground metal piping, a concrete-encased electrode does *not* require a supplemental electrode. (See *250.53(D)(2)*.)

Although the concrete-encased electrode shown in Figure 8-10 is installed horizontally in the foundation wall, these electrodes could also be installed vertically in a foundation wall. The wall however, must be in *direct contact* with the earth. (See *250.52(A)(3)*.)

Grounding and Bonding in Commercial and Industrial Facilities

Electrical wiring systems for commercial buildings and industrial facilities can vary considerably. A small, one-room newsstand with a few luminaires and a couple of convenience outlets is quite different from a high-rise apartment building, hospital, or manufacturing facility. However, *grounding* is grounding and *bonding* is bonding. Grounding, whether residential or commercial, is one of the most important aspects of a sound electrical system. Remember, when properly installed, the ground provides a *direct* path for current to flow back to the *earth.*

Article 250 applies not only to residential wiring but to commercial wiring as well. Depending on the use of the building, the *NEC* or local ordinances may require a different wiring method than would be required in single-family dwellings or even smaller buildings. Commercial buildings require an increased size of service and feeders, and different sizes and types of service-entrance equipment and panelboards. For example, larger commercial buildings utilize a 480/277-volt Y-connected service entrance.

Metal water piping systems in direct contact with the earth are generally used as the primary grounding electrode in commercial installations. As in residential dwellings, the grounding electrode conductor must be made within the first 5 ft (1.52 m) of where the metal underground water piping enters the building. (See *250.52(A)(1)* and *250.54*.)

When metal underground water piping is used as the primary grounding electrode, it must be supplemented by at least one additional grounding electrode. (See *250.53(D)(2)*.) A driven rod can be used as the supplemental grounding electrode, as permitted by *250.52(A)(5)* and *250.54*.

Concrete-Encased Electrode

Concrete-encased electrodes have become popular with the increasing use of nonmetallic water mains. Many communities have mandated using a concrete-encased electrode as the primary electric service electrode because of its proven performance record of providing an excellent connection to earth. Advantages of the concrete-encased electrode include the following:

- No supplemental grounding electrode is required.
- No need to check for the maximum 25-ohm requirement because the permanent moisture under a concrete foundation ensures a low-impedance direct connection to earth.

When using a concrete-encased grounding electrode, be sure that the footing or foundation is in *direct* contact with the earth. Make certain that there is no vapor barrier underneath the footing or foundation.

TRADE TIP   Always work closely with the concrete and rebar contractor. It is necessary to bring one end (*stub-up*) of reinforcing bar (called *rebar*) or the bare 4 AWG copper conductor upward out of the concrete slab or footing at a location near the likely location of the electrical main service. This will provide an easy connection point for the grounding electrode conductor. (See *250.52(A)(3)*.)

Additional Ground Rod Requirements

The most commonly used ground rods are copper-coated steel. Copper provides an excellent connection between the rod and the ground clamp. The steel gives it strength to withstand being driven into the ground. Galvanized and stainless steel rods are also available. *Aluminum* rods are *not permitted.* Ground rods:

- Must be at least ½ in.(13 mm) in diameter.
- Must be installed below the permanent moisture level.
- Must *not be less than* 8 ft (2.5 m) in length (8 ft and 10 ft are most commonly used).
- Must be driven to a depth so that at *least* 8 ft (2.5 m) is in contact with the soil.

If solid rock is encountered, drive the rod at an angle not greater than 45° from vertical, or lay it in a trench that is at *least* 2½ ft (750 mm) deep.

Drive the rod so the upper end is flush with or just below ground level. If the upper end is exposed, the ground rod, the ground clamp, and the grounding electrode conductor must be protected from physical damage.

If *more than one* rod is needed, keep the rods at least 6 ft (1.8 m) apart. Driving the rods closer together does *not* significantly change the resistance values of a single rod. A better technique, and an easy one to remember, is to space multiple rods the *length* of the rods. For example, when driving two 8 ft (2.5 m) ground rods, space them 8 ft (2.5 m) apart.

Ground Clamps

Ground clamps used for bonding and grounding *must be listed* for the purpose. The use of solder to make up bonding and grounding connections is *not* acceptable. Under high levels of fault current, the solder would probably melt, resulting in loss of integrity of the bonding or grounding path. (See *250.8.*)

TRADE TIP   Grounding clamps that are attached to rebar and buried in concrete *must* be rated appropriately.

Several types of ground clamps are illustrated in **Figures 8-11** through **8-14**. These clamps and their attachment to the grounding electrode must conform to *250.70*.

FIGURE 8-11 Typical ground clamps used in residential systems.

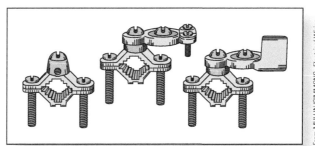

FIGURE 8-12 Armored grounding conductor connected with ground clamp to water pipe.

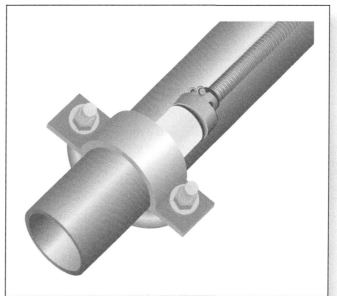

FIGURE 8-13 Ground clamp of the type used to bond (jumper) around water meter.

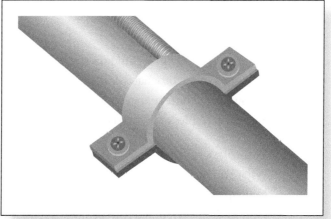

FIGURE 8-14 **FIGURE 8-14**   Ground clamp of the type used to attach ground wire to well casings.

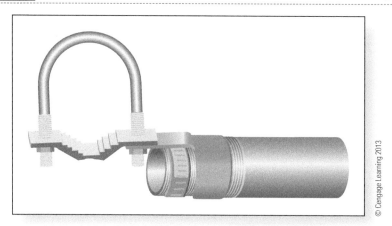

© Cengage Learning 2013

LIFE SKILLS

Service grounding is something that you will have to deal with routinely throughout your career as an electrician. Write down three things that you learned from this section that will make you a more capable apprentice once you are in the field.

Self-Check 2

1. What is the purpose of the service equipment?
2. What is the purpose of the grounding electrode conductor?
3. What is another name for the grounded service conductor?
4. Which *NEC* section describes main bonding jumper requirements?
5. What two components do all residential panelboards have?

8-3   Incorrect Grounding or Lack of Grounding

Whether it is an incorrect or improper ground or no ground at all, one thing is for sure: It will be an *unsafe* situation! Read and remember the consequences in the following examples.

Incorrect Grounding

Grounding (and bonding) techniques *must* always be performed while following the *NEC* requirements. When, for whatever reason, grounding is improperly installed or there is a lack of grounding, consequences will result—sometimes very serious consequences. The following are examples of incorrect grounding *and* resulting consequences.

Example 1

What would result if only two conductors were run out to a facility's tower lights? *Consequence:* A low impedance path with the capacity to safely carry the maximum ground-fault current likely to be imposed on it would *not* be provided (*NEC 110.10*).

Therefore, this system would *not* clear a ground-fault and would *not* open the circuit overcurrent protective device. This is *not Code* compliant and is very unsafe.

Example 2

What would result if metal parts became energized and, although they had been grounded to a ground rod, these metal parts were *not* bonded to an effective ground-fault path (*NEC 250.2* and *250.4(A)(5)*)?
Consequence: If the metal parts become energized, and there is no effective ground-fault path (to ground), a ground-fault current could cause serious injury and possible death.

Example 3

Grounding electrode conductors that have the following characteristics:
1. They are other than aluminum or copper (*250.64(A)*).
2. They are *not* protected against physical damage (*250.64(B)*).
3. They are not spliced, or if they are spliced, an irreversible compression-type connector or exothermic welding was *not* used (*250.64(C)*)

Consequence: This would be an unsafe and ungrounded system.

Lack of Grounding

Whether it's an older home with no direct path to earth ground, or a wiring system where the ground connection that has become disconnected, *ungrounded* wiring is *unsafe*! The following are examples of lack of grounding *and* consequences that would result.

Example 1

What about older ungrounded homes?
Consequence: Older homes often have electrical receptacles and fixtures that are ungrounded. Grounding provides a third path for current to travel. So if there is a leak of any sort, it will flow to ground (earth) and not through the body of a person who touches a defective fixture, appliance, or tool. With *any* ungrounded service, you have potential shock, injury, and even death.

Example 2

What would happen in the following situation? An electrical system is grounded with a grounding rod driven at least 8 feet into the ground outside the house, but a branch circuit is left ungrounded (or a separate wire became disconnected that leads to the Neutral bar of the service panel) or a there is a break in the metal sheathing that runs without a break from each outlet to the panel.
Consequence: The *branches* described are ungrounded, and therefore potential shock, injury, or death could result.

Example 3

What would result in the following situation? A building's main electrical ground relied on a connection to a metal water pipe to a well. The building's ground wires were connected to the metal water pipe. However, the metal pipe exiting the building was replaced with a newer plastic water line between the building and the well.

Consequence: The local ground is now completely ineffective (lost). This would be an ungrounded system and will be very unsafe.

Example 4

What would happen if bare aluminum electrical ground wires became corroded entirely through where the wire touched a damp foundation wall?
Consequence: The electrical ground becomes completely ineffective, and this service is unsafe.

Example 5

What would result if the ground wire between the electrical panel and a building's water pipe or grounding electrode became separated, loose, corroded, incorrectly spliced, or lost entirely?
Consequence: The electrical ground becomes completely ineffective, and this service is unsafe.

With a group of two or three classmates, pick one or two of the examples in this chapter. Then discuss how the potentially dangerous scenario could possibly be remedied or avoided.

Self-Check 3

1. Why must grounding and bonding techniques follow *NEC* requirements?

8-4 ■ *NEC* Requirements—Bonding of Wiring Devices to Outlet Boxes

The *NEC* is very clear about where bonding must be used. This *Code* also states in no uncertain terms how a bonding jumper must be installed. The following highlights the *Code*'s requirements in these areas and emphasizes the importance of electrical continuity.

NEC References and Interpretations

Electrical continuity
a complete, continuous, and connected path that keeps the electrical potential difference at zero.

NEC 250.96(A) requires that bonding be done around connections of metal raceways, cable trays, cable armor, cable sheath, enclosures, frames, fittings, and other metal non-current-carrying parts used as equipment grounding conductors where necessary. Bonding is provided in order to ensure that these systems have electrical continuity and the current-carrying capacity to safely conduct the fault current likely to be imposed on them. Electrical continuity (in this case) refers to a complete, continuous, and connected path in order to keep the electrical potential difference at zero. Whether or not an equipment grounding conductor is run within the raceway, bonding is still required. This ensures the raceway will *not* become energized due to a line-to-enclosure fault without having the capacity and capability of clearing the fault by allowing sufficient current to operate the overcurrent protective device on the line side of the fault.

Figure 8-15 illustrates possible breaks in electrical continuity and potential loss of the ground-fault current return path.

Referring to Figure 8-15, note that arrows point to connections or links in the bonding process. Remember that the *effective ground-fault current path* is required to:

- Be permanent and electrically continuous.
- Have the capacity to conduct safely any fault currents.
- Have sufficiently low impedance to limit the voltage to ground and to facilitate the operation of the circuit-protective devices. (See *250.4(A)(5)*.)

TRADE TIP   Remember *every connection* is important! It only takes one loose locknut or broken fitting to break the direct connection and lose continuity in the fault current path.

NEC 250.96(A) also refers to conditions where a nonconducting coating might interrupt the required continuity of the ground-fault path, and it states that such coatings must be removed.

Locknuts can generally penetrate painted enclosures to establish a good electrical connection. Locknuts are tightened by hand and can be further tightened ¼ turn by means of a screwdriver and hammer.

FIGURE 8-15   Bonding to maintain continuity.

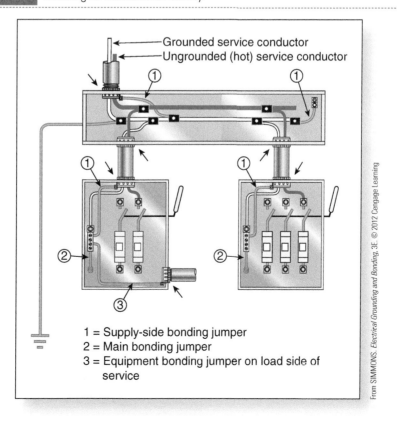

Grounded service conductor
Ungrounded (hot) service conductor

1 = Supply-side bonding jumper
2 = Main bonding jumper
3 = Equipment bonding jumper on load side of service

From SIMMONS. *Electrical Grounding and Bonding, 3E.* © 2012 Cengage Learning

---

TRADE TIP After tightening a locknut used on a painted enclosure, remove the locknut and scrape the paint off or install a bonding bushing if there is any question about the adequacy of the connection.

When bonding receptacles, use an equipment bonding jumper to connect the grounding terminal of a grounding type receptacle to a grounded box. (See *250.146*). **Figure 8-16** shows a self-grounding receptacle. This type receptacle is recognized by the metal clip around the screw.

The receptacle shown in Figure 8-16 is also hospital grade (because it has a green dot on the face).

Four *exceptions* to the general rule requiring the bonding jumper are provided in *250.146(A)* through *(D)* as paraphrased below.

(A) When the box is mounted on the surface and *direct* metal-to-metal contact is made between the receptacle yoke and the box.

(B) When contact devices or yokes designed and *listed* as providing bonding with mounting screws to establish the grounding circuit between the device yoke and flush-type boxes are used. These devices are commonly referred to as self-grounding receptacles. (*Note:* Refer back to Figure 8-16 to see this type of receptacle).

(C) When floor boxes that are *listed* as providing satisfactory grounding continuity are used.

(D) When a receptacle having an isolated equipment grounding terminal required for reduction of electrical noise or electromagnetic interference is used.

FIGURE 8-16 Self-grounding receptacle.

From ZACHARIASON, *Electrical Materials*, 1E. © 2008 Cengage Learning

Equipment Bonding Jumper Installation

The equipment bonding jumper is permitted to be installed either inside or outside of a raceway or enclosure. (See *250.102(E)*.)

Where the jumper is installed on the outside, the length is generally limited to not more than 6 ft (1.8 m). In addition, the bonding jumper is required to be routed with the raceway. This is vital to keep the impedance of the equipment bonding jumper as low as possible. **Figure 8-17** illustrates the installation of equipment bonding conductors external to enclosures.

An equipment bonding jumper longer than 6 ft (1.8 m) is permitted at outside pole locations for the purpose of bonding or grounding *isolated* sections of metal raceways or elbows that are installed in exposed risers of metal conduit or other metal raceway. **Figure 8-18** shows that a bonding jumper installation at outside pole locations is permitted to be longer than 6 ft (1.8 m) in length.

FIGURE 8-17   Installation of equipment bonding jumper.

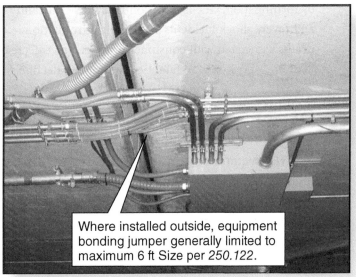

Where installed outside, equipment bonding jumper generally limited to maximum 6 ft Size per *250.122*.

From SIMMONS: *Electrical Grounding and Bonding*, 3E. © 2012 Cengage Learning

FIGURE 8-18   Bonding jumper installation at outside pole locations.

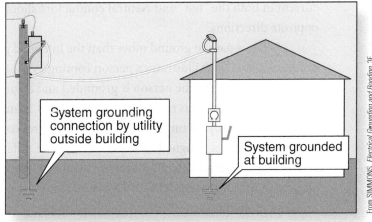

System grounding connection by utility outside building

System grounded at building

From SIMMONS: *Electrical Grounding and Bonding*, 3E. © 2012 Cengage Learning

LIFE SKILLS

One of your goals should be to know the major *NEC* requirements that you will be encountering when you're in the field. Consider making space in a notebook to organize an *"NEC* Requirements to Remember" list.

Self-Check 4

1. What is the importance of bonding?

2. What is meant by electrical continuity?

A ground fault can be considered a "fault" to "ground." This occurs when either of the current-carrying conductors on a circuit becomes inadvertently connected to ground. Examples of this would be: insulation failures in the wiring of buildings or damaged appliances, tools, extension cords, misuse of devices, wiring errors, etc. Some refer to this condition as a "short" to ground. A ground-fault circuit interrupter (GFCI) is designed to sense when a short or fault to ground occurs and "interrupts" the circuit by turning it off. GFCIs have contributed significantly to the reduction of electrocution and severe shock incidents since their introduction in the early 1970s. Electrocution deaths associated with consumer products have decreased from 270 in 1990 to 180 in 2001 and shock-related fatalities continue to decline.

8-5 Ground-Fault Circuit Interrupter (GFCI)

Ground-Fault Circuit Interrupter (GFCI) a device intended for the protection of personnel that functions to de-energize a circuit or portion thereof within an established period of time when current to ground exceeds the values established for a Class A device.

The *NEC* requires ground-fault circuit interrupter (GFCI) protection at various locations. GFCIs are used to prevent people from being electrocuted. A GFCI as defined by *Article 100* is a device intended for the protection of personnel that functions to de-energize a circuit or portion thereof within an established period of time when current to ground exceeds the values established for a Class A device. An *NEC* Informational Note states that a Class A GFCI trips when the current to ground has a value in the range of 4 milliamps to 6 milliamps.

GFCIs operate by sensing the amount of current flow on both the ungrounded ("hot") and grounded (Neutral) conductors supplying power to a device. Current in both conductors should be equal but opposite in polarity. **Figure 8-19** illustrates that current in both the "hot" and Neutral conductors should be the same but flowing in opposite directions.

When a path to ground other than the intended path is established, a ground fault occurs. **Figure 8-20** illustrates a person coming in contact with a defective coffee pot.

In Figure 8-20, the person is grounded and a current of 0.1 ampere is flowing through the person. This results in a current of 10.1 amperes flowing though the "hot" conductor, but only 10 amperes through the Neutral conductor.

The GFCI is designed to detect this current difference and to protect personnel by *opening* the circuit when it detects a current difference of approximately 0.005 ampere (5 milliamperes). *NEC 210.8* lists places where ground-fault protection is required in dwellings.

FIGURE 8-19 Current in "hot" and Neutral conductors.

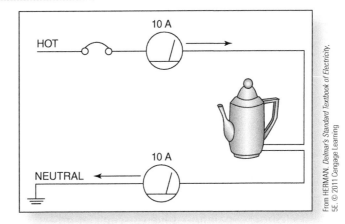

FIGURE 8-20 A ground fault occurs when an unintended path to ground is established.

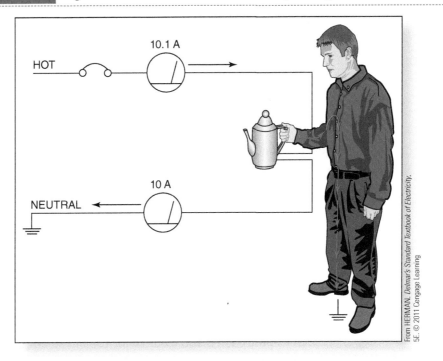

TRADE TIP The effect of an electric current passing through a human body depends on several factors: current, frequency, voltage, body contact resistance, internal body resistance, path of current, duration of contact, and environmental conditions (i.e., dry or humid). Remember: NO SHOCK IS A GOOD SHOCK! Table 8-1 (following) is shown ONLY to give reference as to HOW even very low levels of current affect the human body. Refer to Table 8-1.

Note that in the example illustrated in Figure 8-20, this person would most probably experience severe shock and in just a few seconds, ventricular fibrillation.

With ground-fault protection, the person in Figure 8-20 would have been protected. It's little wonder that the *NEC* has listed code requirements for ground-fault circuit interrupters. GFCIs protect against shock and save lives!

TABLE 8-1 Effect of electric shock.

	Current in Milliamperes @ 60 hertz	
	Men	**Women**
• Cannot be felt	0.4	0.3
• A little tingling—mild sensation	1.1	0.7
• Shock—not painful—can still let go	1.8	1.2
• Shock—painful—can still let go	9.0	6.0
• Shock—painful—just about to point where you can't let go—called "threshold"—you may be thrown clear	16.0	10.5
• Shock—painful—severe—can't let go— muscles immobilize—breathing stops	23.0	15.0
• Ventricular fibrillation (usually fatal)		
• Length of time: 0.03 sec.	1000	1000
• Length of time: 3.0 sec.	100	100

From MULLIN/SIMMONS. *Electrical Wiring Residential*, 17E. © 2012 Cengage Learning

Code Requirements for Ground-Fault Circuit Interrupters

Required ground-fault circuit interrupter protection is stated in *210.8*. The following summarizes this information. In dwellings, ground-fault circuit interrupter (GFCI) protection must be provided for all 15- and 20-amp, 125-volt receptacle outlets installed in:

- Bathrooms
- Garages
- Crawl spaces
- Unfinished basements
- Locations within 6 ft (1.8 m) of wet-bar, utility, or laundry sinks

GFCI protection is also required for 15- and 20-amp, 125 volt receptacles in boat-houses and for boat hoists, outdoors, and for receptacles serving the countertop in kitchens.

LIFE SKILLS

On the jobsite there will be times where you are required to answer direct questions from the customer. Suppose a customer asks why a 20-amp, 125-volt outlet in their crawl space needs GCFI protection, how would you answer?

Self-Check 5

1. What is the purpose of a ground-fault circuit interrupter?
2. What does the *NEC* state regarding GFCIs?

Summary

- *Article 250* of the *NEC* requires that alternating current (ac) systems of 50 to 1,000 volts that supply premises wiring systems where the system can be grounded so that the maximum voltage to ground on the ungrounded conductors does not exceed 150 volts shall be grounded.

- All residential electrical systems rated at 120/240 volts must be grounded.

- Grounding is the act of connecting something to the ground (earth) generally by means of a conductor such as a wire or rod, so it has zero electrical potential.

- Bonding is the connection of two or more conductive objects to establish electrical continuity and conductivity.

- Definitions of and key terms relating to grounding and bonding are found in *Article 100*.

- A service entrance provides a way for the electrical system to get power from the electric utility company.

- Grounding (earthing) takes place at the service entrance.

- A service is the conductors and equipment required for delivering energy from the electric utility to the wiring system of the premises.

- Service equipment is the necessary equipment connected to the load end of the service conductors supplying a building and intended to be the main control and cutoff supply.

- The supplemental grounding electrode is a grounding electrode used to back up a metal water pipe grounding electrode.

- For a grounded system, an unspliced main bonding jumper must be used to connect the equipment grounding conductor and the service-disconnect enclosure to the grounded conductor of the system within the enclosure for each service disconnect.

- The grounding electrode conductor is required to be connected to the grounded (Neutral) service conductor.

- A metal underground water piping system 10 ft (3.0 m) or longer, in direct contact with the earth, is acceptable as the primary grounding electrode as per *250.52(A)(1)*.

- When metal underground water piping is used as the primary grounding electrode, it must be supplemented by at least one additional grounding electrode as per *250.53(D)(2)*.

- Grounding, whether residential or commercial is one of the most important aspects of a sound electrical system.

- The concrete-encased electrode does not require a supplemental ground or the need to check the ground resistance.

- The most commonly used ground rods are copper-coated steel.

- Ground clamps used for bonding and grounding must be listed for the purpose.

- Incorrect grounding or a lack of grounding is unsafe; shock, injury, even death, could result.

- Bonding is provided in order to ensure that connected, metal systems have electrical continuity and the current-carrying capacity to safely conduct the fault current likely to be imposed on them.
- When bonding receptacles, a bonding jumper is required to connect the grounding terminal of a grounding type receptacle to a grounded box, as per *250.146*.
- The equipment bonding jumper is permitted to be installed either inside or outside of a raceway or enclosure, as per *250.102(E)*.
- A ground-fault circuit interrupter (GFCI) is a device intended for the protection of personnel that functions to de-energize a circuit or portion thereof within an established period of time when current to ground exceeds 4 milliamps.
- *NEC 210.8* states that in dwellings, ground-fault circuit-interrupter (GFCI) protection must be provided for all 15- and 20-amp, 125-volt receptacle outlets installed in bathrooms, garages, crawl spaces, unfinished basements, and locations within 6 ft (1.8 m) of wet-bar, utility, or laundry sinks.

Review Questions

True/False

1. All residential electrical systems rated at 120/240 volts must be grounded. (True, False)

2. Any intentional connection between an electric circuit and the earth is known as bonding. (True, False)

3. A reliable conductor used to ensure the required electrical conductivity between metal parts is called a bonding jumper. (True, False)

4. The grounding electrode conductor is not connected to the service equipment. (True, False)

5. A developmental grounding electrode is a grounding electrode used to back up a metal water pipe grounding electrode. (True, False)

Multiple Choice

6. In a residential wiring system, the grounded service conductor is generally called the _____.
   A. Safety conductor
   B. "Hot" conductor
   C. Neutral conductor
   D. Electrode conductor

7. What connects the neutral service conductor to the service-entrance main disconnect enclosure and the equipment grounding conductors?
   A. Equipment bonding jumper
   B. Main bonding jumper
   C. System bonding jumper
   D. Feeder bonding jumper

8. Where in the *NEC* will the requirements for establishing a grounding electrode system be found?
   A. *Article 250, Part III*
   B. *Article 100, Part III*
   C. *Article 350, Part III*
   D. *Article 150, Part III*

9. When a metal underground water piping system is used as a primary grounding electrode, it must be _____.
   A. 10 ft or longer and in direct contact with the earth
   B. Less than 10 ft and in direct contact with the earth
   C. 10 ft or longer, but with no direct contact with earth
   D. Less than 10 ft, but with no direct contact with earth.

10. What does the *NEC* state about using ground rods, pipes, or plates when the resistance to ground of the first ground rod, pipe, or plate is 25 ohms or more?
    A. It cannot be used under any circumstances.
    B. At least one additional ground rod must be added.
    C. Resistance to ground must be increased.
    D. At least four additional rods must be added.

Fill in the Blank

11. _____ electrodes have become popular with the increasing use of nonmetallic water mains.

12. The most commonly used ground rods are _____ steel.

13. When describing bonding, electrical _____ refers to a complete, continuous, and connected path in order to keep the electrical potential difference at zero.

14. In the bonding process, the effective ground-fault current path is required to have the _____ to conduct safely any fault currents.

15. A GFCI is designed to detect a current difference in the "hot" and neutral conductors and _____ the circuit when this current difference exceeds approximately 5 milliamperes.

True/False

16. The effect of an electric current passing through a human body does not depend on the path of current. (True, False)

17. *NEC Section 210.8* does not list unfinished basements as an area required to be protected by a ground-fault circuit interrupter. (True, False)

18. A grounding electrode conductor must be made on the first 5 ft where metal underground water piping enters a building. (True, False)

19. A concrete-encased electrode does not require a supplemental grounding electrode. (True, False)

20. An older home with no direct path to earth ground is considered ungrounded and is unsafe. (True, False)

Chapter 9

Tools

Patricia Brack is happy in her work as a journeyman electrician because the field is so diverse. She enjoys her work, she says, because "every job is different," and she thus can look forward to "the challenge of a new task every day." She also notes, "The aspect of creating something from beginning to end is very satisfying."

Brack sees the beginning of a job as "the most rewarding task with the most challenges." She welcomes the opportunity to formulate a "game plan" that will "get all that power to where the customer wants it, in a neat and workman-like manner."

Brack feels the curriculum she followed in her program at the IEC has prepared her to complete her work in a professional way: "I was always amazed when I was in the field and taking some measurements with the voltmeter/ampmeter and the classroom lesson dawned upon me," she remarks.

Chapter Outline

THE IMPORTANCE OF CARING FOR HAND TOOLS AND
THEIR PROPER USE

BASIC HAND TOOLS

POWER TOOLS

SPECIALTY TOOLS

ELECTRICIAN TOOL KITS

Key Terms

Arbor

Awl

Bevel

Chisel

Chuck

Double-insulated

EMT

Hexagonal

Metallic tubing

Plumb

Polarized plug

Reciprocating

Sheathed cable

Sheathing

T®-Stripper Wire Stripper

Chapter Objectives

After completing this chapter, you will be able to:

1. Explain the importance of caring for basic hand and power tools and their proper use.

2. Identify primary hand and power tools used in the electrical trade.

3. Demonstrate safe and effective use of basic hand and power tools.

Life Skills Covered

 Critical Thinking   Cooperation and Teamwork

Life Skill Goals

Throughout your career as an electrician, you will come to rely upon a wide variety of tools. Whether a physical tool, like a wrench or a hammer, or a written tool, like this text or the *NEC*, mastering your tools is key to mastering your trade. Utilizing the life skills for this chapter, Critical Thinking and Cooperation and Teamwork, will help get you started.

As you go through this chapter, talk to your classmates about its contents. Discuss which tools are best for a particular job. Compare your preferences when it comes to brand and other features. If you're not experienced with a particular tool that is mentioned, find a classmate who is and ask questions. Your peers are every bit as useful as any tool in your box. Make sure you're taking advantage!

This chapter will provide and emphasize guidelines for hand and power tool safety. It will help you to identify basic hand and power tools and demonstrate *safe and effective use. Demonstrations* and *lab exercises* will provide hands-on training in tool usage. This chapter will help you learn about tools and ways to avoid tool-related injuries.

Introduction

You may have heard the expression that hand tools are extensions of our hands. You certainly will agree that if we misuse our hands, we experience pain. Misusing hand tools will increase the possibility of injury to ourselves or people working around us.

Sometimes real-life examples make more of an impression than words in a book. A California construction worker slipped and fell on a nail gun, which shot six nails into his face, neck, and skull. A Houston-area remodeler accidently shot a ½-inch nail through his chest. Believe it or not, these contractors are the lucky ones because they survived. One of every five *fatal* work injuries occurs on a construction site. According to accident facts compiled from the Bureau of Labor Statistics, U.S. Department of Labor, and from the book *Accident Facts*, published by the National Safety Council, hand tools are involved in 6% of all compensated work injuries. This figure increases to 14% for noncompensated work injuries. Learn the proper care and use of hand and power tools. *Work smart* and *avoid injuries!*

9-1　The Importance of Caring for Hand Tools and Their Proper Use

Causes of hand tool injuries can generally be traced to some type of improper use or the lack of proper maintenance of a hand tool. It is important for electricians, tradesmen, and all tool users to be aware of guidelines for the care and safe use of hand tools. The following information should be considered whenever you are using hand tools:

- Wear personal protective equipment (PPE) as required.

- Use the right tool for the job. Serious damage and injury may result from using a tool that is *not* designed for the job.

- Be sure to use the correct *size* tool.

- Use a tool only *after* you have been trained in basic safety and proper use of the tool.

- Inspect tools frequently to be sure they are in good condition, and keep them that way.

- Keep tools clean and dry. (Clean hand tools work better).
- Lubricate tools when necessary. (Lubrication allows tools with hinged joints to work easily).
- Repair damaged tools promptly, and dispose of broken or damaged tools that cannot be repaired.
- Dispose of razor blades and utility knife blades in a puncture-resistant sharps container.
- Pay attention when using tools. (Only *trained and qualified* individuals should work on electrical circuits).
- Store tools properly in a safe place when not in use.
- Keep all cutting tools sharp so the tool will move smoothly without bending or skipping.

TRADE TIP More injuries result from dull cutting tools than from sharp ones. This is because *dull* cutting tools require *much more* force to do their job. With more force being applied, a tool more easily slips and could cause injury to you or damage to equipment.

- Maintain a good grip on the tool, and stand in a balanced position to avoid sudden slips.
- Ensure that your work area is clean, dry, well lit, and free of obstructions.

LIFE SKILLS

As an electrician, you will rely heavily on your tools. To get the most out of them, you must take care of them. Together with a partner, come up with two lists: five tool care "do's" and five tool care "don'ts." Then, find another group and compare notes.

Self-Check 1

1. Why is the proper care and safe use of hand tools important?
2. List four helpful tips for properly using tools.

9-2 Basic Hand Tools

There are many tools used in the electrical trade. Hand tools fit easily into a tool pouch worn on an electrician's hip. The following information describes the most common tools used by electricians.

Wrenches

Wrenches are commonly used hand tools. Their main function is holding and turning nuts, bolts, cap screws, plugs, and various threaded parts. Quality wrenches are designed to keep leverage and intended load in safe balance.

Standard wrench types, with both American standard inch and metric openings, are available for almost every operation and service.

Open-End Wrenches

The most widely used open-end wrenches have openings at a 15° angle, which permits complete rotation of hex nuts, with a 30° swing by flopping the wrench. They are available in both single- and double-head patterns; double-head patterns have different openings in each head. **Figure 9-1** shows examples of open-end wrenches.

Adjustable Wrenches

Regular pattern adjustable wrenches are available in lengths from 4 to 24 inches. Adjustable wrenches are designed to provide a wide range of capacity in a single tool and are a convenient service wrench. They are *not* intended to replace fixed opening wrenches for production or general service work.

An adjustable wrench, sometimes called a *crescent* wrench, is used to tighten couplings and connectors, tighten pressure-type wire connectors, and remove and hold nuts and bolts. The jaws of a crescent wrench are smooth and fit well around bolt heads and nuts. **Figure 9-2** shows a crescent wrench.

The crescent wrench has one fixed jaw and one movable jaw. A worm gear is turned to adjust the jaws to any number of head sizes. Common sizes in residential electrical include an 8- and a 10-inch size.

FIGURE 9-1 Open-end wrenches.

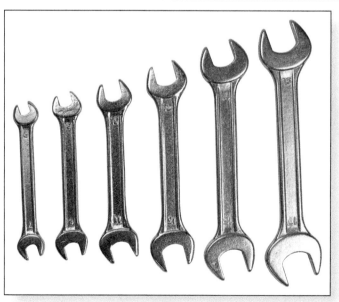

FIGURE 9-2 Crescent wrench.

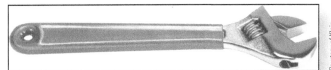

Pliers

Pliers of various types are used by just about anyone who uses tools. There are many types and sizes. Each pliers is designed for a specific use, although their versatility makes some pliers adaptable for many jobs. The most common types of pliers, also considered pouch tools, are listed here:

- Lineman
- Long-nose
- Diagonal cutting

Lineman

Bevel refers to any edge cut at an angle to a flat surface.

There are two head patterns available: standard, also known as **bevel** nose, and New England, referred to as the round nose. Bevel refers to any edge cut at an angle to a flat surface. A lineman pliers, sometimes called *side-cutter* pliers, is one of the most widely used of electricians' hand tools. These pliers are used to cut cables, conductors, and small screws. They are also used to form large conductors and to pull and hold conductors. Always use pliers that are large enough for the job. Generally, the handles should be around 9 inches in length so that a minimum of hand pressure is required to cut the conductor or cable. In most cases, only one hand is needed to make the cut, although with large cable two hands may be required. Lineman pliers are available in a variety of styles and sizes. The 9¼-inch handle with the New England nose is the tool of choice for residential electrical work. These pliers are shown in **Figure 9-3**.

FIGURE 9-3 Lineman pliers.

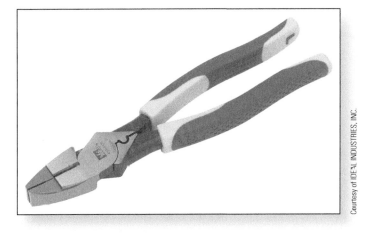

Pliers used for electrical work are coated with vinyl for better comfort and to provide a secure grip.

Long-Nose Pliers

Long-nose pliers, often referred to as *needlenose* pliers, are used to form small conductors, cut conductors, and hold and pull conductors. **Figure 9-4** shows an example of the long-nose pliers.

The narrow head of the long-nose, or, as it is often called, the needlenose pliers, makes it easier to work in tight areas. Electricians generally use long-nose pliers with at least an 8-inch handle.

Diagonal Cutting Pliers

Diagonal cutting pliers, sometimes referred to as *dikes*, are used to cut cables and conductors in areas with limited working space. **Figure 9-5** illustrates diagonal cutting pliers with 8-inch, high-leverage handles and an angled head.

FIGURE 9-4 Long-nose pliers.

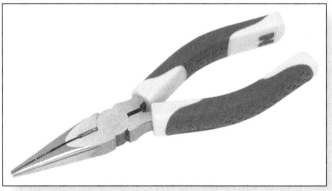

Courtesy of IDEAL INDUSTRIES, INC.

FIGURE 9-5 Diagonal cutting pliers.

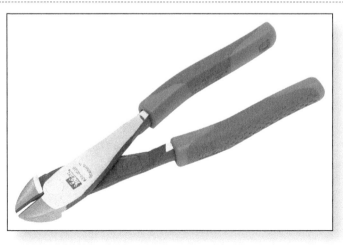

Courtesy of IDEAL INDUSTRIES, INC.

Wire Strippers

Wire strippers are primarily used to strip (remove) insulation from conductors. This tool is also used to cut and form conductors. Wire strippers come in several styles, and the following are the two styles often used by electricians:

- T®-Stripper Wire Stripper
- Cable ripper

T®-Stripper Wire Stripper

In residential electrical work, the most often used wire stripper is the nonadjustable type called a T®-Stripper Wire Stripper. **Figure 9-6** shows a T®-Stripper Wire Stripper–style wire stripper.

The T®-Stripper Wire Stripper is designed to strip the insulation from several different wire sizes without having to be adjusted for each size. Wire sizes most often encountered in residential electricity are from 10 through 18 AWG. The T®-Stripper Wire Stripper is the best tool to strip insulation from wires of this size.

Cable Ripper

Cable rippers are used to remove the outside sheathing from a nonmetallic sheathed cable. Sheathing is a protective covering that wraps or surrounds something, and a sheathed cable is one protected by a nonconductive covering, such as vinyl. **Figure 9-7** shows a cable ripper.

Cable rippers are designed to slit the cable sheathing. A knife or cutting pliers is then used to cut the sheathing off the cable.

Screwdrivers

Screwdrivers are designed in several styles, and the selection of a screwdriver depends on the type of screw that it needs to fit. **Figure 9-8** illustrates several screwdriver tip styles.

T®-Stripper Wire Stripper a wire stripper designed to strip insulation from several different wire sizes without having to be adjusted for each size.

Sheathing a protective covering that wraps or surrounds something.

Sheathed cable a cable protected by a nonconductive covering, such as vinyl.

FIGURE 9-6 T®-Stripper Wire Stripper.

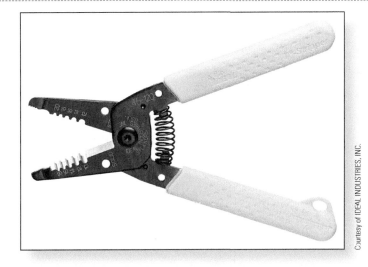

Courtesy of IDEAL INDUSTRIES, INC.

FIGURE 9-7 Cable ripper.

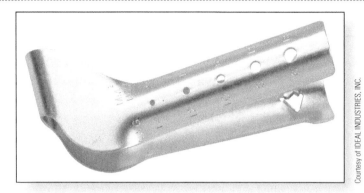

Courtesy of IDEAL INDUSTRIES, INC.

FIGURE 9-8 Screwdriver tip styles.

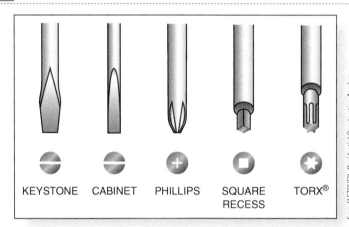

KEYSTONE CABINET PHILLIPS SQUARE RECESS TORX®

From FLETCHER. *Residential Construction Academy,* 3E. © 2012 Cengage Learning

In electrical work, the most common types of screwdrivers used are the Keystone and Phillips screwdrivers. The Keystone is used to install and remove slot-head screws. This screwdriver is also used to tighten and loosen slot-head lugs. Phillips-tip screwdrivers are used to remove and install Phillips-head screws and to tighten and loosen Phillips-head lugs. **Figures 9-9** and **9-10** illustrate Keystone-tip and Phillips-tip screwdrivers.

A very handy screwdriver used to tighten and loosen screws when there is limited working space is the *stubby* screwdriver. Stubby screwdrivers are designed in both Keystone and Phillips styles. **Figures 9-11(A)** and **(B)** show Keystone and Phillips stubby screwdrivers.

FIGURE 9-9 A Keystone-tip screwdriver.

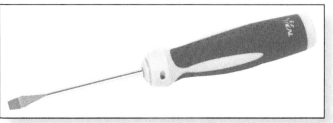

Courtesy of IDEAL INDUSTRIES, INC.

FIGURE 9-10 A Phillips-tip screwdriver.

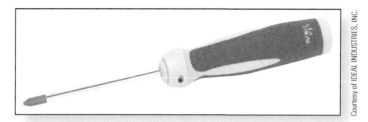

FIGURE 9-11 (A) Keystone stubby and (B) Phillips stubby.

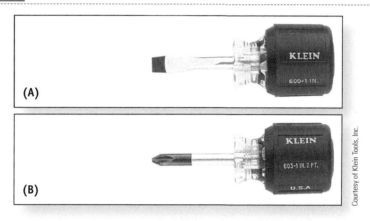

(A)

(B)

Cordless Screwdriver

Cordless screwdrivers use rechargeable batteries instead of an electrical cord to power a strong reversible motor attached to a shaft. **Figure 9-12** shows a cordless screwdriver.

FIGURE 9-12 A cordless screwdriver.

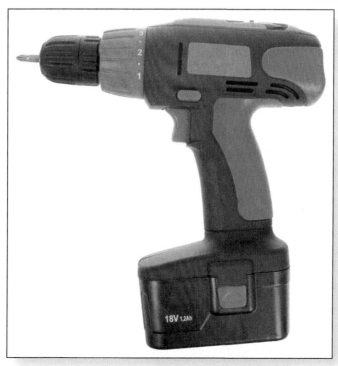

Cordless screwdrivers typically feature a number of interchangeable bits, including flat blade, Phillips, and socket attachments. Advantages of using a cordless (power) screwdriver are to prevent hand exhaustion and create more torque than standard screwdrivers.

TRADE TIP Most cordless drills have an adjustable clutch feature that allows the amount of turning force to be increased or decreased. This allows the cordless drills to act like a power screwdriver. Electricians often use this feature for installing receptacles and switches in device boxes and for other tasks that normally involve using a manual screwdriver.

Using a cordless (power) screwdriver is much faster than using a manual screwdriver. It is also much easier on the installing electrician.

Knife

A valuable tool often used by residential electricians is the electrician's knife. The electrician's knife can be used to open cardboard boxes containing electrical equipment and to strip insulation from large conductors and cables. One commonly used electrician's knife includes both a cutting blade and a screwdriver blade. Two other popular electrician's knives are the *hawkbill* knife that has a curved blade and the *utility* knife with a retractable blade. **Figure 9-13** illustrates three commonly used electrician's knives.

Hammer

The electrician's hammer is a very useful tool for the residential electrician. A hammer is used to drive and pull nails or staples, pry boxes loose, break wallboard, and strike **awls** and **chisels**. An awl is a small tool for marking surfaces or for punching small holes. A chisel is an edge tool with a flat steel blade with a cutting edge. **Figure 9-14** shows an 18-ounce electrician's hammer.

Awl a small tool for marking surfaces or for punching small holes.

Chisel an edge tool with a flat steel blade with a cutting edge.

An electrician's hammer should have long, straight claws to simplify the removal of electrical equipment. The handle for the electrician's hammer should be strong and shock-absorbent. The most popular hammers are designed with fiberglass handles. Referring again to Figure 9-14, note the fiberglass handle and comfortable neoprene grip. The 18- and 20-ounce hammers are the most commonly used by residential electricians.

Hacksaw

A hacksaw is generally used to cut some types of conduit and is also used to cut larger conductors and cables. **Figure 9-15** shows a heavy-duty hacksaw that uses 12-inch blades.

Hacksaws are currently designed with rugged frames that are lightweight but provide ample rigidity that gives excellent control when cutting. The best all-around hacksaw blade for electricians has 24 teeth per inch. Hacksaw blades for electrical work are also available in configurations of 18 and 32 teeth per inch.

FIGURE 9-13 (A) Electrician's knife, (B) hawkbill knife, and (C) utility knife.

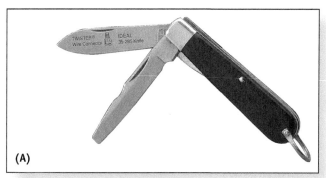

(A)

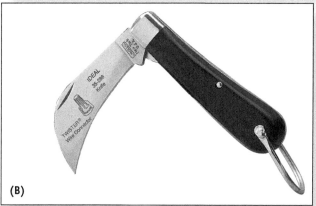

(B)

(C)

Courtesy of IDEAL INDUSTRIES, INC.

FIGURE 9-14 Electrician's hammer.

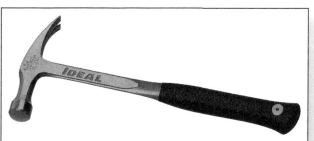

Courtesy of IDEAL INDUSTRIES, INC.

FIGURE 9-15 Hacksaw.

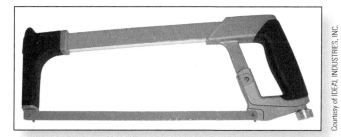

Courtesy of IDEAL INDUSTRIES, INC.

Tape Measure

The tape measure is one of the most useful tools in an electrician's toolbox. It is an instrument consisting of a narrow strip (cloth or metal) marked in inches or centimeters and is used for measuring. **Figure 9-16** shows a picture of a 25-foot-long tape measure.

A tape measure is generally used to take measurements to determine the correct location of electrical equipment. Tape measures come in standard 12-, 16-, 20-, and 25-foot lengths. The most commonly used tape measure is the 25-foot length with 1-inch-wide tape. If the tape is too narrow, it may bend or *break down* when it is extended.

Folding Rule

A folding rule is basically a ruler that folds into itself, by hinges at intervals, and which can be extended to measure. **Figure 9-17** illustrates a 6-foot folding rule.

A folding rule typically opens up to a maximum length of 6 feet and works well when measuring short distances and when making the same size cut, or marking out boxes for receptacles. The folding rule is often used for print reading in order to determine box locations on blueprints and confirm depth and setout of electrical boxes. This rule is hinged every six inches, making it useful for making templates for pipe bending. However, electricians generally use tape measures more often than folding rules.

FIGURE 9-16 Tape measure.

FIGURE 9-17 A folding rule.

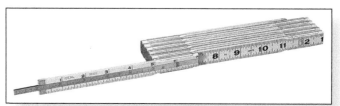

LIFE SKILLS

To work efficiently and effectively on a jobsite, you should be familiar with all of the basic tools listed in this section. On a scale of 1–5, rate how comfortable you are using each of the tools. Then, commit to improving on your areas of weakness (Cooperation and Teamwork).

Self-Check 2

1. What are pouch tools?
2. List three types of pliers and uses for each.
3. List two types of screwdrivers. What is a stubby, and when is it used?
4. Where might an electrician use a hammer?
5. What length is the most commonly used tape measure?

9-3 Power Tools

There are a variety of electric power tools used by electricians. Tradesmen in all industries use power tools.

> **TRADE TIP** When using a power tool, *don't become overconfident and don't take shortcuts!* Read and follow the instructions for a given power tool. The majority of hand injuries in the electrical trade come from trying to use a "two-handed" tool with only one hand. This practice usually leads to serious personal injury and in some cases loss of limb.

Double-insulated a form of electrical protection featuring two separate insulation systems to help protect against electrical shock from internal malfunctions.

Electric power tools include both those powered by 120-volt ac and those powered by low-voltage dc electricity. Ac-powered tools with a two-prong plug are **double-insulated**, a form of electrical protection featuring two separate insulation systems to help protect against electrical shock from internal malfunctions. All other power tools have a three-prong grounding attachment plug.

Cordless power tools are powered by dc delivered by rechargeable batteries. Many of the guidelines listed for the care and proper use of hand tools (found on page 214) apply to power tools as well. A good rule to live by when working with power tools is this: *Use a tool only after you have been trained in basic safety and proper use of the tool!*

The following are guidelines that should be considered when using power tools:

- Always wear personal protective equipment (PPE) as required.
- Do not operate power tools in explosive atmospheres such as in the presence of flammable liquids like gasoline. Sparks created by power tools could ignite gasoline fumes.
- Keep bystanders away from the work area. Flying debris could cause injuries.
- Be sure that grounded tools are plugged into a properly installed grounded receptacle outlet.

> **TRADE TIP**   **NEVER** remove the grounding prong from a grounding-type attachment plug. This is very unsafe and could result in a lethal shock.

- Double-insulated power tools use a polarized attachment plug. A polarized plug is a type of plug that has one prong longer than the other so it can only be inserted into an outlet in one way. Be sure that it is plugged into a correctly installed polarized receptacle.
- Do not operate electric power tools in wet conditions.
- Do not misuse the cord on a power tool. *Never* carry a tool by its cord.

> **TRADE TIP**   *On a residential construction site,* **ALL** 120-volt power tools with cords *must* be plugged into ground-fault circuit-interrupter (**GFCI**)–protected receptacles.

- When using a power tool outside, be sure to use an extension cord marked with "W-A" or "W." These cords are designed for outside use.
- *Never* use a power tool when tired or when taking medications that cause drowsiness.
- *Never* operate a power tool wearing loose clothing or jewelry.
- Make certain that the power tool is switched OFF *before* plugging it into an outlet.
- Be sure that all chuck keys or other tightening wrenches are removed before you apply power.
- Be sure to have firm footing when using a power tool.
- *Always* use the proper power tool for the job.
- *Always* store power tools in a dry and clean location away from children and other untrained people.
- *Never* use a damaged or defective tool.

Power Drills

A power drill with an appropriate bit is designed to bore holes for the installation of wire runs, conduits, and other electrical equipment in wood, metal, plastic, or other material. Residential electricians frequently use power drills. Like most power tools, power drill models are designed with power cord and plug or cordless with rechargeable batteries. The main types of power drills are as follows:

- Pistol-grip
- Hammer
- Cordless

Pistol-Grip Drill

Drills with pistol grips are the most common type in use today. Pistol-grip drills are small, relatively lightweight, and easy to use. This type of drill looks like and is held like a pistol. Most of the time, this type drill is simply referred to as a variable-speed drill. **Figure 9-18** shows a variable-speed reversing drill with keyed chuck.

Chuck the part of the drill that holds the drill bit securely in place.

Pistol-grip (variable) drills are available in three common **chuck** sizes: ¼, ⅜, and ½ inch. A chuck is the part of the drill that holds the drill bit securely in place. A chuck key or wrench is used to tighten and loosen the drill bits in place. Some of the newer types are designed to operate without a chuck key. They use a *keyless* chuck.

The most commonly used drills in the electrical trade are the ⅜- and ½-inch sizes. For drilling small holes at higher speeds, use a ⅜-inch drill. If you are boring larger holes at lower speeds, the ½-inch drill is the best choice. The speed of a drill in revolutions per minute (rpm) is related to the size of the chuck. The larger the chuck size, the slower the speed of the drill. The torque, or turning force, of a drill is directly proportional to the chuck size. Therefore, more torque is available with a drill that has a larger chuck. This type drill not only resembles a pistol, it also has a *trigger*. The trigger switch is used to control the speed of the drill. If you squeeze the trigger, the drill turns. The harder you squeeze, the faster the drills turns. Most drills in use today are reversible. This means that the drill can turn clockwise, and when switched, counterclockwise.

A drill bit is the tool that is attached to a drill that actually does the hole boring. Pistol-grip drills can be used with a variety of drill bits. The following details drill bits and their uses.

- A twist bit is designed to drill wood or plastic at high speed and metal at a lower speed.
- A flat-bladed spade bit, sometimes called a *speed-bore* bit, is used to drill holes in wood at high speed.
- A masonry bit is used to drill holes in concrete, brick, and other masonry surfaces.
- An auger bit is used for drilling wood at a relatively slow speed.

FIGURE 9-18   Pistol-grip drill.

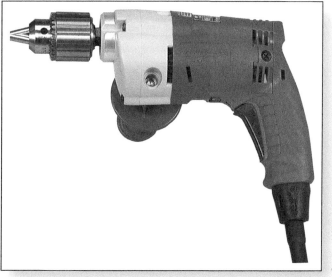

Courtesy of Milwaukee Electric Tool Corporation

Figure 9-19 shows some common bits, and Figure 9-20 is an example of an auger bit.

Hammer Drill

The hammer drill is similar to a standard variable-speed drill, with the exception that it is provided with a hammer action for drilling masonry or concrete walls and floors. Figure 9-21 illustrates a hammer drill.

FIGURE 9-19 Common drill bits used with pistol-grip drill.

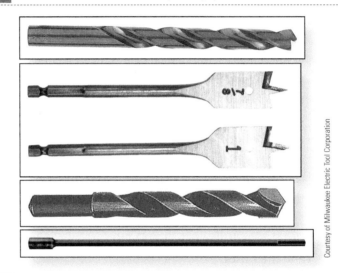

Courtesy of Milwaukee Electric Tool Corporation

FIGURE 9-20 An auger bit.

Courtesy of Milwaukee Electric Tool Corporation

FIGURE 9-21 A hammer drill.

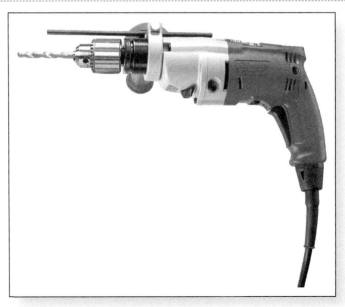

Courtesy of Milwaukee Electric Tool Corporation

Masonry bits used to bore holes are used with hammer drills. When using this type of drill, note that while the drill is turning the masonry bit, it is also moving the bit in a reciprocating, or hammering, motion. Some hammer drills are designed so that it can switch back and forth from a *drill only* mode to a *hammer drill* mode. Pistol-style hammer drills are designed with ⅜- or ½-inch chucks.

Cordless Drill

Cordless drills have become very popular with cordless models of pistol-grip and hammer drills being available. **Figure 9-22** shows a ½-inch cordless drill with an 18-volt power supply (rechargeable battery).

The source of electrical power for these cordless drills is generally the nickel metal hydride rechargeable battery. The voltage supplied to these cordless drills is typically 12, 14.4, 18, or 24 volts. However, there are manufacturers that have designed a 28-volt cordless drill that with the higher voltage provide more power. Typically, a cordless drill kit comes with the drill, a battery charger, and at least one battery.

Power Saws

Power saws are used by electricians on a regular basis. Power saws are used to cut plywood backboards in order to mount electrical equipment and for cutting building framing members during the installation of electrical wiring. Two popular power saws are:

- Circular
- Reciprocating

FIGURE 9-22 A cordless drill.

Courtesy of Milwaukee Electric Tool Corporation

FIGURE 9-23 A circular saw.

Courtesy of Milwaukee Electric Tool Corporation

Circular Saw

A circular saw is an electric saw that turns a round, flat blade to cut wood, metal or plastic depending on the blade selected. Circular saws have a handle with on/off trigger switch, an **arbor** nut to hold the blade in place, and guards to protect the operator from touching the spinning blade. An arbor is a mandrel, or a tool component that can be used to grip other moving tool components. **Figure 9-23** shows a circular saw.

Arbor a mandrel, or a tool component, that can be used to grip other moving tool components.

The circular saw is often called a *skilsaw*. This name comes from the first portable electric saw, which was made by a company named Skil. Most circular saws currently use a cord-and-plug connection. However, several manufacturers have introduced powerful cordless models. The circular saw is not generally used by electricians, but every electrician should be familiar with this tool and use it *only when properly trained.*

Reciprocating Saw

A **reciprocating** saw is a type of saw in which the cutting action is achieved through a push-and-pull reciprocating motion of the blade. Reciprocating refers to the moving back and forth of the blade on this saw. Recall that the circular saw blade moves in a round or circular rotation. **Figure 9-24** shows an example of a reciprocating saw.

Reciprocating moving back and forth.

The reciprocating saw is also known as a recipro saw, saber saw, or Sawzall (a trademark of the Milwaukee Electric Tool Company.) The saw shown in Figure 9-24 is a Sawzall. The reciprocating saw has a large blade and a handle designed to allow the saw to be used comfortably on vertical surfaces. This saw has a foot at the base of the blade. The user rests this foot against the surface being cut so that the tendency of the blade to push away or pull toward the cut as the blade travels through its cycle is countered.

Reciprocating saws can make a straight or curved cut in many materials, including wood and metal. These saws generally have a cord and plug, like circular saws, but manufacturers are also producing cordless models.

FIGURE 9-24 A Sawzall reciprocating saw.

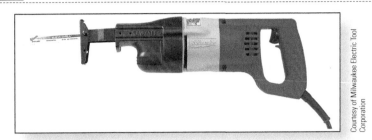

Courtesy of Milwaukee Electric Tool Corporation

Electricians frequently use reciprocating saws to cut off conduit, wood, and fasteners flush to the surface. Reciprocating saws work well for demolition work and for tasks such as cutting fittings in an existing pipe run, cutting cable trays, or cutting any other raceway too big for the bandsaw. These saws are ideal for working in areas with limited space.

As with any power tool, proper safety procedures should be followed, and personal protection equipment should be worn. Here are some additional guidelines to follow when using a reciprocating saw:

- Always have three teeth in a cut. Less than three teeth will result in the blade getting hung up on the material being cut. More than three teeth will slow down the cutting process.
- Turning the blade 180° will make it easier to make a flush cut. This will prevent the head of the saw from getting in the way.
- Splintering can be limited by putting the finished side of the wood down when cutting. Covering the cutting line with masking tape also helps.
- If a blade should break when cutting a thin material, switch to a blade with more and finer teeth.
- The best way to start a cut that is away from the edge of a material is to drill a starter hole larger than the widest part of the blade.
- When cutting *metal*, be sure to cut at a lower speed and use a good-quality oil.

Portable Bandsaw

Metallic tubing or conduit is an electrical piping system used for protection and routing of electrical wiring.

There are not many tools that can cut metallic tubing and other types of conduit or other stock more easily than a bandsaw. Metallic tubing or conduit is an electrical piping system used for protection and routing of electrical wiring. When it is not feasible to bring the stock to the machine, portable bandsaws offer the option of taking the machine to the work. A portable bandsaw kit generally includes a carrying case, starter blades, and a cord or battery with charger used with the cordless models. **Figure 9-25** shows an electrician using a portable bandsaw.

Cutting blades are changeable to fit the job. Fine-tooth blades are used for smaller and more delicate materials, and larger serrations, or notched edges, are used for heavier stock. Portable bandsaws cut metal, concrete, plastic, and wood. These saws can cut items measuring up to 4¾ inches in diameter, which should handle most field applications. Electricians use bandsaws to cut conduit and other raceways, threaded

FIGURE 9-25 A portable bandsaw being used at a jobsite.

rod, angle iron, strut, ceramic tile, concrete, sometimes large-gauge wire, and a variety of other materials found on a construction site. A bandsaw is one of the most versatile and powerful tools in any tool kit.

When using the bandsaw, it is very important to choose the proper blade. The saw must not grab and catch on the material being cut. Always let the weight of the saw do most of the work. These saws usually weigh 13 to 19 pounds and are designed for two-hand gripping for a natural downward cutting motion when the cut object is locked into a stand-up vice.

The following are basic bandsaw safety tips:

- Always pay close attention to the job at hand.
- Keep *all* guards in place and of proper length.
- Handle the stock with care.
- *Never* leave the bandsaw on.
- Keep bandsaw area clean.
- Keep hands in a safe position.
- Wear safety equipment.
- Never wear loose clothing.

Although power tools are often the most efficient method for getting a job done, they can also be one of the most dangerous. With a small group of classmates, review the power tool usage guidelines from the beginning of this section, and discuss why each item on the list is necessary.

Self-Check 3

1. What is a double-insulated tool, and why is double-insulation important?
2. List four guidelines that should be considered when using power tools.
3. List two types of power drills and applications of each.
4. What are other names for the reciprocating saw, and what are a few applications for this saw?
5. Where might an electrician use a bandsaw? List three safety tips that should be followed when using a bandsaw.

9-4 Specialty Tools

Some tools are important but are not used on a regular basis and are not typically found in a tool pouch. These tools are referred to as specialty tools. They are often used for very specific jobs performed by electricians. As with *all* tools, proper care and safe use must be practiced.

Knockout Punch

A knockout punch is used to cut holes for installing cable or conduit in metal boxes, equipment, and appliances. Conduit is a tube or pipe that provides a protected passageway for running wire or cable. **Figure 9-26(A)** shows a manual knockout set, and **Figure 9-26(B)** shows a hydraulic knockout set.

Using the knockout punch for cutting holes generally requires an electrician to drill a hole through the enclosure at the spot where the hole needs to be made. The drilled hole should be just large enough for the threaded stud of the knockout set to fit easily through. The knockout set is then used to make a hole that will match the trade size of the conduit or cable connector that is used.

A manual knockout set is generally used to make trade-size holes from ½ inch through 1¼ inches. Hydraulic knockout sets are also designed for these sizes and for sizes up to 6 inches. It is important for electricians to be familiar with both manual and hydraulic knockout tools.

Keyhole Saw

A keyhole saw is a long, narrow saw used for cutting small, often awkward features in various materials such as wood, hardboard, plastic, and metal. Keyhole saw blades are typically 10- to 12-inches long. They may be fixed or retractable. Keyhole saws work

FIGURE 9-26 (A) A manual knockout set and (B) a hydraulic knockout set.

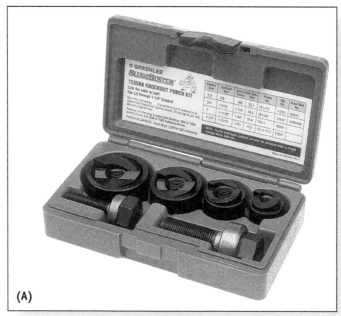

(A)

(B)

well for making openings for conduit and for cutting holes in wallboard to install electrical boxes. The keyhole saw is sometimes referred to as a pad saw, jab saw, or compass saw. **Figure 9-27** shows an example of a keyhole saw.

Because keyhole saws cut on the pull stroke and have very sharp teeth, they make clean cuts, even in hard-to-get-at areas.

Fish Tape

A fish tape, also known as a *draw wire* or *draw tape*, is a tool used by electricians to route new wiring through walls, ceiling cavities, and electrical conduit. **Figure 9-28(A)** and **(B)** show fish tapes.

FIGURE 9-27 A keyhole saw.

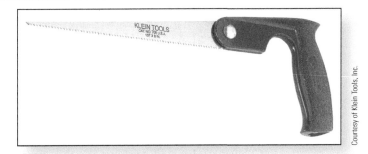

Courtesy of Klein Tools, Inc.

FIGURE 9-28 (A) A steel fish tape and (B) a fiberglass fish tape.

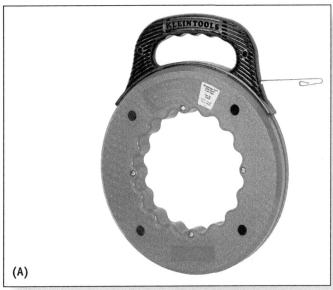

(A)

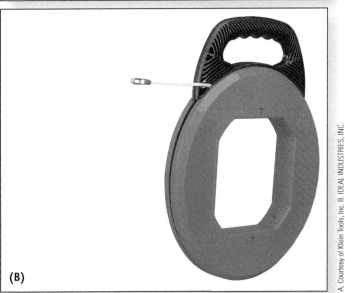

(B)

A. Courtesy of Klein Tools, Inc. B. IDEAL INDUSTRIES, INC.

Made of a narrow band of spring steel, fiberglass, or nylon, by careful manipulation the tape can be *fished*, or guided, through the confined spaces within walls, ceilings, or conduit. Once guided through, the new wiring can be pulled by attaching it to the end of the fish tape and pulling the tape back from where it came.

Modern fish tapes are housed in a metal or nonmetallic case called the reel. This reel is used to play out the fish tape to the length needed, and then *reeled* back in.

EMT (Electrical Metallic Tubing) Bender

EMT sometimes called thinwall, is commonly used instead of galvanized rigid conduit (GRC), as it is less costly and lighter.

An **EMT** bender is used to bend electrical metallic tubing up to 1.25 inches. **Figure 9-29** shows an example of an EMT bender. EMT, sometimes called thinwall, is commonly used instead of galvanized rigid conduit (GRC), as it is less costly and lighter.

EMT, electrical metallic tubing, is a thin, small-diameter wall pipe that is commonly used to house electrical wiring as it runs from one point to another in a residential or commercial building. EMT and conduit, although rigid and durable, can be bent and shaped to match the contour of walls, pipes, and other objects encountered in a structure. Bending conduit is simplified with the use of an EMT conduit bender. EMT benders are designed with a long handle, typically 38 inches long, made of iron for durability or aluminum to be lightweight. The EMT benders also have a rounded head into which the conduit is inserted and bent. Special markings on the bender head help the electrician make the desired bend accurately.

Level

Plumb perfectly vertical; the surface of the item being leveled is at a right angle to the floor or platform the electrician is working from.

A level is a tool used to ensure that the surface or angle is level or plumbed properly. *Plumb* means perfectly vertical; the surface of the item you are leveling is at a right angle to the floor or platform you are working from. Electricians use levels to plumb conduit, equipment, and appliances. The level of choice for electricians is the *torpedo* level, which is sometimes referred to as a spirit or bubble level. The torpedo level uses a glass vial or bubble to determine the evenness of the slope or angle. **Figure 9-30** illustrates a torpedo level.

The body of torpedo levels is rectangular in shape, with ends of the level typically narrower than the middle section. All torpedo levels have at least one vial in the middle section that is filled with some type of spirits, and *multipurpose* torpedo levels have up to three different vials. A 9-inch torpedo level with an aluminum frame and a magnetic edge on one side is shown in Figure 9-30. A magnetic torpedo level is

FIGURE 9-29   EMT bender.

FIGURE 9-30 A torpedo level.

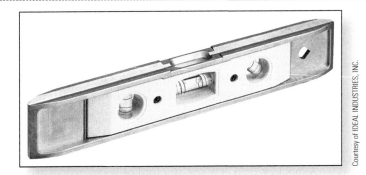

Courtesy of IDEAL INDUSTRIES, INC.

very useful because it will stay on metal electrical equipment while the equipment is being installed.

Chisel

A chisel is a tool with a characteristically shaped cutting edge, or blade, on its end, for carving or cutting a hard material such as wood, stone, or metal. The most common chisels used are the wood chisel and the cold chisel. **Figure 9-31** shows examples of chisels.

Electricians often use wood chisels to cut recesses for dimmer switches to be installed in or around the moldings framing a door. Chisels are also used in skirting outlets and cutting rebates, or stepped rectangular recesses, in studs or headers to accommodate cables. Cold chisels can be used to cut out rivets or seized nuts, cut sheet metal, chase masonry, or chip parts of a brick.

FIGURE 9-31 (A) A metal chisel and (B) A wood chisel.

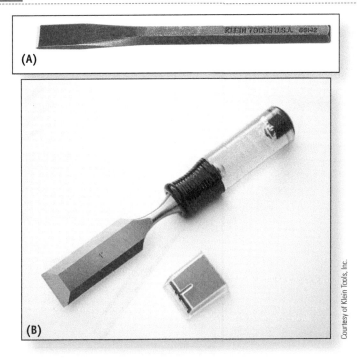

Courtesy of Klein Tools, Inc.

Cable Cutter

A cable cutter is designed to enable easy cutting of larger size cables and conductors. **Figure 9-32** shows an example of a cable cutter.

A lineman pliers is typically used to cut most cables and conductors used in residential wiring. However, cables with conductors larger than 10 gauge and single conductors larger than 2 AWG are more easily cut to size by using a cable cutter.

Hex Key Set

Hexagonal refers to six sides or edges.

A hex key, or Allen key, is a tool of **hexagonal** cross section used to drive bolts and screws that have a hexagonal socket in the head. Hexagonal refers to six sides or edges. This type wrench is used to install or remove recessed type bolts and screws. Hex keys come in fixed sets or individually. An Allen wrench set is displayed in **Figure 9-33**.

Some features of hex keys are listed here:

- The tool is simple, small, and light.
- The contact surfaces of the screw or bolt are protected from external damage.

FIGURE 9-32 A cable cutter.

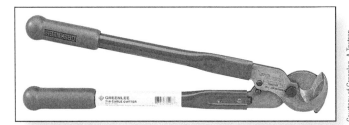

Courtesy of Greenlee, A Textron Company

FIGURE 9-33 An Allen wrench set.

Courtesy of IDEAL INDUSTRIES, INC.

- There are six contact surfaces between bolt and driver.
- The tool can be used with a headless screw.
- The screw can be inserted into its hole, using the key.
- Either end of the tool can be used to take advantage of reach or torque (turning force).

Fuse Puller

A fuse puller is a tool constructed of a strong nonconductive material, such as high-impact nylon, that is used to remove cartridge fuses from fuse boxes or other electrical enclosures. **Figure 9-34** shows different sizes of fuse pullers.

Some fuse puller models are hinged so that one end accommodates a certain range of fuse sizes, and the other end accommodates another range of fuse sizes.

Rotary BX Cutter

A rotary BX cutter is a tool that is used to strip the outside armor sheathing from Type AC cable, Type MC cable, and Flex cables. A rotary cutter is displayed in **Figure 9-35**.

The rotary cutter uses a rotating handle that turns a small cutting wheel that is designed to cut armor without damage to internal conductors. These cutters have an

FIGURE 9-34 Fuse pullers.

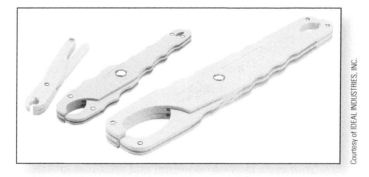

Courtesy of IDEAL INDUSTRIES, INC.

FIGURE 9-35 A rotary cutter.

Courtesy of IDEAL INDUSTRIES, INC.

adjustable cutting depth and slice through casings, or outer coverings, with only a few turns. When installing a residential electrical system using a metal-sheathed cable, this tool will save time!

LIFE SKILLS

Specialty tools are important because there are one or two specific jobs that can be performed more effectively with them than with a standard hand or power tool. If you had to select your own "specialty," the one or two things that you think you will do best at as an apprentice, what would they be?

Self-Check 4

1. Where might an electrician use a knockout punch?
2. What are two other names for a keyhole saw? List two applications of the keyhole saw.
3. What is a fish tape, and why is it used?
4. What is EMT, and why is an EMT bender used?
5. Where might an electrician user a cable cutter?

9-5 Electrician Tool Kits

With *safety first* in mind, selecting and using the proper tool is vital. As discussed throughout this chapter, improper or unsafe use of a tool *can be lethal*! You are receiving training in the proper care and safe use of tools; in addition, there are agencies that detail requirements for hand and power tools. Know these requirements and follow them. Be aware of other tool usage guidelines that will help you do a competent job and keep you *safe*! Learn what tools you will need on the job, and keep your tools in good condition and your tool kit safe.

Standards

OSHA Standard 1926.302 specifically covers the requirements for hand and power tools. Tools used by electricians must be compliant with OSHA 1910.335 and NFPA 70E standards. In addition, the American National Standards Institute (ANSI) also has standards relating to tools.

Most tool manufacturers have good information on tools from specifications to prices. Another source of information on tools is the Web site of the Hand Tools Institute at www.hti.org. There are a number of safety education articles available. The institute provides access to a 90-plus-page publication, "Guide to Hand Tools," that includes topics for selecting, properly using, and maintaining tools, as well as possible hazards involved with tool usage.

On the jobsite, you will be called upon to use portable electric power tools. Electricity available is usually in the form of temporary power, covered by *Article 590* of the *NEC*.

NEC 590.6(A) requires that *all 125-volt, single-phase, 15-, 20-, and 30-ampere receptacle outlets that are not a part of the permanent wiring of the building or structure and that are in use by personnel shall have ground-fault circuit interrupter protection for personnel.*[*]

Do *not* ignore or defeat this requirement! You should always carry and use as part of your tool kit a portable GFCI of the types shown in **Figure 9-36**. This inexpensive yet invaluable device *could save your life*!

Tool Kits

There are many reputable tool manufacturers and each one offers a variety of tool kits suited for work in the electrical trade. You must make a well-informed decision on not only what tools you will need for the job, but what tool manufacturer best suits your needs. The following are two examples of electrician's tool kits.

12-Piece Electrician's Tool Set

- Four pliers with dipped grip handles (9¼-inch lineman with fish tape puller; 8-inch diagonal; 8-inch long-nose; 10-inch tongue-and-groove)
- T®-Stripper Wire Stripper
- Four cushion grip screwdrivers (#1 × 3 Phillips; #2 × 4 Phillips; 3/16 × 4 slotted cabinet tip; ¼ × 4 heavy-duty slotted round shank)

FIGURE 9-36 Portable plug-in cord sets with built-in GFCI protection.

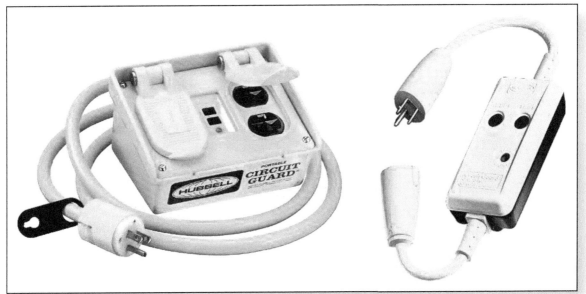

From MULLIN/SIMMONS. *Electrical Wiring Residential*, 17E. © 2012 Cengage Learning

---

[*] Reprinted with permission from NFPA 70-2011.

- 25-foot tape measure
- Electrician's leather tool pouch
- Web belt

13-Piece Journeyman Electrician's Kit

- Three pliers with smart-grip handles (9¼-inch lineman with fish tape puller; 8-inch long-nose; 10-inch tongue-and-groove)
- T®-Stripper Wire Stripper
- Four cushion grip screwdrivers (#1 × 3 Phillips; #2 × 4 Phillips; 3/16-inch × 4 slotted cabinet tip; ¼-inch × 4 slotted round shank)
- Hawkbill knife
- 25-foot tape measure
- Medium-size fuse puller
- Torpedo level
- Journeyman electrician's tote

These kits are only meant to give you an idea of what is offered in kit sets. Remember, get the tools you need. Good tools are a *good investment*!

You may be called upon to use a tool that was not described in this chapter. Find out about the tool *before* you use it. Read the manual or ask for help. Keep yourself up-to-date on new tools for electricians. As an electrician, it will be *your* responsibility to stay up on technological changes and trends. You are responsible to know *what* tool to use and *how* to use it. You and your tools will be working together for a long time. Know your tools and take care of them!

LIFE SKILLS

Now that you know the basic elements of an electrician's tool kit, it's time to create your own. Pretend that money is not an issue and build your "dream" tool kit. What tools would you include? Which brands would you select? Why? When you are done, compare your kit with your classmates'.

Self-Check 5

1. List three agencies that cover requirements or standards for hand and power tools.
2. Why is a portable plug-in cord set with built-in GFCI protection important to an electrician?
3. What should you do if you are asked to use an unfamiliar tool?

Summary

- Causes of hand tool injuries typically can be traced to some type of improper use or the lack of proper maintenance of a hand tool.
- Be aware of and follow guidelines for the care and safe use of hand tools and power tools.
- Hand tools that fit easily into an electrical tool pouch are often referred to as pouch tools.
- Wrenches are one of the most widely used hand tools. Their main function is holding and turning nuts, bolts, cap screws, plugs, and various threaded parts.
- The most commonly used pliers by electricians are lineman, long-nose, and diagonal cutters.
- In electrical work, the most common type screwdrivers are the Keystone and Phillips.
- Cordless screwdrivers use rechargeable batteries instead of an electrical power cord. Using a cordless driver is faster and easier on the tool user.
- A hammer is used to drive and pull nails, pry boxes loose, break wallboard, and strike awls and chisels.
- The tape measure is an instrument consisting of a narrow strip (cloth or metal) marked in inches or centimeters, and is used for measuring.
- There are a variety of electric power tools used by electricians.
- Electric power tools include both those powered by 120-volt ac and those powered by low-voltage dc electricity.
- Specialty tools are important but are not used on a regular basis and are not typically found in a tool pouch.
- OSHA, NFPA, and ANSI are agencies that list requirements and standards for hand and power tools.

Review Questions

True/False

1. Causes of hand tool injuries can generally be traced to some type of improper use or the lack of proper maintenance of a hand tool.   (True, False)

2. Repair damaged tools when time permits, and always keep broken or damaged tools that cannot be repaired.   (True, False)

3. The main purpose of pliers is holding and turning nuts, bolts, cap screws, plugs, and various threaded parts.   (True, False)

4. A crescent wrench is used to tighten couplings and connectors, tighten pressure-type wire connectors, and remove and hold nuts and bolts.   (True, False)

5. The 9¼-inch handle with the New England nose is the pliers of choice for residential electrical work.   (True, False)

Multiple Choice

6. The long-nose pliers is often referred to as _____.
  A. Dikes
  B. Side cutters
  C. Needlenose
  D. Lineman

7. The T®-Stripper Wire Stripper is the best tool to strip insulation from wire sizes that range from _____.
  A. 1 through 4 AWG
  B. 5 through 8 AWG
  C. 10 through 18 AWG
  D. 24 through 36 AWG

8. Which of the following is *not* a common screwdriver tip?
  A. Keystone
  B. Hex
  C. Cabinet
  D. Phillips

9. Which of the following is *not* commonly used as an electrician's knife?
  A. Hawkbill
  B. Electrician's
  C. Jackknife
  D. Utility

10. Which of the following is *not* a standard tape measure length?
  A. 10-foot
  B. 16-foot
  C. 20-foot
  D. 25-foot

Fill in the Blank

11. _____ is a form of electrical protection featuring two separate insulation systems.

12. Power drills with _____ grips are the most common type in use today.

13. A(n) _____ drill bit is designed to drill wood or plastic at high speed and metal at a lower speed.

14. Masonry bits used to bore holes are used with _____ drills.

15. A(n) _____ saw is a type of saw in which cutting action is achieved through a push-and-pull, back-and-forth motion of the blade.

16. A(n) _____ punch is used to cut holes for installing cable or conduit in metal boxes, equipment, and appliances.

17. A(n) _____ saw is a long, narrow saw used for cutting small, often awkward features in various materials.

18. A(n) _____ tape is a tool used by electricians to route new wiring through walls, ceiling cavities, and electrical conduit.

19. A(n) _____ bender is used to bend electrical metallic tubing up to 1.25 inches.

20. Most electricians use the _____ level, sometimes called a bubble level.

21. A(n) _____ is a tool with a characteristically shaped cutting edge on its end, for carving or cutting a hard material such as wood.

22. Single conductors larger than _____ AWG are more easily cut to size by using a cable cutter.

23. An Allen key is a tool of _____ cross section, used to drive bolts and screws that have a socket in the head.

24. A rotary _____ cutter is a tool that is used to strip the outside armor sheathing from Type AC cable.

25. Tools used by electricians must be compliant with OSHA _____ and NFPA 70E standards.

Chapter 10

Wiring Overview

Career Profile

Trenton Johnston, a project manager at Mosher Enterprises, Inc., finds the electrical trade exciting and stimulating: "Everything about my industry is constantly changing, and it keeps me motivated to take on new goals and projects," he enthuses. He also notes, "As the electrical trade changes from year to year, the career path for those of us involved is always expanding."

As a project manager, Johnston starts his day following up on customer and job concerns. When his crews arrive, he updates their supervisors on activities for the day. He also follows up on the status of all jobs that he is managing and makes sure that he is ahead of schedule and the needs of the customer. "My job is to make our company money, but at the same time the customers have to be happy so that they come back to the company in the future," he explains.

Of his apprenticeship program at the Northern New Mexico IEC, Johnston asserts, "My education keeps me connected to the guys in the field. I understand on a daily basis what the workers are going through and what they need to be successful. It also has allowed me to be thorough in my problem solving so that my jobs are profitable and my value in the company continues."

When he has the chance to advise his apprentices and journeymen, Johnston tells them, "Every job has its own unique challenges, and many things cannot be controlled. You have to be patient, have good communication skills, and be able to manage difficult or stressful situations." He also emphasizes that "they have to always be willing to learn something new. No one in this industry will ever know everything, and regardless of your job title you can still learn new/better/faster ways of doing things."

Chapter Outline

SPECIFICATIONS USED IN MAKING ELECTRICAL INSTALLATIONS

SYMBOLS AND NOTATIONS USED IN ELECTRICAL DRAWINGS AND PLANS

NATIONALLY RECOGNIZED TESTING LABORATORIES

FUSES AND CIRCUIT BREAKERS AND CURRENT RATINGS

Key Terms

Circuit breaker

Detailed drawing

Electrical drawing

Elevation drawing

Floor plan

Fuse

Interrupting rating

Label

Legend

Non-time-delay fuse

Plot plan

Schedule

Sectional drawing

Section line

Service cables

Specifications

Time-delay fuse

Chapter Objectives

After completing this chapter, you will be able to:

1. Describe how specifications are used in making electrical installations.
2. Identify symbols and notations used in electrical drawings, and explain how they are used.
3. List nationally recognized testing laboratories (NRTLs), and describe their responsibility for establishing electrical standards and ensuring that materials meet those standards.
4. Describe basic types of fuses and circuit breakers.
5. Define the term *interrupting rating*.

Life Skills Covered

 Goal Setting

 Critical Thinking

Life Skill Goals

For an aspiring electrician, the importance of understanding this chapter's content cannot be understated. You can't build a career without the proper foundation, and for an electrician the basics of wiring are a cornerstone. The life skills for this chapter, Goal Setting and Critical Thinking, go hand in hand. Your goal should be to ponder and eventually understand the content.

Set a goal to learn the chapter so well that you can answer each unit's Self-Check questions without consulting your notes. Be sure to consider not only *what* the answer is, but also *why* the answer is correct. If you can master the basics of wiring, you will be ahead of the pack when you reach the more advanced lessons later in this course.

This chapter describes a building plan and emphasizes electrical specifications, drawings, and prints indicated on these plans. You learn about these specifications and the many symbols and notations found on electrical drawings. You also learn that in addition to plans and specifications, understanding and adhering to related *NEC* requirements is crucial to becoming a competent electrician.

Introduction

Before a house or any structure is built, a plan must be developed. An architect prepares a residential building plan. This plan includes a set of drawings that shows the necessary details and instructions that skilled trades workers use as a guide to build the house.

This chapter describes how electricians know what products are safe, by introducing nationally recognized testing laboratories (NRTLs) and their role in safety testing products. Safety is reinforced in this chapter by describing basic types of fuses and circuit breakers, often referred to as the "safety valves" of an electrical circuit. Learn and appreciate the importance of NRTLs and overcurrent protection devices so that you use *safe* products and understand how to select and replace fuses and circuit breakers.

10-1 Specifications Used in Making Electrical Installations

Before we describe specifications, which are the part of a building plan that provides more *specific* details about the construction of a building, let's describe a building plan and learn about its main parts.

Building Plan

A building plan may be called the *prints, blueprints, drawings,* or *construction drawings.* What a building plan is called is *not* important. What is important is the ability of an electrician to know the various parts of a typical building plan and how to read and interpret the information found on them. The main parts of a residential building plan are as follows:

- Plot plan
- Floor plans
- Elevation drawings
- Detail drawings
- Electrical drawings

Following is a description of each part of a building plan, with *electrical* drawings broken down further into electrical symbols, schedules, and specifications.

Plot Plan

Plot plan an architectural drawing that is generally the first sheet in a set of building plans. It shows all the major features and structures on a piece of property and is drawn as if you were looking down on the property from a considerable height.

A **plot plan** is an architectural drawing and is generally the first sheet in a set of building plans. It shows all the major features and structures on a piece of property. The plot plan is a "top-down" orientation, which means it is drawn with a view as if you were looking down on the property from a considerable height. An electrician uses a plot plan to obtain some important information. For example, if a backyard swimming pool is planned or driveway lighting is to be installed, the plot plan can be used to determine where the underground wiring trench should be dug. **Figure 10-1** shows a typical plot plan.

Floor Plans

Floor plan a diagram, or drawing, usually to scale, that shows the relationships among rooms, spaces, and other physical features at one level of a structure.

A **floor plan** is a diagram, or drawing, usually to scale, that shows the relationships among rooms, spaces, and other physical features at one level of a structure. These details are shown from a view directly above the structure or house. In a floor plan, it appears as if the roof were removed and you were looking straight down at the floors and cut-off walls. Typically, each floor and the basement have a floor plan.

Room sizes and wall lengths are generally specified by dimensions drawn between the walls. Details of fixtures like sinks, water heaters, furnaces, and so on, are also included in floor plans, along with notes to specify finishes, construction methods, and symbols for electrical items. Notes and symbols listed on floor plans are interpreted by electricians as they install wiring and electrical equipment. An example of a floor plan is shown in **Figure 10-2**.

FIGURE 10-1 A plot plan.

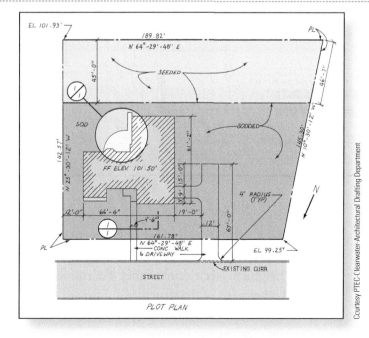

PLOT PLAN

Courtesy PTEC-Clearwater-Architectural Drafting Department

FIGURE 10-2 A floor plan.

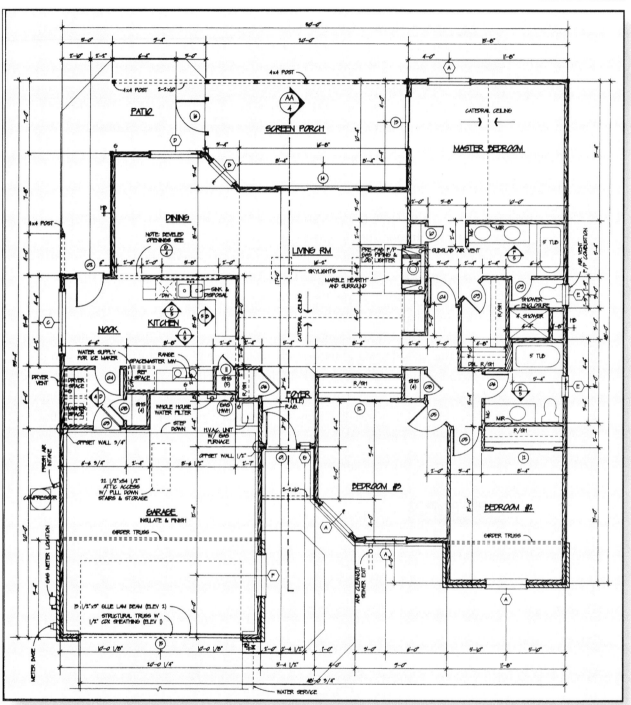

Elevation drawing a one-dimensional (flat) projection that shows the side of a structure or house as seen head on and facing a certain direction.

Elevation Drawings

An **elevation drawing** is a one-dimensional (flat) projection that shows the side of a structure or house, as seen head on and facing a certain direction. Elevation drawings are helpful to electricians as they determine the heights of windows, doors, porches, and other parts of the structure. Information such as this is not found on floor plans and is needed where installing electrical devices like outside luminaires, doorbells, and outside receptacles. **Figure 10-3** shows an elevation drawing.

FIGURE 10-3 An elevation drawing.

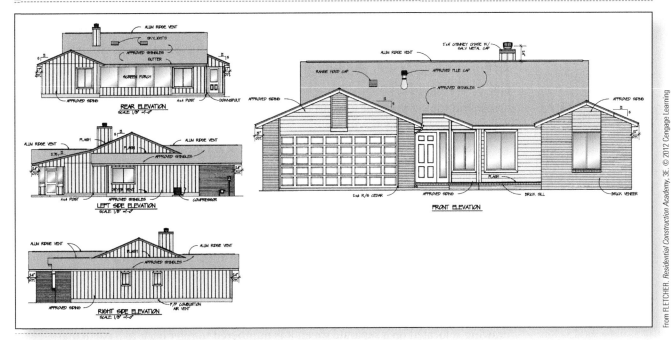

REAR ELEVATION
SCALE 1/8" =1'-0"

LEFT SIDE ELEVATION
SCALE 1/8" =1'-0"

RIGHT SIDE ELEVATION
SCALE 1/8" =1'-0"

FRONT ELEVATION

From FLETCHER. *Residential Construction Academy*, 3E. © 2012 Cengage Learning

Sectional drawing a part of the structure that shows a cross-sectional view of a specific part of the dwelling. It is a view that enables you to see the inside of a structure or house.

Sectional Drawings

A **sectional drawing** is a part of the structure that shows a cross-sectional view of a specific part of the dwelling. It is a view that enables you to see the inside of a structure or house. You are able to see information not shown on a floor plan or elevation drawing. In order to produce a sectional drawing, you need to imagine you have taken a saw and cut off the side of a house. When the side of the house is removed, you have a cutaway view of the rooms and structural members of that part of the house. **Figure 10-4** shows a typical sectional drawing.

FIGURE 10-4 A sectional drawing.

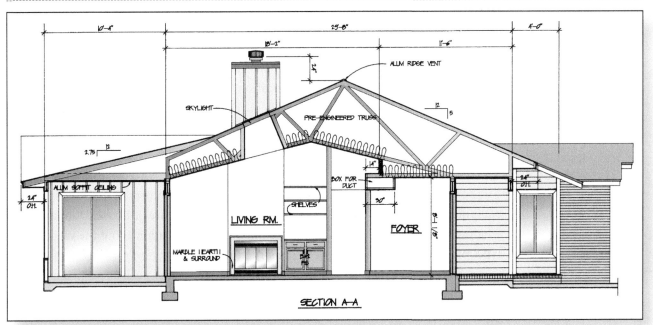

SECTION A-A

From FLETCHER. *Residential Construction Academy*, 3E. © 2012 Cengage Learning

The point on a floor plan or elevation drawing that is denoted by the sectional drawing is shown with a dashed line with arrows on the ends and is called a section line. A section line is a broad line consisting of long dashes followed by two short dashes; at each end of the line are arrows that show the direction in which the cross section is being viewed. Building plans generally have several sectional drawings, and to distinguish between them, letters or numbers are located at the end of the arrows on the section lines. As an example, a typical section drawing may be labeled "Section C-C."

Sectional drawings provide useful information for electricians. An electrician uses a wall section drawing to determine how to run cable through walls or floors and to show how thick the wood is for drilling.

Detail Drawings

A detailed drawing is a separate large-scale drawing that shows very specific details of a small part or particular section of a structure. The large scale or enlarged view makes details much easier to see than in a sectional drawing. Detailed drawings are generally listed on the same building plan sheet where the structure feature appears. If a detailed drawing is located on a separate sheet, it is numbered to refer back to a particular location on the building plan. A detailed drawing is illustrated in **Figure 10-5.**

Electrical Drawings

An electrical drawing is a part of the building plan that shows information about the electrical supply and distribution for the structure's electrical system. Electrical drawings present in detail what is required of an electrician to complete the installation of the electrical system. Electrical symbols, a type of shorthand that denotes equipment and devices, show the electrician which electrical items are required and where they are to be located. Symbols are used to provide a less cluttered electrical plan. **Figure 10-6** (p. 254) illustrates the use of symbols and notations on a floor plan.

Electrical symbols and notations are covered in great detail in the next section. A typical electrical floor plan is shown in **Figure 10-7** (p. 254).

Electrical plans are important to electrical contractors, as they use this information to estimate the amount of material and labor needed. This information is used to project a total cost for the electrical system installation and also in the bidding process. Electrical plans provide an excellent record of the electrical system and can be reviewed if technical issues surface.

Schedules

Schedules are tables used on building plans that provide information about specific equipment or materials used in the construction of a structure or house. Door and window schedules list the sizes and other important information about the various types of doors and windows used in the structure.

Schedules provide specific information for electricians, such as particular electrical equipment that needs to be installed in the structure. As an example,

FIGURE 10-5 A detailed drawing.

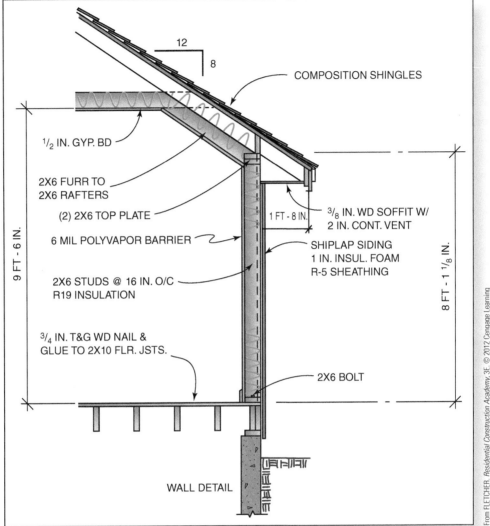

COMPOSITION SHINGLES

12
8

1/2 IN. GYP. BD

2X6 FURR TO
2X6 RAFTERS

(2) 2X6 TOP PLATE

6 MIL POLYVAPOR BARRIER

2X6 STUDS @ 16 IN. O/C
R19 INSULATION

3/4 IN. T&G WD NAIL &
GLUE TO 2X10 FLR. JSTS.

9 FT - 6 IN.

1 FT - 8 IN.

3/8 IN. WD SOFFIT W/
2 IN. CONT. VENT

SHIPLAP SIDING
1 IN. INSUL. FOAM
R-5 SHEATHING

8 FT - 1 1/8 IN.

2X6 BOLT

WALL DETAIL

From FLETCHER. *Residential Construction Academy*, 3E. © 2012 Cengage Learning

panelboard schedules are used on residential and commercial prints. They identify and describe the panel. These schedules also list current, voltages, and loads. Another example is a luminaire schedule that lists and describes the various types of luminaires used in the structure. By referring to this schedule, an electrician also determines the type and the amount of lamps used with each luminaire. **Figure 10-8** shows an example of a luminaire schedule.

Specifications

Specifications the part of a building plan that provides more *specific* details about the construction of a building.

Building plan **specifications** are those parts of the building plan that provide more *specific* details about the construction of a structure. Although specifications provide detailed information to all construction trades involved with the structure, understanding and interpreting electrical specifications is a requirement for an electrician. Electrical specifications, or *specs* as they are often called, list specific

FIGURE 10-6 Use of symbols and notations on a floor plan.

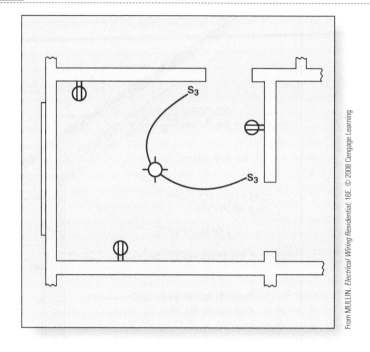

FIGURE 10-7 A floor plan.

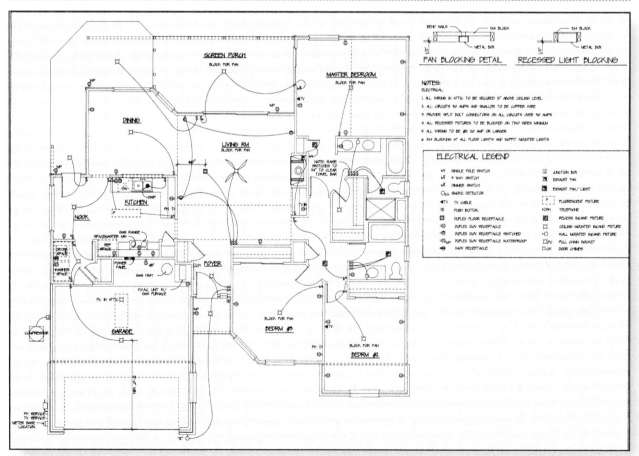

From FLETCHER. Residential Construction Academy, 3E. © 2012 Cengage Learning

FIGURE 10-8 A luminaire schedule.

Symbol	Number	Manufacturer and Catalog Number	Mounting	Lamps Per Fixture
A	2	Lightolier 10234	Surface	2 40-watt T-8 CWX Fluorescent
B	4	Lightolier 1234	Surface	4 40-watt T-8 WWX Fluorescent
C	1	Progress 32-486	Surface	1 60-watt medium base LED
D	1	Progress 63-8992	Pendant	5 60-watt medium base incandescent
E	3	Lithonia 12002-10	Recessed	1 75-watt medium base reflector
F	3	Hunter Paddle Fan 1-3625-77	Surface	3 25-watt medium base CFL
G	2	Nutone Fan/Light Model 162	Recessed	1 60-watt medium base incandescent

manufacturer's catalog numbers and other information. This information is necessary so that electrical items are the correct size, type, and proper rating. There may be unique specifications listed; for example, for federally funded project installations, all materials and parts to be used must be made in the United States of America. Specifications are often used by electrical contractors as they estimate installation costs and prepare bids. These contractors know the importance of reading *all* of the specs. Missing certain details could be a *costly* mistake.

NEC Requirements Regarding Services

The *NEC* defines a service as the conductors and equipment for delivering electrical energy from the serving electric utility to the wiring system of the premises served. There are two types of service entrances used to deliver electrical energy to a residential wiring system:

- Overhead service
- Underground service

Overhead Service

The overhead service is the type most often installed in residential wiring. It is less expensive and takes less time to install than an underground service. An overhead service entrance includes the service conductors between the terminals of the service equipment main disconnect and a point outside the home where they are connected to overhead wiring that is connected to the utility's electrical system. The overhead wiring is placed high enough to protect it from physical damage and to keep it away from people. **Figure 10-9** shows an example of a typical overhead service entrance.

FIGURE 10-9 An overhead service entrance.

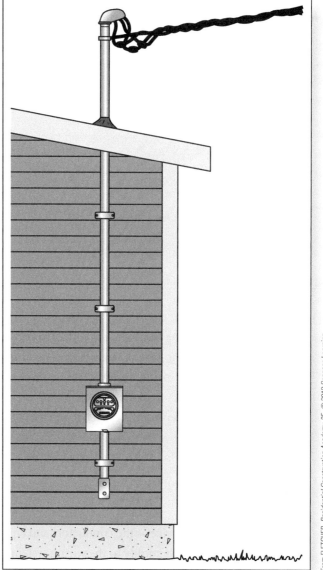

From FLETCHER, *Residential Construction Academy*, 3E. © 2012 Cengage Learning

Underground Service

An underground service entrance is often installed as an alternative to an overhead service. The service conductors between the terminals of the service main disconnect and the point of connection to the utility wiring are buried in the ground at a depth that protects the conductors from physical damage and also prevents accidental human contact with the conductors. A typical underground service entrance is shown in **Figure 10-10**.

Service conductors that are made up in the form of a cable are called **service cables**. These service cables are installed to supply the electrical power from the utility company to the building's electrical system. This cable is most often run from the utility's service point to the meter base, and then from the meter base to the main panel. Major electrical appliances such as electric ranges are sometimes connected using this cable as a branch circuit.

Service cables conductors that are made up in the form of a cable to supply the electrical power from the utility company to the building's electrical system.

FIGURE 10-10 An underground service entrance.

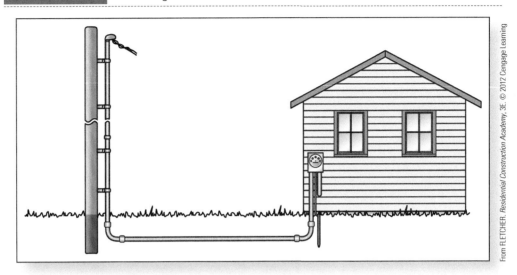

From FLETCHER. *Residential Construction Academy*, 3E. © 2012 Cengage Learning

Service entrance (SE) cable types are covered in *Article 338* of the *NEC*. This cable type is defined as a single conductor or multiconductor assembly provided with or without an overall covering. Most residential installations use aluminum rather than copper conductors in service-entrance cables because of the larger size requirements and expense of copper.

The three types of service cables are as follows:

- SEU is the most commonly used for service-entrance installations. It contains three conductors enclosed in a flame-retardant, moisture-resistant outer covering. **Figure 10-11** shows a typical SEU service-entrance cable.

- SER is similar to SEU but has four conductors wrapped in a round configuration. If service-entrance cable is chosen as the wiring method, SER cable is used to wire a new electric range or electric dryer. Type SER is also used as a feeder from the main service panel to a subpanel. Type SER service-entrance cable is shown in **Figure 10-12**.

- USE is a three-conductor cable used for underground service-entrance installations. It has a moisture-resistant covering, but because these cables are buried, it is not flame retardant. Type USE is also used for underground feeder or branch circuits. **Figure 10-13** shows a USE service-entrance cable.

FIGURE 10-11 Service-entrance cable, Type SEU.

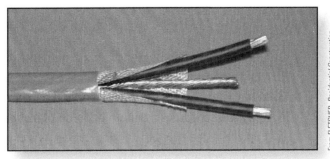

From FLETCHER. *Residential Construction Academy*, 3E. © 2012 Cengage Learning

FIGURE 10-12 Service-entrance cable, Type SER.

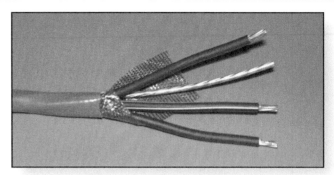

From FLETCHER. *Residential Construction Academy*, 3E. © 2012 Cengage Learning

FIGURE 10-13 Service-entrance cable, Type USE.

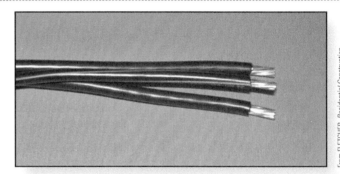

From FLETCHER. *Residential Construction Academy*, 3E. © 2012 Cengage Learning

Because the service entrance is one of the most important parts of a residential electrical system, electricians must prepare and plan for the installation and know advantages and disadvantages of both overhead and underground service entrances. *Article 230 of the NEC* covers many of the requirements for the installation of service entrances.

LIFE SKILLS

Before any electrician tackles a project, he or she needs to have a plan. Before you dive into this chapter, create your own plan for how you master the content.

Self-Check 1

1. List the five main parts of a building plan.
2. Why might the sectional drawing be important to an electrician?
3. What is an electrical drawing?
4. Why are schedules found in building plans important to an electrician?
5. List the three types of service cables. Which one is most commonly used for service-entrance installations?

10-2 Symbols and Notations Used in Electrical Drawings and Plans

Recall that the electrical drawings present in detail what is required of an electrician to complete the installation of the electrical system. Electrical symbols show the electrician which electrical items are required and where they are to be located. Electricians must be able to recognize every symbol shown on the plan and understand how to interpret cabling and wiring diagrams.

"Symbols for Electrical Construction Drawings," a standard published by the American National Standards Institute (ANSI), shows electrical symbols used on electrical drawings. Lighting outlet symbols are shown in **Figure 10-14** (p. 260). **Figure 10-15** (p. 261) shows common electrical symbols for receptacle outlets, and **Figure 10-16** (p. 262) shows common symbols used to show switch types. **Figure 10-17** (p. 263) shows a number of other symbols used to represent common pieces of electrical equipment.

Electrical drawings generally have ANSI electrical symbols representing devices, equipment, and so forth. However, some plans use symbols that are not standard. Where symbols other than ANSI standards are used, a legend is usually included in the plans. A legend is the part of a building plan that describes the various symbols and abbreviations used on the plan.

Electrical wiring is represented on electrical plans by using symbols. On a plan, a curved, dashed line that goes from a switch symbol to a lighting outlet symbol indicates that the outlet is controlled by that switch. This is illustrated in **Figure 10-18** (p. 264).

This line is always curved to eliminate confusion about whether it is a hidden line or a switch leg. A hidden line is also dashed, but it is always drawn straight. In a cabling diagram, a curved line is used to represent the wiring method used. As an example, a home run is drawn as a curved solid line with an arrow on its end. A home run indicates the wiring from that point goes all the way to the load center. A number next to the *home run* symbol indicates its circuit number. In cable diagrams, slashes on the curved solid line are sometimes used to indicate how many conductors there are in the cable. A properly drawn cable diagram saves time and makes the installation of the wiring much easier. Cable diagrams also help in troubleshooting a circuit if a problem should develop in the future. The difference between a cabling diagram and a wiring diagram is illustrated in **Figure 10-19** (p. 264).

Referring to Figure 10-19(A), note that 2-wire cable is being used to wire this room. An electrician knows that this cable is to be run between electrical boxes. In this diagram, the *home run is clearly listed.* This cabling diagram indicates to the electrician that the cable is to be run from the receptacle box to the load center.

Legend that part of a building plan that describes the various symbols and abbreviations used on the plan.

TRADE TIP   Cabling diagrams do *not* show actual connections to be made. A wiring diagram, as shown in Figure 10-19(B), shows exactly where the conductors terminate. Wiring diagrams are helpful to electricians where wiring circuits and equipment. However, with experience, wiring diagrams are not used as much.

FIGURE 10-14 Lighting outlet symbols.

OUTLETS	CEILING	WALL
SURFACE-MOUNTED INCANDESCENT OR LED	○ ⊕ ⊗	⊢○ ⊢⊕ ⊢⊗
LAMP HOLDER WITH PULL SWITCH	○ PS Ⓢ	⊢○ PS ⊢Ⓢ
RECESSED INCANDESCENT OR LED	⊡ Ⓡ ⊘	⊢⊡ ⊢Ⓡ ⊢⊘
SURFACE-MOUNTED FLUORESCENT	▭ ▭○	▭ ▭○
RECESSED FLUORESCENT	▱ ▭○R	▱ ▭○R
SURFACE OR PENDANT CONTINUOUS-ROW FLUORESCENT	▭▭▭ / ○▭▭	
RECESSED CONTINUOUS-ROW FLUORESCENT	▱▱▱ / ○R▭▭	
BARE LAMP FLUORESCENT STRIP	⊢─┼─┼─┤	
TRACK LIGHTING	⊤ ○ ○	⊤ ○ ○
BLANKED OUTLET	Ⓑ	⊢Ⓑ
OUTLET CONTROLLED BY LOW-VOLTAGE SWITCHING WHEN RELAY IS INSTALLED IN OUTLET BOX	Ⓛ	⊢Ⓛ

LIFE SKILLS

Symbols and notations are frequently used in the electrical field to save space and make recognition easier. Are there any symbols or shorthand that you can use in your notes to help you study more efficiently?

FIGURE 10-15 Receptacle outlet symbols.

RECEPTACLE OUTLETS		
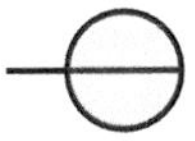 ⊖	SINGLE-RECEPTACLE OUTLET	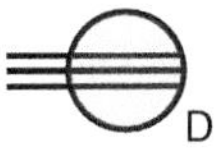 ⊜D ELECTRIC CLOTHES DRYER OUTLET
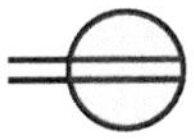 ⊜	DUPLEX-RECEPTACLE OUTLET	 Ⓕ FAN OUTLET
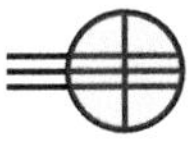 ⊕	TRIPLEX-RECEPTACLE OUTLET	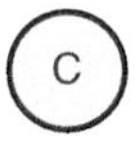 Ⓒ CLOCK OUTLET
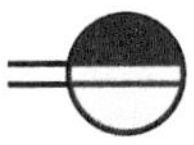 ⊖	DUPLEX-RECEPTACLE OUTLET, SPLIT CIRCUIT	 ⊙ FLOOR OUTLET
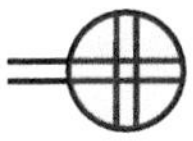 ⊕	DOUBLE-DUPLEX RECEPTACLE (QUADPLEX)	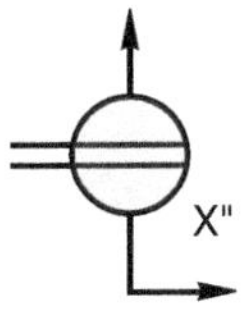 X" MULTIOUTLET ASSEMBLY; ARROW SHOWS LIMIT OF INSTALLATION. APPROPRIATE SYMBOL INDICATES TYPE OF OUTLET, SPACING OF OUTLETS INDICATED BY "X" INCHES.
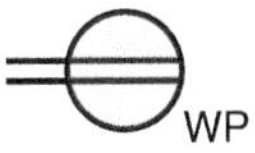 ⊜WP	WEATHERPROOF RECEPTACLE OUTLET	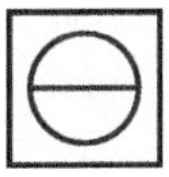 ⊟ FLOOR SINGLE-RECEPTACLE OUTLET
 ⊜GFCI	GROUND FAULT CIRCUIT INTERRUPTER RECEPTACLE OUTLET	 ⊟ FLOOR DUPLEX-RECEPTACLE OUTLET
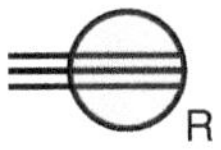 ⊜R	RANGE OUTLET	 ◩ FLOOR SPECIAL-PURPOSE OUTLET
	SPECIAL-PURPOSE OUTLET (SUBSCRIPT LETTERS INDICATE SPECIAL VARIATIONS. DW = DISHWASHER. A, B, C, D, ETC., ARE LETTERS KEYED TO EXPLANATION ON DRAWINGS OR IN SPECIFICATIONS).	

From FLETCHER. *Residential Construction Academy*, 3E. © 2012 Cengage Learning

FIGURE 10-16 Electrical switch symbols.

SWITCH SYMBOLS	
S OR S_1	SINGLE-POLE SWITCH
S_2	DOUBLE-POLE SWITCH
S_3	THREE-WAY SWITCH
S_4	FOUR-WAY SWITCH
S_D	DOOR SWITCH
S_{DS}	DIMMER SWITCH
S_G	GLOW SWITCH TOGGLE— GLOWS IN OFF POSITION
S_K	KEY-OPERATED SWITCH
S_{KP}	KEY SWITCH WITH PILOT LIGHT
S_{LV}	LOW-VOLTAGE SWITCH
S_{LM}	LOW-VOLTAGE MASTER SWITCH
S_{MC}	MOMENTARY-CONTACT SWITCH
 ⬦M	OCCUPANCY SENSOR—WALL MOUNTED WITH OFF-AUTO OVERRIDE SWITCH
 ⬦M P	OCCUPANCY SENSOR—CEILING MOUNTED "P" INDICATES MULTIPLE SWITCHES WIRE-IN PARALLEL
S_P	SWITCH WITH PILOT LIGHT ON WHEN SWITCH IS ON
S_T	TIMER SWITCH
S_R	VARIABLE-SPEED SWITCH
S_{WP}	WEATHERPROOF SWITCH

From FLETCHER. *Residential Construction Academy*, 3E. © 2012 Cengage Learning

FIGURE 10-17 Miscellaneous electrical symbols.

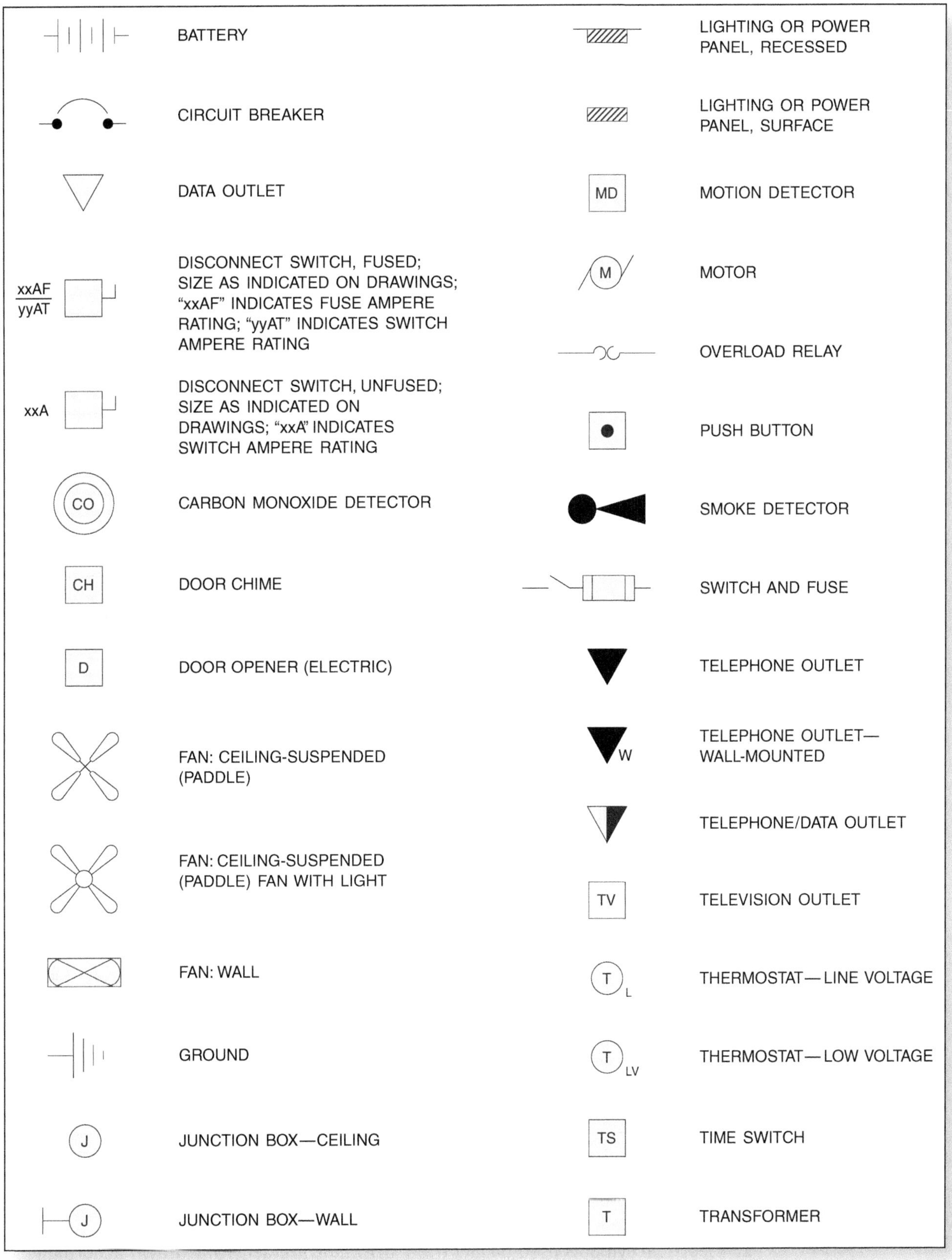

Symbol	Description	Symbol	Description
	BATTERY		LIGHTING OR POWER PANEL, RECESSED
	CIRCUIT BREAKER		LIGHTING OR POWER PANEL, SURFACE
	DATA OUTLET	MD	MOTION DETECTOR
xxAF/yyAT	DISCONNECT SWITCH, FUSED; SIZE AS INDICATED ON DRAWINGS; "xxAF" INDICATES FUSE AMPERE RATING; "yyAT" INDICATES SWITCH AMPERE RATING	M	MOTOR
			OVERLOAD RELAY
xxA	DISCONNECT SWITCH, UNFUSED; SIZE AS INDICATED ON DRAWINGS; "xxA" INDICATES SWITCH AMPERE RATING		PUSH BUTTON
CO	CARBON MONOXIDE DETECTOR		SMOKE DETECTOR
CH	DOOR CHIME		SWITCH AND FUSE
D	DOOR OPENER (ELECTRIC)		TELEPHONE OUTLET
	FAN: CEILING-SUSPENDED (PADDLE)	W	TELEPHONE OUTLET—WALL-MOUNTED
	FAN: CEILING-SUSPENDED (PADDLE) FAN WITH LIGHT		TELEPHONE/DATA OUTLET
		TV	TELEVISION OUTLET
	FAN: WALL	T L	THERMOSTAT—LINE VOLTAGE
	GROUND	T LV	THERMOSTAT—LOW VOLTAGE
J	JUNCTION BOX—CEILING	TS	TIME SWITCH
J	JUNCTION BOX—WALL	T	TRANSFORMER

From FLETCHER. *Residential Construction Academy*, 3E. © 2012 Cengage Learning

FIGURE 10-18   Electrical wiring symbols.

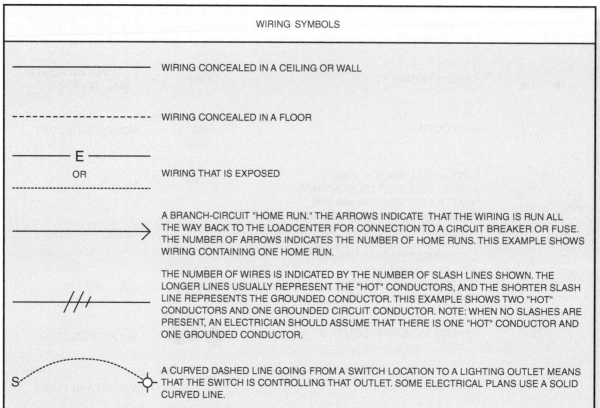

FIGURE 10-19   Cabling diagram versus a wiring diagram.

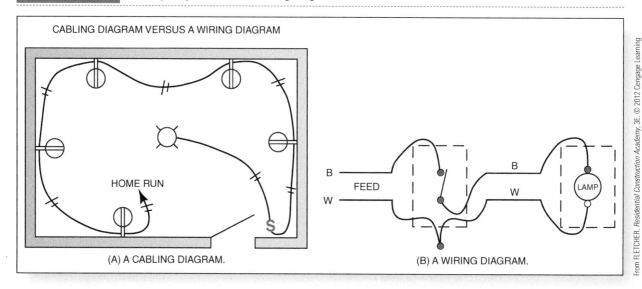

    Self-Check 2

1. Electrical symbols used on electrical drawings are generally drawn according to what standard? Who published this standard?

2. What is a legend, and why might one be used on an electrical plan?

3. In electrical wiring diagrams, what do the slashes on a line represent?

4. What is the major difference between a cabling diagram and a wiring diagram?

10-3 . Nationally Recognized Testing Laboratories

Hardware and materials used by residential electricians must be approved. The *National Electrical Code (NEC)* tells us in *Article 90* that it is the authority having jurisdiction (AHJ) who has the job of approving the electrical materials used in an electrical system. The AHJ is usually the local or state electrical inspector in your area. Electricians are required by the *NEC* in *110.3(B)* to use and install equipment and materials according to the instructions that come with their listing or labeling. AHJs often base their approval of electrical equipment on whether it is listed and labeled by a nationally recognized testing laboratory (NRTL).

Purpose of an NRTL

Manufacturers, consumers, regulatory agencies, and others recognize how important a nationally recognized testing laboratory is in reducing safety risks. Third-party testing is necessary unless you have the required test equipment and knowledge of how to test a product for safety. NRTLs have the knowledge, resources, and test instruments to test and evaluate products for safety.

An NRTL performs tests based on a specific nationally recognized safety standard. Once the product has been tested and is compliant with the safety standard, the product is determined to be free from reasonably foreseeable risk of fire, electric shock, and related hazards. It is then listed and has a listing marking label. Listing means that an NRTL has tested a device and it has met the NRTL's requirements for safety, fire, and shock hazard. A label is an identifying mark of a specific testing laboratory. This label verifies that the product complies with appropriate standards. Electricians, AHJs, and others know that if the product is installed and used according to installation instructions, it will operate in a safe manner.

Label an identifying mark of a specific nationally recognized testing laboratory (NRTL).

Example NRTLs

As already stated, the primary purpose of an NRTL is to *reduce* safety risks. The following describes several NRTLs.

Underwriters Laboratories

One of the best known of the NRTLs is Underwriters Laboratories (UL). Most electrical product manufacturers submit their products to UL, where the equipment is put through a variety of tests. These tests determine whether the product performs safely while operating under normal and abnormal conditions, to meet published standards. After UL testing and evaluating, when it is determined that the product is in compliance with the specific standards it was tested for, the manufacturer is permitted to put the UL label on its product. The product is then *listed* in a UL Directory. **Figure 10-20** shows a typical placement of the Underwriters Laboratories mark on HVAC units.

Underwriters Laboratories lists products that conform to a specific safety standard. This does *not* mean that UL approves a product. The UL mark indicates that selected

FIGURE 10-20 These HVAC units show a typical placement of the UL logo.

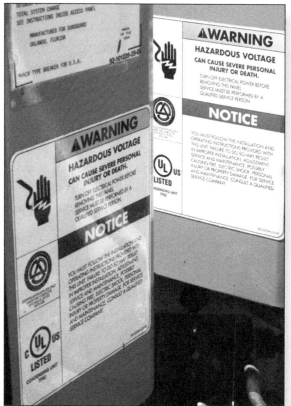

© Cengage Learning 2013

samples of the product have been subjected to testing and evaluations using nationally recognized safety standards with regard to fire, electric shock, and related safety hazards.

In the area of residential wiring, three useful UL publications are as follows:

- Electrical Construction Equipment Directory ("Green Book")
- Electrical Appliance and Utilization Equipment Directory ("Orange Book")
- Guide Information for Electrical Equipment Directory ("White Book")

The Green and Orange Directories list technical information regarding a certain product, and names and addresses are listed for manufacturers, along with manufacturers' identification numbers. As an electrician, you will find the "White Book" most useful when you are looking for specific requirements, permitted uses, and limitations for a particular product.

These UL directories can be obtained by contacting:

Underwriters Laboratories, Inc.
333 Pfingsten Road
Northbrook, IL 60062-2096
Phone number: 847-272-8800
Web site: http://www.ul.com

CSA International

CSA International is part of CSA Group, which also includes the Canadian Standards Association. CSA provides worldwide service and its marks appear on billions of

FIGURE 10-21 The CSA International label.

Courtesy of CSA International (csa-international.org)

qualified products around the world. For almost a century, CSA has tested and certified products for manufacturers. CSA offers:

- In-depth technical expertise
- Testing and certification to U.S. and Canadian standards including the Canadian Electrical Code (CEC)
- Testing and certification for the International Electrotechnical Commission (IEC) standards

An illustration of this label is shown in **Figure 10-21**.

CSA International can be contacted by using the following information:

CSA International
8501 East Pleasant Valley Road
Cleveland, Ohio, USA, 44131-5575
Phone number: 1-866-463-1785 or 1-216-328-8113
Web site: http://www.csa-international.org

Intertek Testing Services

Intertek Testing Services (ITS), formerly known as Electrical Testing Laboratories (ETL), is a nationally recognized testing laboratory. Its ETL SEMKO division tests, evaluates, labels, lists, and provides follow-up service for the safety testing of electrical products in accordance with nationally recognized safety standards.

CSA and ITS labels and listings are often used as a basis for the AHJ approval of equipment required by the *NEC*.

Intertek Testing Services can be reached by contacting:

Intertek Testing Services, NA, Inc.
ETL SEMKO
3933 US Route 11
Cortland, NY 13045
Phone number: 607-753-6711
Web site: http://www.intertek.com

National Electrical Manufacturers Association (NEMA)

The National Electrical Manufacturers Association (NEMA) is an organization that establishes certain construction standards for the manufacture of electrical equipment.

NEMA represents approximately 500 manufacturers of electrical products. Electrical standards, which in many cases are very similar to UL and other nationally recognized standards, are developed by NEMA. There are NEMA representatives on the *NEC* code-making panels (CMPs).

Information provided by NEMA can be obtained by contacting:

National Electrical Manufacturers Association
1300 North 17th St., Suite 1847
Rosslyn, VA 22209
Phone number: 703-841-3200
Web site: http://www.nema.org

LIFE SKILLS

Why do you think it is important that NRLTs are national organizations instead of state or local?

Self-Check 3

1. Why is it better to have a third-party agency such as an NRTL do product safety testing?
2. What is the most recognizable NRTL?
3. List the three UL publications useful in the area of residential wiring. Which of these publications is most useful?
4. What is NEMA? How can this association be reached?

10-4 ▪ Fuses and Circuit Breakers and Current Ratings

Article 240 of the *NEC* covers overcurrent protection. This article states that overcurrent protection for conductors and equipment is provided to open the circuit if the current reaches a value that causes an excessive or dangerous temperature in conductors or conductor insulation. Circuit breakers and fuses provide overcurrent protection for residential main services, branch circuits, and feeders. These devices are the *safety valves* of an electrical circuit.

Fuses

Fuse an overcurrent protection device that opens a circuit where the fusible link is melted away by the extreme heat caused by an overcurrent.

Fuses are a reliable and economic form of overcurrent protection. *NEC 240.50* through *240.54* provides the requirements for plug fuses, fuse holders, and adapters. A fuse is an overcurrent protection device that opens a circuit where the fusible link is melted away by the extreme heat caused by an overcurrent. There are two styles of fuses that are generally used in a residential installation:

- Plug
- Cartridge

Time-delay fuse
a fuse with a time-delay configuration, or time-current characteristic, that is designed with a heat sink next to the fusible element.

Both types are available in a **time-delay fuse** or **non-time-delay fuse** design. A fuse with a time-delay configuration or time-current characteristic is one that is designed with a heat sink next to the fusible element. The heat sink *absorbs* heat before the fusible link melts and provides the fuse with a time delay. A non-time-delay fuse has one fusible element, or link. Where an overcurrent condition exists, the link melts, offering protection to connected circuits.

TRADE TIP   Non-time-delay fuses are not used to protect, for example, motor circuits because of the high starting inrush of current of motors.

Non-time-delay fuse a fuse with one fusible element, or link. Where an overcurrent condition exists, the link melts, offering protection to connected circuits.

Plug Fuses

There are two types of plug fuses:

- Edison base
- Type S

Edison-Base Fuse

The Edison-base plug fuse is used *only* to replace existing Edison-base fuses. **Figure 10-22** shows examples of Edison-base fuses.

Edison-base fuses cannot be used in *new* installations. The reason for this is that all sizes of Edison-base plug fuses have the same base size and fit the same fuse holder. Over fusing could result with this type fuse, for example, if a person put a 30-ampere Edison-base fuse in a fuse holder that required a 15-ampere fuse.

Type S Fuse

All new installations that require plug fuses must use Type S, or safety, fuses as overcurrent protection devices. This type of fuse can only fit and work in a fuse holder with the same amperage rating as the fuse. **Figure 10-23(A)** shows a 15-ampere-rated (blue in color) Type S plug fuse and **(B)** adapter.

As seen in Figure 10-23, Type S fuses are color coded. The 15-amp fuse in Figure 10-23 is blue; a 20-amp is orange; and a 30-amp is green in color. Type S fuses are considered tamper resistant. As an example, a 30-ampere Type S fuse cannot be put into a 15-ampere Type S adapter and work. This rejection base design is to prevent over-fusing of circuits where using plug fuses.

FIGURE 10-22   Edison-base plug fuses.

Courtesy of Cooper Bussmann, Inc.

FIGURE 10-23 (A) Type-S plug fuse and (B) adapter.

Courtesy of Cooper Bussmann, Inc.

Cartridge Fuses

Cartridge fuses are used occasionally in residential wiring. There are two styles of cartridge fuses:

- Ferrule
- Blade-type

Ferrule

The ferrule style is designed as a cylindrical tube of insulating material, such as cardboard, with metal caps on each end. Examples of ferrule-type cartridge fuses are shown in **Figure 10-24**. Ferrule-type cartridge fuses are available with ampere ratings of 15, 20, 30, 40, 50, and 60.

FIGURE 10-24 Ferrule-type cartridge fuses.

Courtesy of Cooper Bussmann, Inc.

Blade-Type

Blade-type cartridge fuses are available in fuses that are rated 60 to 600 amps. This type cartridge fuse has a cylindrical body of insulating material with protruding metal blades on each end. **Figure 10-25** shows a 250-volt, 100-ampere blade-type fuse.

Circuit Breakers

The most popular type of overcurrent protection device in residential wiring is the **circuit breaker**. A circuit breaker is an automatic overcurrent device that trips into an open position where an overcurrent is detected and therefore stops the circuit current. Three circuit conditions cause a circuit breaker to trip:

- Overload
- Short circuit
- Ground fault

The thermal/magnetic-type circuit breaker is used in residential wiring. Thermal tripping is caused by an overload. Circuit breakers used in residential wiring are available in single-pole for use on 120-volt circuits and two-pole for use on 240-volt circuits. **Figure 10-26** shows a single-pole circuit breaker on the left side of the figure and a two-pole breaker on the right side.

Circuit breaker
an automatic overcurrent device that trips where an overcurrent is detected into an open position and stops the current flow in an electrical circuit.

FIGURE 10-25 A blade-type fuse.

Courtesy of Cooper Bussmann, Inc.

FIGURE 10-26 Molded-case circuit breakers.

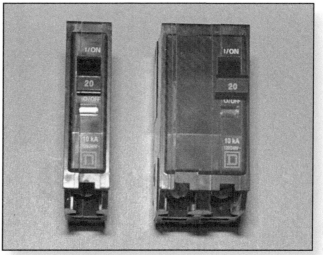

From FLETCHER. *Residential Construction Academy*, 3E. © 2012 Cengage Learning

Circuit breakers used for residential circuits are rated in voltage and current. They are available in 15, 20, 30, 40, and 50 amperes. There are also larger sizes available, used as the main service-entrance disconnecting means, that are rated at 100 or 200 amperes. Residential circuit breakers generally have a voltage rating of 120/240 volts. The slash (/) between the lower and higher voltage rating indicates that the circuit breaker has been tested for use on a circuit with the higher voltage between "hot" conductors (240 V) and with the lower voltage from a "hot" conductor to ground (120 V).

Circuit breakers can also be designed to work in ground-fault circuit interrupter (GFCI) and arc-fault circuit interrupter (AFCI) models. Regular overcurrent protection is provided with these designs along with GFCI and AFCI protection to the whole circuit. These types of circuit breakers are recognized by the push-to-test button on the front and the length of white insulated wire attached to them. **Figure 10-27** shows a GFCI and an AFCI circuit breaker.

Interrupting Rating

Interrupting rating the highest current rated voltage that a device is intended to interrupt under standard test conditions.

All fuses and circuit breakers intended to interrupt the circuit at fault levels must have an adequate interrupting rating wherever they are used in the electrical system, as stated in *110.9* of the *NEC*. Fuses or circuit breakers that have inadequate interrupting ratings could cause a fuse to blow (open) or a circuit breaker to trip (off) while attempting to clear a short circuit. The *NEC* defines the term *interrupting rating* as the highest current rated voltage that a device is intended to interrupt under standard test conditions. It is important that the test conditions match the actual installation needs. Interrupting ratings and short-circuit ratings should *not* be confused.

Short-circuit current ratings are marked on the equipment. Keep in mind, the main purpose for overcurrent protection is to *open* the circuit before conductors or conductor insulation is damaged where an overcurrent condition occurs. Fuses and circuit breakers must be selected to ensure that the short-circuit current rating of any

FIGURE 10-27 A GFCI and an AFCI circuit breaker.

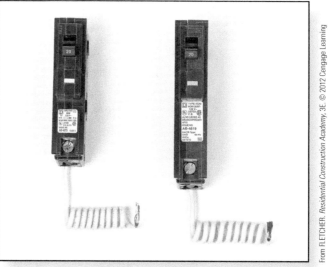

From FLETCHER. *Residential Construction Academy*, 3E. © 2012 Cengage Learning

electrical system component is not exceeded should a short circuit or a high-level ground fault occur. Sufficient interrupting ratings of OCPDs does *not* ensure adequate short-circuit protection for the system components.

Most circuit breakers have an interrupting rating of either 5000 or 10,000 amps. The interrupting rating of fuses can be as high as 300,000 amps. For circuit breakers, if the interrupting rating is other than 5000 amps, it must be marked on the breaker. Fuses must have the interrupting rating marked on their casings.. The interrupting rating of a fuse is defined as the maximum amount of current the device can handle without failing. The interrupting rating of a device must be at least as high as the available fault current. The available fault current is the maximum amount of current that can flow through the circuit if a fault occurs.

LIFE SKILLS

This section features three subsections: Fuses, Circuit Breakers, and Interrupting Rating. Write a paragraph explaining how the three are related.

Self-Check 4

1. What does *Article 240* of the *NEC* state about overcurrent protection for conductors and equipment?
2. What are the two main overcurrent protection devices?
3. If a new installation requires plug fuses, what type must be used?
4. What are the two types of cartridge fuses?
5. List the three conditions that cause a circuit breaker to trip.

Summary

- An electrician should know the various parts of a building plan and how to read and interpret the information found on them.
- The main parts of a residential building plan are as follows: plot plan, floor plans, elevation drawings, detail drawings, and electrical drawings.
- An electrical drawing is a part of the building plan that shows information about the electrical supply and distribution for the structure's electrical system.
- Schedules are tables in building plans that provide information about specific equipment or materials used in the construction of a structure or house. Schedules provide specific information for electricians, such as particular electrical equipment that must be installed in the structure.
- Specifications, such as specific manufacturer's catalog numbers and other information necessary so that electrical items are the correct size, type, and proper rating, are important to an electrician.

- Service conductors that are made up in the form of a cable are called service cables. These service cables are installed to supply the electrical power from the utility company to the building's electrical system.

- SE cable types are covered in *Article 338* of the *NEC*. This cable type is defined as a single conductor or multiconductor assembly provided with or without an overall covering.

- The three types of service-entrance cables are Types SEU, SER, and USE.

- Electrical symbols show the electrician which electrical items are required and where they are to be located.

- The American National Standards Institute (ANSI) has published a standard titled "Symbols for Electrical Construction Drawings," which shows the electrical symbols used on electrical drawings.

- The *NEC* tells us in *Article 90* that it is the authority having jurisdiction (AHJ) who has the job of approving the electrical materials used in an electrical system.

- Many AHJs base their approval of electrical equipment on whether it is listed and labeled by a nationally recognized testing laboratory (NRTL).

- NRTLs are third-party testing agencies that have the knowledge, resources, and test instruments to test and evaluate products for safety.

- Three of the most recognizable NRTLs are Underwriters Laboratories (UL), CSA International (formerly Canadian Standards Association), and Intertek Testing Services (ITS).

- *Article 240* of the *NEC* covers overcurrent protection and states that overcurrent protection for conductors and equipment is provided to open the circuit if the current reaches a value that causes an excessive or dangerous temperature in conductors or conductor insulation.

- Overcurrent protection for residential main services, branch circuits, and feeders is provided by circuit breakers or fuses.

- A fuse is an overcurrent protection device that opens a circuit where the fusible link is melted away by the extreme heat caused by an overcurrent.

- The two styles of fuses that are generally used in a residential installation are plug and cartridge.

- A circuit breaker, the most often used type of overcurrent protection device in residential wiring, is an automatic overcurrent device that trips into an open position where an overcurrent is detected, and stops the current flow in an electrical circuit.

- The three circuit conditions that cause a circuit breaker to trip are overload, short circuit, and ground fault.

- *NEC 110.9* states that all fuses and circuit breakers intended to interrupt the circuit at fault levels must have an adequate interrupting rating wherever they are used in the electrical system.

- The interrupting rating is the highest current rated voltage that a device is intended to interrupt under standard test conditions.

Review Questions

True/False

1. Specifications are the part of a building plan that provides more *specific* details about the construction of a building. (True, False)

2. A plot plan is a diagram, or drawing, usually to scale, showing the relationship between rooms, spaces, and other physical features at one level of a structure. (True, False)

3. An elevation drawing is a one-dimensional projection that shows the side of a structure as seen head on and facing a certain direction. (True, False)

4. A sectional drawing is a separate large-scale drawing that shows very specific details of a small part or particular section of a structure. (True, False)

5. Electrical drawings present in detail what is required of an electrician to complete the installation of the electrical system. (True, False)

Multiple Choice

6. On a building plan, the tables used to provide information about specific equipment or materials used in the construction of a structure or house are called _____.
   A. Specifications
   B. Legends
   C. Schedules
   D. Drawings

7. The conductors and equipment for delivering electric energy from the serving electric utility to the wiring system are referred to as the _____.
   A. Front end wiring
   B. Overhead wiring
   C. Service
   D. Branch circuit

8. Which of the following is *not* a type of service-entrance cable?
   A. SEU
   B. SCR
   C. SER
   D. USE

9. Which organization published a standard titled "Symbols for Electrical Construction Drawings"?
   A. OSHA
   B. NEMA
   C. ANSI
   D. AAMI

10. The part of a building plan that describes the various symbols and abbreviations used on the plan is referred to as the _____.

    A. Legend

    B. Table

    C. Specifications

    D. Symbol diagram

Fill in the Blank

11. _____ on a curved line are sometimes used in a cable diagram to indicate how many conductors are in the cable.

12. A(n) _____ diagram shows exactly where the conductors terminate, for example, in an electrical box.

13. Once an NRTL tests a product and the product is considered to be free from reasonably foreseeable risk of fire, electric shock, and related hazards, the product is then _____.

14. The UL _____ indicates that representative samples of the product have been tested and evaluated to nationally recognized safety standards with regard to fire, electric shock, and related hazards.

15. CSA International is the _____ counterpart of Underwriters Laboratories, Inc., in the United States.

16. *Article 240* of the *NEC* covers _____ protection.

17. A(n) _____ is an overcurrent protection device that opens a circuit where the fusible link is melted away by the extreme heat caused by an overcurrent.

18. _____ fuses are not used to protect, for example, motor circuits because of the high starting inrush of current of motors.

19. New installations cannot use _____ fuses.

20. _____ cartridge fuses are available in fuses that are rated larger than 60 amps.

21. Circuit breakers are the most often used type of overcurrent _____ device in residential wiring.

22. _____ rating is the highest current rated voltage that a device is intended to interrupt under standard test conditions.

23. A(n) _____ plan is generally the first sheet in a set of building plans and shows all the major features and structures on a piece of property.

24. A(n) _____ drawing is a part of the structure that shows a cross-sectional view of a specific part of the dwelling.

25. The _____ service is the service type most often installed in residential wiring.

ACTIVITY 10-1   Wiring Activity

As you have learned, three-way switches are used to control a light or lighting circuit from more than one location. **Figure 10-28** represents a typical room layout in a dwelling unit. Using this figure for reference, complete the connections on the following cabling diagram (**Figure 10-29**).

FIGURE 10-28 Typical room layout.

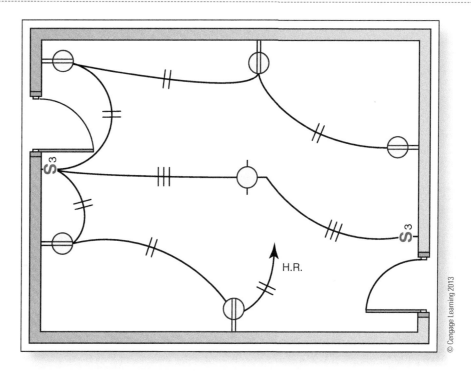

© Cengage Learning 2013

FIGURE 10-29 Cabling diagram.

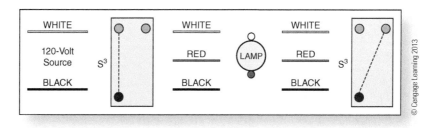

© Cengage Learning 2013

Chapter 11

Wiring Devices

Chapter Outline

RECEPTACLES—MARKINGS AND OPERATION

SWITCHES—MARKINGS AND OPERATION

DIMMER CONTROL DEVICE

FUSES—PURPOSE AND OPERATION

CIRCUIT BREAKERS—PURPOSE AND OPERATION

GROUND-FAULT CIRCUIT INTERRUPTER (GFCI) AND ARC-FAULT
CIRCUIT INTERRUPTER—INSTALLATION AND OPERATION

Key Terms

Ambient temperature

Ballast

Dimmer switch

Double-pole switch

Duplex receptacle

Four-way switch

Multiwire circuit

Neutral

Outlet

Receptacle

Single receptacle

Split-wired receptacle

Three-way switch

Chapter Objectives

After completing this chapter, you will be able to:

1. Identify markings on a single receptacle and duplex receptacle, and describe the operation of each device.

2. Describe the operation and markings of single-pole, three-way, and four-way toggle switches.

3. Explain the functions of a dimmer control device.

4. Describe the operation of a fuse.

5. Describe the operation of a circuit breaker.

6. Describe installation and operation of ground-fault circuit interrupters (GFCI) and arc-fault circuit interrupters (AFCI).

Life Skills Covered

 Critical Thinking

 Goal Setting

Life Skill Goals

This chapter contains a lot of information, and its first life skill, Critical Thinking, is aimed at helping you absorb it all. When you're on a jobsite, it is crucial that you be able to analyze information and then draw conclusions about what it means. As you go through this chapter, read it carefully and take notes. This chapter helps you to *understand* the material rather than just memorize it. When you get out into the field, you will be glad that you took the time to learn this information. The second life skill covered is Goal Setting. We discuss several goals that you should keep in mind as you read this chapter. However, you should be coming up with goals far beyond the scope of this book. Before you read the chapter, think about your larger career goals. How will a base knowledge of wiring devices help you reach them?

As an electrician, you must remember that no matter what the device is, *all* wiring methods must be in accordance with the *National Electrical Code*. Anything else is unacceptable! Learning the purpose, function, and operation of wiring devices sets the foundation for actually making the correct wiring connections as per the *NEC*.

Introduction

There are many things an electrician must learn and remember, but two of the most important are *SAFETY FIRST* and *IMPROPER WIRING CAN CAUSE FIRES, PERSONAL INJURIES, AND ELECTROCUTIONS!*

This chapter describes some of the most commonly used *wiring devices*, including receptacles, switches, and dimmer controls. It also covers the purpose and operation of overcurrent, ground-fault, and arc-fault protection devices: fuses, circuit breakers, ground-fault circuit interrupters, and arc-fault circuit interrupters.

11-1 Receptacles—Markings and Operation

This section covers conductor identification and describes several types of receptacles. The markings on and parts of the most common receptacle, the duplex, are described in detail.

Conductor Identification

For an electrician to know which wires go to which terminals, whether the device is a receptacle, a switch, or a circuit breaker, he or she must know and understand conductor color coding. *Articles 200* and *210* of the *NEC* cover this material.

Grounded Conductor

Multiwire circuit a circuit in residential wiring that consists of two ungrounded conductors that have 240 volts between them and a grounded conductor that has 120 volts between it and each ungrounded conductor.

The *NEC* requires that the grounded (identified) conductor have an outer finish that is either continuous white or gray for alternating-current circuits. This conductor may have an outer finish (not green) that has three continuous white stripes along the conductor's entire length. The grounded conductor is also called a neutral conductor in multiwire circuits. A multiwire circuit in residential wiring consists of two ungrounded conductors that have 240 volts between them and a grounded conductor that has 120 volts between it and each ungrounded conductor.

Ungrounded Conductor

Ungrounded conductors can be any color or pattern other than green, white, gray, or with three continuous white stripes. This conductor is referred to as the "hot" conductor. If this "hot" conductor and a grounded conductor or grounded surface such as a metal water pipe come in contact with each other, a short circuit results.

Typically, an ungrounded conductor has an outer finish that is colored black, red, blue, green, yellow, brown, orange, or gray. Black is the only color used for conductors that are 1 AWG or larger.

Three Continuous White Stripes

Where working with residential wiring, it is unlikely that you will work with a conductor that has three continuous white stripes. This is a method of identifying conductors done by cable manufacturers and is generally found on larger size conductors.

Grounded Neutral Conductor

The white grounded conductor is commonly referred to as the *neutral*. You will learn, however, later in your studies, that this conductor is not always a neutral conductor.

The following are *NEC* requirements for grounded neutral conductors:

- *310.2(A)*—The grounded neutral conductor for branch circuits and feeders must be insulated.
- *230.41*—For residential services, the grounded neutral conductor is permitted to be insulated or bare.
- *250.20(B)(1)*—For residential wiring, the 120/240-volt electrical system is grounded by the electric utility, at its transformer, and again by the electrician, at the main service. (The system must be *grounded* so the maximum voltage to ground on the ungrounded conductors does not exceed 150 volts).
- *250.26(2)*—The neutral conductor is required to be grounded for single-phase, 3-wire systems.

Electricians usually use the term *neutral* whenever they refer to a white grounded circuit conductor. The correct definition of neutral as listed in the *NEC* is as follows: *The conductor connected to the neutral point of a system that is intended to carry current under normal conditions.**

Receptacles

A receptacle is a device installed in an electrical box that allows an electrician to access current from the wiring system and deliver it through a cord and attachment plug to a piece of equipment. Many people, including some electricians, *incorrectly* refer to a receptacle as an outlet, but this is an improper term. An outlet is the point on the

Neutral the conductor connected to the neutral point of a system that is intended to carry current under normal conditions.

Receptacle a device installed in an electrical box that allows an electrician to access current from the wiring system and deliver it through a cord and attachment plug to a piece of equipment.

Outlet the point on the wiring system at which current is taken to supply electrical equipment.

---

*Reprinted with permission from NFPA 70-2011.

wiring system at which current is taken to supply electrical equipment. **Figure 11-1** shows a receptacle *outlet* (top), as well as three *outlets:* one with a single receptacle, one with a duplex receptacle, and one with two duplex receptacles.

In *Article 100*, the *NEC* defines a receptacle *outlet* as the branch-circuit wiring and the box where one or more receptacle *devices* are to be installed. The *NEC* goes on to define a *receptacle* as a strap (yoke) with one, two, or three contact devices.

FIGURE 11-1 Receptacle outlets and receptacles.

A RECEPTACLE OUTLET IN WHICH ONE OR MORE RECEPTACLES WILL BE INSTALLED.

A RECEPTACLE OUTLET WITH A SINGLE RECEPTACLE. (ONE CONTACT DEVICE)

A RECEPTACLE OUTLET WITH A MULTIPLE (DUPLEX) RECEPTACLE. THIS IS TWO RECEPTACLES. (TWO CONTACT DEVICES)

A RECEPTACLE OUTLET WITH TWO MULTIPLE (DUPLEX) RECEPTACLES. THIS IS FOUR RECEPTACLES. (FOUR CONTACT DEVICES)

Single Receptacle

A **single receptacle** is shown in Figure 11-1. This is a single contact device with no other contact device on the same yoke. Receptacles such as these are sometimes used in residential wiring, for example, a receptacle installed for a mid-size to large air-conditioning unit, which is generally a single receptacle with a 20-ampere, 125-volt rating. The most common type of receptacle used in residential wiring is a duplex receptacle rated for 15 amperes, using 14-gauge wire, or rated for 20 amperes for use with 12-gauge wire at 125 volts.

Duplex Receptacle

A **duplex receptacle** is also shown in Figure 11-1, and it is the most common receptacle type used in residential wiring; it has two receptacles on one strap; each receptacle is capable of providing power to a cord-and-plug-connected electrical load. The smaller "hot" conductor on the plug enters the *shorter* contact slot on the receptacle. The larger *grounded* conductor enters the longer contact slot on the receptacle. A U-shaped ground slot receives the round grounding conductor, if there is one. On one side of the receptacle, silver screws connect with the ground contact slots. On the other side, brass- or bronze-colored screws connect with the "hot" contact slots. A green screw in the receptacle connects with the green grounding conductor. **Figure 11-2** shows the markings on and the parts of a duplex receptacle.

The markings on and the parts of a duplex receptacle, as shown in Figure 11-2, are important for electricians to know. The following features are typically found on the *front* of a receptacle:

- Connecting tabs, to connect the top and bottom halves of a duplex receptacle—They can be removed to provide different wiring configurations

FIGURE 11-2 A duplex receptacle.

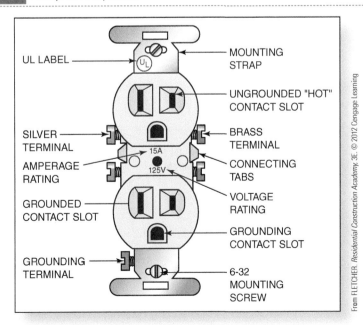

with split-wired receptacles. A split-wired receptacle is a duplex receptacle wired so that the top outlet is "hot" all the time, and the bottom outlet is switch controlled.

- Mounting strap, to attach the receptacle to a device box—Modern receptacles are held in place by screws through cardboard or plastic mounting straps.
- Ratings—Amperage and voltage ratings are written on the receptacle.
- Label—A label from an NRTL is on the receptacle. This is required for approval by the AHJ.

Figure 11-3 shows the back of a duplex receptacle. Information listed should be checked *before* installation. The following features, as shown in Figure 11-3, are generally found on the *back* of a duplex receptacle:

- Push-in terminals—Sometimes called *backstabs*, these are used where electricians strip a conductor and push it into the hole rather than terminating on the screws.
- Strip gauge—This gauge is used to let the electrician know how much insulation needs to be stripped off the conductor where using the push-in fittings.
- Conductor size—This tells the electrician what the maximum conductor size is for this device. Most duplex receptacles are rated for 14 or 12 AWG conductors.
- Conductor material markings—These markings indicate what conductor material is okay to use with the device. "CU" indicates that only copper conductors can be used. "Cu-Clad" indicates that only copper-clad aluminum can be used, and "CO/ALR" indicates that either copper or copper-clad aluminum can be used. Aluminum is also allowed to be used with receptacles marked "CO/ALR."

FIGURE 11-3 The back of a duplex receptacle.

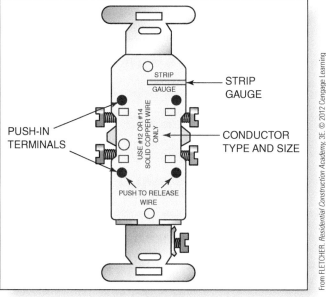

LIFE SKILLS

To safely install a terminal, you must know the color coding associated with wiring each particular terminal. Set a goal to learn the color of wiring for each of the terminal types listed in this chapter. Try making flashcards to help you study.

Self-Check 1

1. What does the *NEC* require for ac circuits regarding the conductor's outer finish?
2. What is the difference between a receptacle and an outlet?
3. Describe a duplex receptacle.
4. List three features generally found on the back of a duplex receptacle.

11-2 ▪ Switches—Markings and Operation

Lighting outlets installed in residential wiring are controlled by using switches. These devices are often called a toggle switch, a light switch, or a snap switch. The following types of switches are most commonly used in residential wiring:

- Single-pole
- Double-pole
- Three-way
- Four-way

Single-Pole Switch

The single-pole switch is the most commonly used switch in residential wiring. The single-pole switch is used in 120-volt circuits to control a lighting outlet or other outlets from only one location. This type switch is typically installed in a bathroom and located next to the door. It allows a person to turn on a luminaire when the person enters the room. This same type switch could be used to control a bathroom ventilation fan from one location. The basic parts of a single-pole switch are shown in **Figure 11-4**.

The following describes the main parts of a single-pole switch:

- Switch toggle—Places the switch in ON or OFF position when moved up or down.
- Screw terminals—Attach the lighting circuit wiring to the switch. On single-pole switches, the two terminal screws are the same color, usually bronze.
- Grounding screw terminal—Attaches the circuit-grounding conductor to the switch. It is green.
- Mounting ears—Secure the switch in a device box with screws.

FIGURE 11-4 The parts of a single-pole switch.

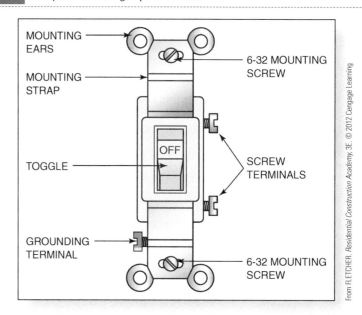

MOUNTING EARS

MOUNTING STRAP

6-32 MOUNTING SCREW

TOGGLE

OFF

SCREW TERMINALS

GROUNDING TERMINAL

6-32 MOUNTING SCREW

From FLETCHER. *Residential Construction Academy*, 3E. © 2012 Cengage Learning

TRADE TIP Where installing single-pole switches, make certain that where the toggle is up, the writing on it reads "ON." If you have it installed upside down, it reads "NO."

Single-pole switches, along with all other switch types, have current and voltage ratings listed on them. Switches may also have push-in terminals similar to those on receptacles.

Double-Pole Switch

Double-pole switch a switch type used to control two separate 120-volt circuits or one 240-volt circuit from one location.

A **double-pole switch** is a switch type used to control two separate 120-volt circuits or one 240-volt circuit from one location. **Figure 11-5** shows a double-pole switch.

The design features of a double-pole switch are similar to those of a single-pole switch. However, the double-pole switch has four terminal screws instead of two. The two top screws are the *line* terminals and are the same color. The two bottom screws are *load* terminals and are the same color, but a different color from the top screws. A double-pole switch is commonly used to control 240-volt appliances like clothes dryers. The double-pole switch works like a single-pole switch except that the line and load terminals are each ganged together and operate as one.

Three-Way Switch

Three-way switch a switch type used to control a 120-volt lighting load from two locations.

A **three-way switch** is a switch type used to control a 120-volt lighting load from two locations. An example of where a three-way switch is used is in an upstairs/downstairs circuit with one three-way switch located at the top of the stairway, and

FIGURE 11-5   The parts of a double-pole switch.

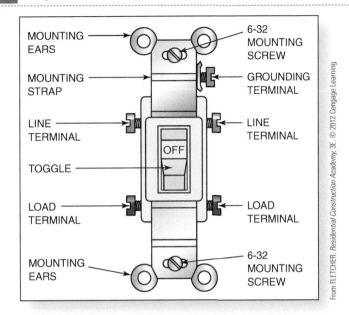

another three-way switch, at the bottom of the stairs. This way, a person can turn the light on or off before going up the stairs or from the top of the stairway.

Three-way switches get their name because they have *three* screw terminals on them. A three-way switch and its main parts are shown in **Figure 11-6**.

FIGURE 11-6   The parts of a three-way switch.

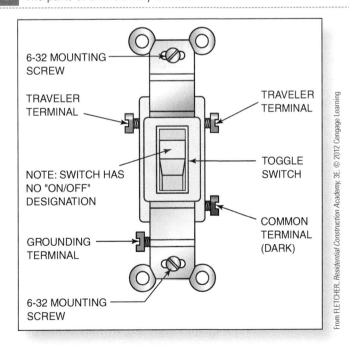

The following features make three-way switches different from single-pole switches:

- Three-way switches have *three* terminals. Two of the terminal screws, called *traveler terminals*, are brass colored and are across from each other on opposite sides of the switch. The other terminal screw is black and is referred to as the common terminal. This terminal is also called the *point* or *hinge* terminal.

TRADE TIP   Your ability to identify the common terminal and the traveler terminals enables you to make the correct connections where wiring three-way switches in a lighting circuit.

- The toggle, unlike a single-pole switch, does *not* have any ON or OFF position listed on it. It does not matter which way the three-way switch is mounted in a device box; there is no up or down mounting orientation on this type switch.

TRADE TIP   Install three-way switches in *pairs*. Run a three-way cable or three wires in a raceway between the two three-way switches for proper connection.

Four-Way Switch

Four-way switch a switch type that, where used in conjunction with two 3-way switches, allows control of a 120-volt lighting load from more than two locations.

A **four-way switch** is a switch type that, where used in conjunction with two 3-way switches, allows control of a 120-volt lighting load from more than two locations. It is commonly used in rooms with three doorways requiring switches. A three-way switch is located at two of the doorways, and a four-way switch, at the third. Where wired and operating properly, a person can turn on or off the room lighting outlet or outlets from any of the three locations.

The four-way switch has two sets of screw terminals (*four* total) on the switch. **Figure 11-7** shows a four-way switch and its parts.

Generally, manufacturers distinguish the two sets of terminals with different colors. Usually, one set is black, and the other set is brass or another lighter color. The two screws in each set are opposite each other on most four-way switches. Four-way switch terminals are often called *traveler terminals*. There is *no common terminal* as there is on a three-way switch. However, similar to three-way switches, there is *no* ON or OFF designation listed on the toggle of a four-way switch. And like the three-way switch, it does not make any difference which way the switch is mounted in a device box.

LIFE SKILLS

In your career, you will always be installing switches, so it is important to know which one to choose. Come up with one application for each of the switch types listed in this chapter, and write a short paragraph explaining why the switch that you chose is the right one for the job.

FIGURE 11-7 The parts of a four-way switch.

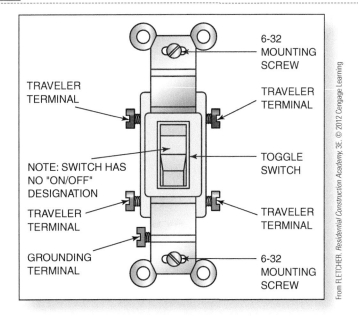

6-32
MOUNTING
SCREW

TRAVELER
TERMINAL

TRAVELER
TERMINAL

TOGGLE
SWITCH

NOTE: SWITCH HAS
NO "ON/OFF"
DESIGNATION

TRAVELER
TERMINAL

TRAVELER
TERMINAL

GROUNDING
TERMINAL

6-32
MOUNTING
SCREW

From FLETCHER. *Residential Construction Academy,* 3E. © 2012 Cengage Learning

Self-Check 2

1. List four types of switches commonly used in residential wiring.
2. What are the differences between a single-pole and double-pole switch?
3. List two examples where three-way switches may be used.
4. What is the difference between a three-way switch and a four-way switch?

11-3 ▪ Dimmer Control Device

Dimmer switch
a switch type that
raises or lowers the
lamp brightness of
a luminaire.

A dimmer switch is a switch type that raises or lowers the lamp brightness of a
luminaire. Dimmers are used in homes to lower the level of light. The two basic types of
dimmers are electronic and autotransformer.

Electronic Dimmers

Electronic dimmer control devices are available in both single-pole and three-way
configurations. A large luminaire located over a dining room table could be controlled
by using a single-pole dimmer switch. Two of the more popular styles of dimmer
switches are shown in **Figure 11-8**.

Electronic dimmers have gained popularity in residential applications. These
dimmers use electronic circuitry to provide dimming capabilities and are usually
rated at 125 volts and 600 watts. These dimmers also come in 1000-watt ratings. The
connected load must not exceed the listed wattage rating of the dimmer.

Electronic dimmers offer several features such as LED locator light, soft-ON and
fade-OFF, and return to previous setting after a power outage.

FIGURE 11-8   Dimmer switches.

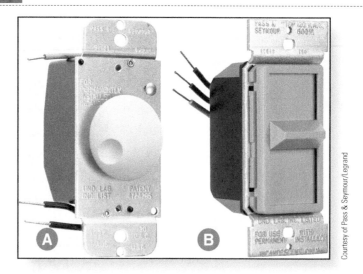

TRADE TIP   Make sure you install a wall box large enough to accommodate the dimmer, conductor, and splices that are in the box. If the wall box is too small, the lack of heat dissipation becomes a problem. Also, forcing the dimmer into the box may cause problems such as short circuits and ground faults.

Autotransformer Dimmers

An autotransformer type of dimmer is physically larger than the electronic dimmer and requires a special wall box. An autotransformer type dimmer is shown in **Figure 11-9**.

The autotransformer type dimmer is not typically used in residential wiring applications unless the load to be controlled is very large. This type dimmer is ordinarily used in commercial applications for the control of large loads.

FIGURE 11-9   Autotransformer-type dimmer control.

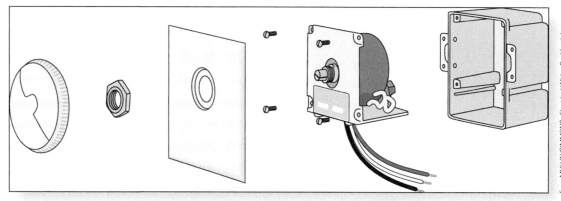

Fluorescent Lamp Dimming

The voltage to the lamp filament is raised or lowered by a dimmer control to dim an *incandescent* lamp. With a lower voltage, the intensity is less. The dimmer is connected *in series* with the incandescent lamp.

Fluorescent lamps, by nature of their design, cannot be dimmed in a simple *series* circuit. To dim fluorescent lamps, special dimmer switches and special dimming **ballasts** are needed. A ballast is a component in a fluorescent luminaire that controls the voltage and current flow to the lamp. The dimmer control allows the dimming ballast to maintain a voltage to the cathodes so that a proper operating temperature is maintained. This also allows the dimming ballast to vary the current in the arc, causing the intensity of light from the fluorescent lamp to vary.

This is a much easier method to provide fluorescent lamp dimming. Some compact fluorescent lamps are available that have an integrated circuit, which is a semiconductor material containing hundreds of thousands of previously separate (discrete) components, designed right into the lamp. This allows the fluorescent lamp to be controlled by a *standard incandescent* lamp dimmer. These *dimmable* compact fluorescent lamps can be used to replace standard incandescent lamps of the medium screw shell base type. *No additional wiring is necessary.* These dimmable fluorescent lamps can be used with dimmers, photocells, occupancy sensors, and electronic timers that are marked as "Incandescent Only."

> **Ballast** a component in a fluorescent luminaire that controls the voltage and current flow to the lamp.

LIFE SKILLS

To reach your goal of fully understanding this chapter's content, you should be aware of issues that could come up in installing a dimmer switch. With a partner, come up with one or two potential problems, and then brainstorm a solution.

Self-Check 3

1. What is the purpose of a dimmer switch?
2. List three features of electronic dimmers.
3. How can fluorescent lamps be dimmed?

11-4 ■ Fuses—Purpose and Operation

A fuse is an overcurrent protection device (OCPD) that opens a circuit where the fusible link is melted away by the extreme heat caused by an overcurrent. Fuses are designed and come in two basic styles: plug and cartridge. The plug-type fuses are available in an Edison-base and as a Type S model. Type S, or safety, fuses are used in all new installations. Both types of plug fuses are shown in **Figure 11-10**.

FIGURE 11-10 Plug fuses.

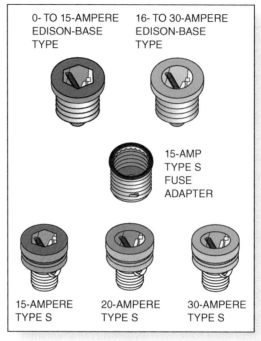

The two available models of cartridge fuses are a ferrule and blade-type. These models are shown in **Figure 11-11**.

Fuses are designed and available with time-delay or non-time-delay characteristics. As described, the main purpose of a fuse is to open in order to protect other circuit elements or wiring in the event of an overcurrent condition.

FIGURE 11-11 Cartridge fuses.

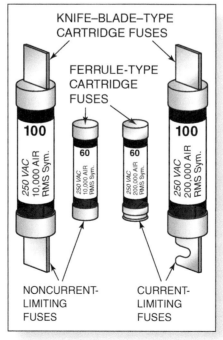

FIGURE 11-12 Three types of plug fuses—single-element, non-time-delay (W); dual-element, time-delay (T); and loaded link, time-delay (TL).

Courtesy of Cooper Bussmann, Inc.

Both plug and cartridge fuses are available with three basic types of time-current characteristics, and they are as follows:

- Non-time-delay—This type of fuse has a single link, or fusible element. One part of the link is *necked down*. Where an overcurrent condition occurs, it opens in the necked down portion, which is the weakest part of the link. Because motor load circuits produce a high *starting* current and then operate with a lower *run* current, non-time-delay fuses are not recommended for this application. A non-time-delay plug fuse is shown and labeled "W" in **Figure 11-12**.

- Time-delay, *dual*-element—One fuse element opens quickly where a short circuit, heavy overload, or ground fault occurs. The other fuse element, in series with the first element, *opens slowly* on overload conditions. Dual-element, time-delay fuses are the recommended protection device for motor circuits because they do *not* open on momentary overloads, such as the initially high starting current of a motor. A time-delay, dual-element plug fuse is shown in Figure 11-12 and is labeled "T."

- Time-delay, loaded link—This type of fuse has a single link, or fusible element, that is *loaded* with a heat sink next to the necked down portion of the link. This *load* acts as a heat sink that absorbs a considerable amount of heat before the necked down portion of the link melts open. This heat sink provides the fuse with time delay. The time-delay, loaded link plug fuse is shown in Figure 11-12 and is labeled "TL."

LIFE SKILLS

To help you best understand the material in the chapter, use your word processing software to make a reference chart for the types of fuses and time-current characteristics listed in the chapter. In one column, list the type of fuse or characteristic. In the other, list its key attributes. Having this chart handy will be a big help when it comes time to sit down and study.

Self-Check 4

1. What is a fuse? What type fuse can only be used in new installations?

2. Describe a time-delay fuse and where one might be used.

11-5 Circuit Breakers—Purpose and Operation

A circuit breaker is a device designed to open and close a circuit manually and to open the circuit automatically on a predetermined overcurrent, without damage to itself where properly applied within its rating. Although fuses are still used to provide overcurrent protection in residential wiring, circuit breakers are used more often.

Circuit breakers are available as a single-pole device for 120-volt applications and as a two-pole device for 240-volt applications. **Figure 11-13** shows three different types of circuit breakers.

Residential installations typically use thermal-magnetic circuit breakers. This type circuit breaker is designed with a bimetallic element. During a *continuous* overload, the bimetallic element moves until it unlatches the inner tripping mechanism of the breaker. Momentary small overloads do *not* cause the element to trip the breaker. If the overload is a high level of current or if there is a short circuit, a magnetic coil in the breaker causes it to interrupt the branch circuit almost instantly. **Figure 11-14** shows a commonly used single-pole and a two-pole circuit breaker.

FIGURE 11-13 Circuit breakers: (A) a single-pole device for 120-volt applications; (B) a two-pole device for 240-volt applications; and (C) twin or dual circuit breakers that can be used for two 120-volt branch circuits.

SINGLE-POLE CIRCUIT BREAKER

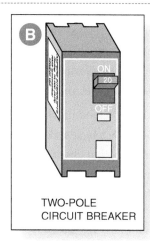
TWO-POLE CIRCUIT BREAKER

TWIN OR DUAL CIRCUIT BREAKER

From FLETCHER, *Residential Construction Academy*, 3E. © 2012 Cengage Learning

FIGURE 11-14 Molded-case circuit breakers.

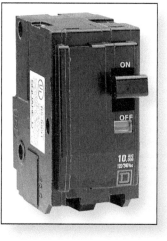

Courtesy of Schneider Electric

Thermal magnetic circuit breakers are temperature sensitive. Because of this characteristic, some circuit breakers are ambient temperature compensated. Ambient temperature is the temperature of the air that surrounds an object on all sides. This compensation partially or completely neutralizes the effect of ambient temperature on the tripping characteristic.

Ambient temperature the temperature of the air that surrounds an object on all sides.

TRADE TIP   *NEC 210.20* and *215.3* permit 100% loading on an overcurrent device *only* if the overcurrent device is listed for 100% loading. At this time, UL has *no* listing of residential-type molded-case circuit breakers that are suitable for 100% loading. It is recommended that a *maximum* loading of 80% or less is good practice.

LIFE SKILLS

Understanding *why* an answer is correct is just as important as knowing the correct answer. In this chapter, we learned that thermal magnetic breakers are most commonly used in residential installations. Why do you think that is the case? Why not another type?

Self-Check 5

1. What type circuit breakers are generally used for residential installations?
2. Explain what happens within a circuit breaker where a continuous overload occurs.

11-6 ▌ Ground-Fault Circuit Interrupter (GFCI) and Arc-Fault Circuit Interrupter—Installation and Operation

A ground-fault circuit interrupter (GFCI) is a device that protects people from dangerous levels of electrical current by measuring the current difference between two conductors of an electric circuit and tripping to an open position if the measured value exceeds approximately 5 milliamperes.

An arc-fault circuit interrupter (AFCI) is a device intended to provide protection from the effects of arc faults by recognizing characteristics unique to arcing and by functioning to de-energize the circuit where an arc fault is detected.

Both of these circuit interrupters are available as circuit breakers or receptacles. Both devices provide protection, but what each protects us from and how each provides this protection differs. The following descriptions highlight installation and operation of a GFCI and an AFCI.

Ground-Fault Circuit Interrupter (GFCI)

Before we take a closer look at those devices that provide ground-fault protection, let's review how a ground-fault circuit interrupter operates, as illustrated in **Figure 11-15**.

You have already learned that some receptacles in a home require ground-fault circuit protection. The two ways that GFCI protection can be provided are by using a GFCI receptacle or a GFCI circuit breaker. With a GFCI circuit breaker, the *whole circuit* becomes protected. However, with a GFCI receptacle, only that receptacle location and any *downstream* receptacles are GFCI protected. The GFCI circuit breaker is more costly than the GFCI receptacle.

The following are available GFCIs:

- 120-volt receptacles
- Single-pole, 120-volt breakers
- Single-pole, 120-volt dual-function, with both GFCI and AFCI features in one breaker
- Two-pole, 240-volt common trip breakers
- Two-pole 120/240-volt common trip dual-function, with both GFCI and AFCI features in one breaker

FIGURE 11-15   Ground-fault circuit interrupter—operation.

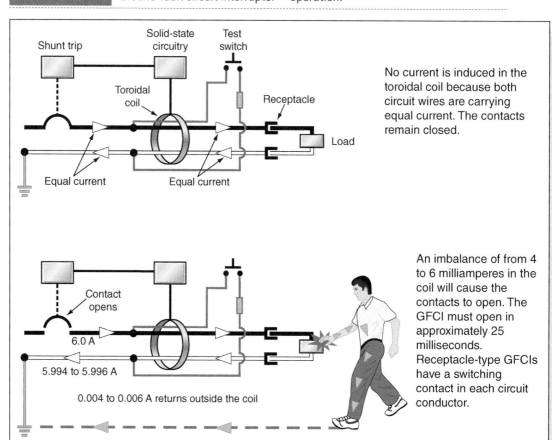

No current is induced in the toroidal coil because both circuit wires are carrying equal current. The contacts remain closed.

An imbalance of from 4 to 6 milliamperes in the coil will cause the contacts to open. The GFCI must open in approximately 25 milliseconds. Receptacle-type GFCIs have a switching contact in each circuit conductor.

From MULLIN/SIMMONS. *Electrical Wiring Residential,* 17E. © 2012 Cengage Learning

- Two-pole, 120/240-volt independent trip, dual-function, with both AFCI and GFCI features in one breaker; suitable for use with a shared neutral on a multiwire branch circuit

- *Faceless*—Strictly for GFCI protection; have only test and reset buttons; do not have a receptacle; mounted in a single device box; and commonly used to protect spas, Jacuzzi tubs, whirlpools, and so on

Figure 11-16 shows a GFCI duplex grounding-type convenience receptacle and a single-pole GFCI circuit breaker.

Receptacle-type GFCIs switch *both* the "hot" and grounded conductors. Refer to **Figure 11-17**. In Figure 11-17, when the test button is pushed, the test current passes through the test button and the sensor and then, bypassing the sensor, back to the opposite circuit conductor. This is how the *unbalance* is created and then monitored by the electronic circuitry to signal the GFCI's contacts to open. Note that because both normal *load* currents pass through the sensor, no unbalance is present.

Ground-fault circuit breakers break *only* the ungrounded ("hot") conductor. The GFCI circuit breaker looks very similar to a standard circuit breaker. **Figure 11-18** shows a single-pole GFCI circuit breaker and a regular single-pole circuit breaker.

There are two differences between a GFCI circuit breaker and a standard breaker. The first is that the GFCI breaker has a white pigtail attached to it, and the second is the push-to-test button located on the front of the breaker. The GFCI breaker is designed with this test button so the breaker can be tested for correct operation once it has been installed and energized.

FIGURE 11-16   A GFCI receptacle and GFCI circuit breaker.

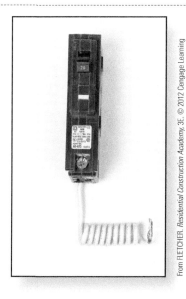

From FLETCHER. *Residential Construction Academy*, 3E. © 2012 Cengage Learning

FIGURE 11-17 Ground-fault circuit-interrupter components and connections.

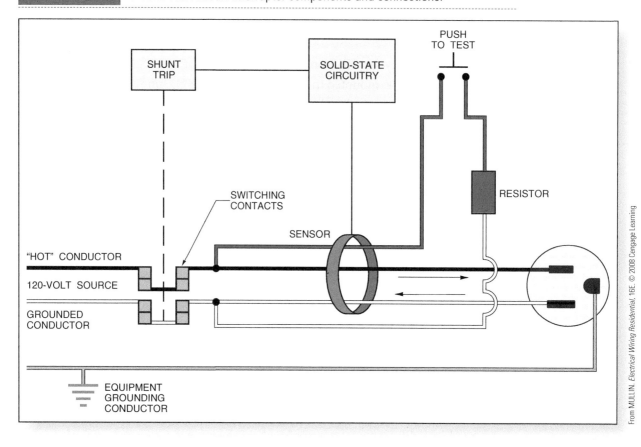

FIGURE 11-18 A GFCI circuit breaker and a standard single-pole circuit breaker.

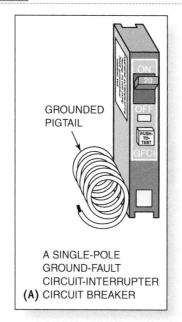

GROUNDED
PIGTAIL

A SINGLE-POLE
GROUND-FAULT
CIRCUIT-INTERRUPTER
(A) CIRCUIT BREAKER

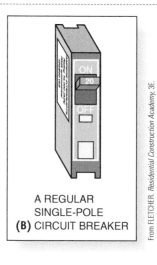

A REGULAR
SINGLE-POLE
(B) CIRCUIT BREAKER

Arc-Fault Circuit Interrupter (AFCI)

According to *Section 210.12*, many 120-volt, 15- and 20-amp branch circuits in newly constructed dwellings must be arc-fault circuit interrupter (AFCI) protected. This comes as a result of a series of studies by the Consumer Product Safety Commission that indicated that a high percentage of the 150,000 residential electrical fires that occur annually in the United States occur because of arc faults.

Arcing in residential wiring applications usually occurs for one of two reasons. The first is an overloaded extension cord or a broken wire in an extension cord or power cord. The overload condition can cause the insulation to break down, creating conditions where arcing between the conductors can occur. Where a conductor is broken, the ends can be close enough that the current can bridge the gap by arcing. In either case, the resulting arc can be at a low enough level that it is not detected by the circuit overcurrent device. The arc, however, is able to ignite flammable materials that it contacts.

The second reason for arcing typically occurs where a plug is partially knocked out of a receptacle, as when tripped over or disturbed by moving furniture. As the plug blades are pulled from the receptacle slot, arcing occurs. The more load on the circuit, the greater the arc. Again, this is a low-level arcing that does *not* necessarily trip the overcurrent device but can still cause a fire.

AFCI devices are designed to trip where they sense rapid fluctuations in the current flow that are typical of arcing conditions. They are set up to recognize the *signature* of dangerous arcs and trip the circuit off where one occurs. AFCIs can distinguish between dangerous arcs and the operational arcs that occur where a plug is inserted or removed from a receptacle or a switch is turned ON or OFF.

NEC requirements tell electricians where to install AFCIs in homes, and UL Standard 1699 specifies the sensing and tripping characteristics of AFCIs.

The following is an overview of the types of AFCIs available:

- Branch/feeder AFCI—a device installed at the origin of a branch circuit or feeder, such as at a load center, to provide protection of the branch circuit wiring, feeder wiring, or both against the unwanted effects of arcing. This device also provides limited protection to branch-circuit extension wiring. It may be a circuit breaker-type device or a device in its own enclosure, mounted at or near a load center. This is the type commonly used in residential wiring.

- Outlet circuit AFCI—an AFCI receptacle. It provides protection for cord sets that are plugged into the receptacle. It is more sensitive than a branch/feeder AFCI. A feed-through AFCI receptacle installed at the first receptacle is permitted by *210.12(B)*. [*Exception:* There is no limitation on the length of the branch-circuit overcurrent device to the first receptacle. However, the branch-circuit wiring method must be RMC, IMC, EMT, or steel armored cable (AC).] Everything downstream from the feed-through AFCI receptacle is AFCI protected.

- Combination AFCI—An AFCI that complies with the requirements for both branch feeder and outlet circuit AFCIs. It is intended to protect downstream branch-circuit wiring, cord sets, and power supply cords.

- Portable AFCI—A plug-in device intended to be connected to a receptacle outlet and provided with one or more outlets. It provides protection to connected cord sets and power-supply cords against the unwanted effects of arcing.

- Cord AFCI—A plug-in device connected to a receptacle outlet to provide protection to the power-supply cord connected to it against the unwanted effects of arcing. The cord may be part of the device. The device has no additional outlets.

LIFE SKILLS

There are numerous types of GFCIs and AFCIs listed in this chapter. It's important that you keep them straight. As you did in the fuse section, use your word processing software to make a reference chart for the AFCI and GFCI listed in the chapter. In one column, list the type of device. In the other, list its key attributes. Having this chart handy is a big help when it comes time to sit down and study. It also helps you build organizational skills that you will use throughout your career.

Self-Check 6

1. What is the purpose of a GFCI?
2. What is the purpose of an AFCI?
3. List four types of GFCIs available.
4. List four types of AFCIs available.

Summary

- The *NEC* requires that the grounded (identified) conductor in ac circuits have an outer finish that is either continuous white or gray, or have an outer finish (not green) that has three continuous white stripes along the conductor's entire length.

- An ungrounded conductor, commonly called the "hot" conductor, must have an outer finish that is a color other than green, white, gray, or with three continuous white stripes.

- The white grounded conductor is commonly referred to as the neutral.

- A receptacle is a device installed in an electrical box that allows an electrician to access current from the wiring system and deliver it through a cord and attachment plug to a piece of equipment.

- An outlet is the point on the wiring system at which current is taken to supply electrical equipment.

- A duplex receptacle is the most common receptacle type used in residential wiring; it has two receptacles on one strap; each receptacle is capable of providing power to a cord-and-plug-connected electrical load.

- The following type switches are most commonly used in residential wiring: single-pole, double-pole, three-way, and four-way.
- A dimmer switch is a switch type that raises or lowers the lamp brightness of a luminaire.
- Some compact fluorescent lamps are available that have an integrated circuit designed right into the lamp, which allows the fluorescent lamp to be controlled by a standard incandescent lamp dimmer.
- A fuse is an overcurrent protection device (OCPD) that opens a circuit where the fusible link is melted away by the extreme heat caused by an overcurrent.
- Fuses are designed and come in two basic styles: plug and cartridge.
- A circuit breaker is a device designed to open and close a circuit manually and to open the circuit automatically on a predetermined overcurrent without damage to itself where properly applied within its rating.
- On a continuous overload, a bimetallic element in the thermal-magnetic circuit breaker moves until it unlatches the inner tripping mechanism of the breaker.
- A ground-fault circuit interrupter (GFCI) is a device that protects people from dangerous levels of electrical current by measuring the current difference between two conductors of an electric circuit and tripping to an open position if the measured value exceeds approximately 5 milliamperes.
- An arc-fault circuit interrupter (AFCI) is a device intended to provide protection from the effects of arc faults by recognizing characteristics unique to arcing and by functioning to de-energize the circuit where an arc fault is detected.
- Ground-fault circuit breakers break *only* the ungrounded ("hot") conductor.

Review Questions

True/False

1. The *NEC* requires that the ungrounded ("hot") conductor have an outer finish that is either continuous white or gray. (True, False)

2. Ungrounded ("hot") conductors that are 1 AWG or larger must be black. (True, False)

3. An outlet is the point on the wiring system at which current is taken to supply electrical equipment. (True, False)

4. The *NEC* defines a receptacle as a strap (yoke) with one, two, or three contact devices. (True, False)

5. A duplex receptacle is the most common receptacle type used in residential wiring. (True, False)

Multiple Choice

6. On a duplex receptacle, what color is the screw used for terminating the circuit bare or green grounding conductor?
A. Silver
B. Brass-colored
C. Black
D. Green

7. Which of the following information or features is *not* found on the back of a duplex receptacle?
A. Conductor size
B. Ratings
C. Strip gauge
D. Push-in terminals

8. What type switch is the most commonly used in residential wiring?
A. Single-pole
B. Double-pole
C. Three-way
D. Four-way

9. What type switch is used to control two separate 120-volt circuits or one 240-volt circuit from one location?
A. Single-pole
B. Double-pole
C. Three-way
D. Four-way

10. What type switch is used to control a 120-volt lighting load from two locations?
A. Single-pole
B. Double-pole
C. Three-way
D. Four-way

Fill in the Blank

11. A(n) _____ switch does not have any ON or OFF position listed on it.

12. A(n) _____ switch is a switch type that, where used in conjunction with two 3-way switches, allows control of a 120-volt lighting load from more than two locations.

13. Four-way switch terminals are also called _____ terminals.

14. A(n) _____ switch is a switch type that raises or lowers the lamp brightness of a luminaire.

15. A(n) _____ compact fluorescent lamp can be used to replace a standard incandescent lamp of the medium screw shell base type.

16. A(n) _____ is an overcurrent protection device that opens a circuit where the fusible link is melted away by the extreme heat caused by an overcurrent.

17. Plug-type fuses are available in an Edison-base and a Type _____ model.

18. Dual-element, _____ fuses are the recommended protection device for motor circuits because they do not open on momentary overloads.

19. A circuit breaker is a device designed to open the circuit _____ on a predetermined overcurrent, without damage to itself where properly applied within its rating.

20. Residential installations typically use _____ circuit breakers.

21. A GFCI is a device that protects people from dangerous levels of electrical current by measuring the current _____ between two conductors of an electric circuit and tripping to an open position if the measured value exceeds 5 milliamps.

22. An AFCI is a device intended to provide protection from the effects of arc faults by recognizing characteristics unique to arcing and by functioning to _____ the circuit where an arc fault is detected.

23. With a GFCI receptacle, only that receptacle location and any _____ receptacles are GFCI protected.

24. Ground-fault circuit breakers break only the _____ conductor.

25. AFCI devices are set up to recognize the _____ of dangerous arcs and trip the circuit off where one occurs.

ACTIVITY 11-1 Receptacle Identification

Matching

Match each lettered item in **Figure 11-19** with its description in the numbered list.

_____ 1. brass "hot" terminal screw

_____ 2. UL label

_____ 3. grounding contact slot

_____ 4. ungrounded "hot" contact slot

_____ 5. silver "grounded conductor" terminal screw

_____ 6. amperage rating

_____ 7. mounting strap

_____ 8. grounding terminal screw

_____ 9. grounded contact slot

_____ 10. 6-32 mounting screw

_____ 11. connecting tabs

_____ 12. voltage rating

FIGURE 11-19 Receptacle.

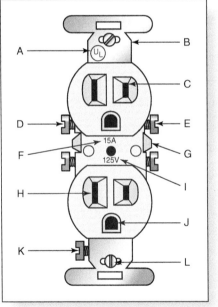

From FLETCHER. *Residential Construction Academy, 3E.* © 2012 Cengage Learning

ACTIVITY 11-2 Single-Pole Switch Identification

Matching

Match each lettered item in **Figure 11-20** with its description in the numbered list.

_____ 1. terminal screws

_____ 2. mounting strap or yoke

_____ 3. toggle

_____ 4. 6-32 mounting screw

_____ 5. grounding terminal screw

_____ 6. mounting ears

_____ 7. 6-32 mounting screw

FIGURE 11-20 Single-pole switch.

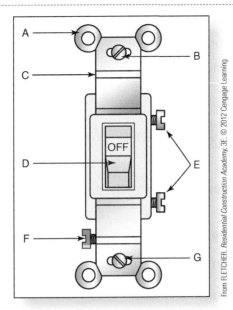

From FLETCHER. *Residential Construction Academy, 3E.* © 2012 Cengage Learning

ACTIVITY 11-3 Double-Pole Switch Identification

Matching

Match each lettered item in **Figure 11-21** with its description in the numbered list.

_____	1. mounting ears		_____	4. grounding terminal screw
_____	2. mounting ears		_____	5. line terminal screw
_____	3. line terminal screw		_____	6. load terminal screw

FIGURE 11-21 Double-pole switch.

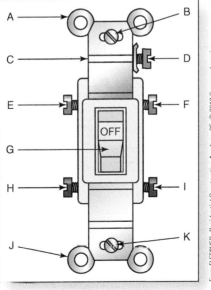

From FLETCHER. *Residential Construction Academy, 3E.* © 2012 Cengage Learning

(Continues)

(CONTINUED)

_____ 7. 6-32 mounting screw   _____ 10. toggle

_____ 8. mounting strap or yoke   _____ 11. 6-32 mounting screw

_____ 9. load terminal screw

ACTIVITY 11-4   Three-Way Switch Identification

Matching

Place each letter in **Figure 11-22** with the appropriate markings or terminology in the numbered list .

_____ 1. Traveler terminal screw   _____ 5. Toggle

_____ 2. Grounding terminal screw   _____ 6. 6-32 mounting screw

_____ 3. Common terminal screw   _____ 7. Traveler terminal screw

_____ 4. 6-32 mounting screw

FIGURE 11-22 Three-way switch.

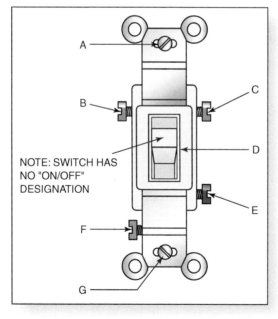

NOTE: SWITCH HAS NO "ON/OFF" DESIGNATION

From FLETCHER, *Residential Construction Academy*, 3E. © 2012 Cengage Learning

ACTIVITY 11-5 Four-Way Switch Identification

Matching

Place each letter in **Figure 11-23** with the appropriate markings or terminology in the numbered list.

_____ 1. Traveler terminal screw

_____ 2. Traveler terminal screw

_____ 3. 6-32 mounting screw

_____ 4. 6-32 mounting screw

_____ 5. Traveler terminal screw

_____ 6. Traveler terminal screw

_____ 7. Grounding terminal screw

_____ 8. Toggle

FIGURE 11-23 Four-way switch.

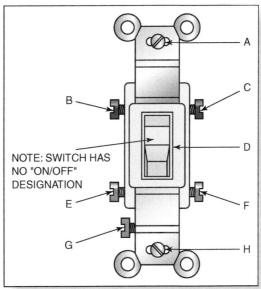

NOTE: SWITCH HAS NO "ON/OFF" DESIGNATION

From FLETCHER. *Residential Construction Academy*, 3E. © 2012 Cengage Learning

Chapter 12

Wiring Methods

Career Profile

In his current position as Human Resources Director, **Greg Anderson, Rex Moore Electrical Contractors & Engineers**, is "responsible for all matters involving workforce planning and development, human resource development, compensation and rewards, employee and labor relations and risk management." He finds this work rewarding because his "duties vary from day to day and season to season." Anderson credits his apprenticeship program with launching his professional success: "I was always good with mechanical tasks and working with my hands. After graduating from high school, I entered the electrical trade through a friend of the family. That led to completing the apprenticeship program, teaching apprenticeship and journey-level training programs, and ultimately [to a job] in executive management for one of the largest contractors in the nation." Greg thus encourages current apprenticeship students to "stick with [the program]." He also advises, "Don't ever think you know it all, but don't be afraid to take the risk of showing up and giving it your all."

Chapter Outline

NEC REQUIREMENTS FOR THE INSTALLATION OF NMC

NEC REQUIREMENTS FOR THE INSTALLATION OF MC CABLE

NEC REQUIREMENTS FOR THE INSTALLATION OF UF CABLE

NEC REQUIREMENTS FOR THE INSTALLATION OF EMT

SWITCHES—WIRING METHODS

NEC REQUIREMENTS FOR REPLACING EXISTING GROUNDED
AND UNGROUNDED RECEPTACLES

FIVE TYPES OF CIRCUIT CONDITIONS

Key Terms

Armored cable, Type AC

Cable

Combination switch

Impedance

Knob-and-tube wiring

Metal clad cable, Type MC

Nonmetallic-sheathed cable, Type NMC

Normal loading

Open circuit

Overload

Raceway

Service mast

Short circuit

Snap switch

Underground feeder cable, Type UF

Chapter Objectives

After completing this chapter, you will be able to:

1. Describe *NEC* requirements for the installation of NMC.
2. Describe *NEC* requirements for the installation of MC cable.
3. Describe *NEC* requirements for the installation of UF cable.
4. Describe *NEC* requirements for the installation of EMT.
5. Given a wiring method, identify the correct wiring connections for single-pole, three-way, and four-way switching as per *NEC* requirements.
6. List *NEC* requirements for replacing existing ungrounded receptacles.
7. Describe the five types of circuit conditions.

Life Skills Covered

 Communication Skills

 Cooperation and Teamwork

Life Skill Goals

A quick scan of this chapter's section heading reveals that wiring methods are all based on *NEC*® requirements. The life skills that are covered in this chapter, Communication Skills and Cooperation and Teamwork, are designed to help you gain an understanding of what is necessary to understand and work from these requirements.

As you go through the chapter, find a partner or two to work with. Get together after class for a few minutes and talk about the different aspects of the various *NEC* requirements. How does the information covered affect what you will be doing in the field?

As an electrician, you must be able to recognize types of cables and know when and where to use them. This chapter describes these cables as well as introduces enclosed channels, referred to as raceways, for holding wires and cables. Wiring three- and four-way switches is presented, along with replacing receptacles, and this chapter describes the five possible circuit conditions. Read and learn that in addition to normal loading of a circuit, other conditions may exist that could cause any of the following: overload, short circuit, ground fault, or open circuit. As an electrician, you are required to recognize these conditions and *know what to do* to correct them!

Raceway an enclosed channel of metal or nonmetallic materials designed expressly for holding wires or cables.

Introduction

Conductors installed for circuits in residential wiring are generally installed as part of a cable. A cable is a factory assembly of two or more insulated conductors that have an outer sheathing that holds everything together; the outside sheathing can be metallic or nonmetallic. The most commonly used cables are: nonmetallic-sheathed, armored cable, metal clad cable, underground feeder cable, and service-entrance. In addition to cable recognition and the ability to know when and where to use different types of cable, you need to know and understand *NEC* installation requirements for commonly used cables.

Cable a factory assembly of two or more insulated conductors that have an outer sheathing that holds everything together; the outside sheathing can be metallic or nonmetallic.

12-1 *NEC* Requirements for the Installation of NMC

Nonmetallic-sheathed cable, Type NMC is typically referred to as Romex by most electricians. Romex is the name first given to NMC by the Rome Wire and Cable Company, and although other companies manufacture NMC, electricians call it Romex. NMC is used more than any other wiring method in residential applications because it is inexpensive to purchase and install.

The *NEC* covers nonmetallic-sheathed cable, Type NMC in *Article 334*. It is defined by the *NEC* as a factory assembly of two or more insulated conductors having an outer sheathing of moisture-resistant, flame-retardant, nonmetallic material. *Article 334* states where NMC *shall be permitted* to be used. Two examples are: in one- and two-family dwellings or cable trays where the cables are identified for the use. This article also lists where NMC *shall not be permitted*. The following are just some of those areas: exposed in dropped or suspended ceilings in other than one- and two-family and multifamily dwellings; as a service-entrance cable; or embedded in poured cement, concrete, or aggregate. **Figure 12-1** shows an example of nonmetallic cable. Note that (A) shows the black ungrounded conductor; (B) designates the bare equipment grounding conductor; and (C) is the white grounded conductor.

Nonmetallic-sheathed cable, Type NMC a factory assembly of two or more insulated conductors having an outer sheathing of moisture-resistant, flame-retardant, nonmetallic material.

FIGURE 12-1   A typical nonmetallic-sheathed cable, Type NMC.

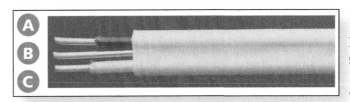

Courtesy of Southwire Company

NMC is available with two or three conductors and comes in sizes from 14 through 2 AWG copper, and 12 through 2 AWG aluminum. Two-wire cable has a black and a white insulated conductor along with a bare grounding conductor. The 3-wire cable has a black, a red, and a white conductor along with a bare equipment grounding conductor. It is used in residential applications for general-purpose branch circuits, small-appliance branch circuits, and individual branch circuits. *Table 310.16* provides the ampacity listing of NMC in the *60° Celsius* column.

UL 719 lists two types of nonmetallic cable, and the *NEC* covers three types of NMC:

- Type NM-B is by far the most commonly used type and has a flame-retardant, moisture-resistant, nonmetallic outer jacket. It can be used in dry locations only. The conductor insulation is rated at 90°C. However, the ampacity of the NM-B is based on 60°C.

- Type NMC-B is not used often in residential work. It has a flame-retardant and moisture-, fungus-, and corrosion-resistant nonmetallic outer jacket and can be used in dry or damp locations. The conductor insulation and ampacity are rated the same as the Type NM-B cable.

- Type NMS-B is used in new homes that have home automation systems using the latest technology. Power conductors, telephone wires, coaxial cable for video, and other data conductors are all contained in the same cable. It has a moisture-resistant, flame-retardant, nonmetallic outer jacket.

LIFE SKILLS

Among electricians, NMC is commonly referred to as Romex instead of by its proper name. Are there any other tools or pieces of equipment that have nicknames? Why is it important to know the real name and the nickname?

Self-Check 1

1. What do most electricians call nonmetallic-sheathed cable, Type NMC?
2. Describe NMC.

12-2. *NEC* Requirements for the Installation of MC Cable

The authority having jurisdiction (AHJ) may not allow nonmetallic-sheathed cable to be used in residential construction in some locations throughout the United States. In those cases, the wiring method most often used is armored or metal clad cable. These cables are shown in **Figure 12-2**. Both of these cables have a metal outer sheathing and provide very high levels of physical protection for the conductors in the cable.

Armored cable has been around for a long time and has a proven track record. Electricians usually refer to it as *BX* cable. This is a trademark owned by General Electric Company that has become a generic term for *any* company's armored cable.

Armored cable, Type AC is covered in *Article 320* of the *NEC*. Type AC cable is defined as a fabricated assembly of insulated conductors in a flexible metallic enclosure. *Article 320* states where AC *shall be permitted* to be used, for example, for feeders and branch circuits both exposed and concealed, in cable trays, and in dry locations. This article also lists where AC *shall not be permitted*. The following are just a few of those areas: where subject to physical damage, in damp or wet locations, or embedded in plaster finish on brick or other masonry in damp or wet locations. Type AC cable comes with two, three, or four conductors. To ensure *electrical continuity* of the outside flexible metal sheathing, a small aluminum bonding wire is included. This bonding wire permits the outside metal sheathing to be used as the grounding conductor. Armored cable is also available with a green insulated grounding conductor. Type AC comes in wire sizes 14 through 1 AWG copper, and 12 through 1 AWG aluminum. Like Romex, Type AC is found in the *60° Celsius* column of *Table 310.16*.

Metal clad cable is very similar in appearance to armored cable. There are, however, several differences. *Article 330* of the *NEC* covers metal clad cable, Type MC and defines it as a factory assembly of one or more insulated circuit conductors enclosed in an armor of interlocking metal tape or a smooth or

Armored cable, Type AC a fabricated assembly of insulated conductors in a flexible metallic enclosure.

Metal clad cable, Type MC a factory assembly of one or more insulated circuit conductors enclosed in an armor of interlocking metal tape or a smooth or corrugated metallic sheath.

FIGURE 12-2   Armored cable (top) and metal clad cable cable (bottom).

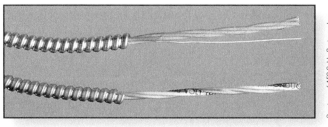

Courtesy of AFC Cable Systems, Inc.

corrugated metallic sheath. *Article 330* states where MC *shall be permitted* to be used, for example, for services, feeders, and branch circuits; for power, lighting, control, and signal circuits; indoors or outdoors; or in any raceway. This article also lists where MC *shall not be permitted:* where subject to physical damage or where exposedto any destructive corrosive conditions. In metal clad cable, the number of conductors is unlimited. Type MC comes in wire sizes from 18 AWG through 2000 kcmil copper, and 12 AWG through 2000 kcmil aluminum. With Type MC, the outer sheathing cannot be used as a grounding conductor unless it is specifically listed as such. A green insulated grounding conductor is always included in the cable assembly. The ampacity for Type MC is found using the *60° Celsius* column of *Table 310.16*; however, for 1/0 AWG and larger conductors, the *75° Celsius* column can be used.

Types AC and MC cables are very similar, and there is no writing to help identify these cable types on the metal sheathing. To help you to distinguish one from the other:

- Look at how the conductors are wrapped. Type AC cable has a light brown paper covering each conductor, whereas Type MC has *no* individual wrapping on the conductors.

- There is a polyester tape over *all* the conductors in Type MC cable, and as you have already learned, Type MC has *no* aluminum bonding wire in it.

TRADE TIP A new style of Type MC cable was introduced in 2006. This cable is manufactured with an internal aluminum bonding wire very similar to Type AC cable. Because it contains the aluminum bonding wire along its entire length, the outer metal sheathing is *listed* as an acceptable grounding means. This new style of Type MC is expected to become the standard style used in residential wiring situations where metal-sheathed cable is used.

LIFE SKILLS

With a partner, discuss the requirements for the installation of MC cable. What are two or three important things to remember?

Self-Check 2

1. What is another name commonly used for armored cable?
2. What wire sizes are available for Type AC cable?
3. List two differences between Type AC and Type MC cable.

12-3 . *NEC* Requirements for the Installation of UF Cable

Underground feeder cable, Type UF a listed factory assembly of one or more insulated conductors with an integral or an overall covering of nonmetallic material suitable for direct burial in the earth.

Article 340 of the *NEC* covers underground feeder cable, Type UF cable. *Article 340* states where UF *shall be permitted* to be used. Some examples are listed here: for underground, including direct burial in the earth; for wiring in wet, dry, or corrosive locations under the recognized wiring methods of the *NEC*; or supported by cable trays, but shall be the multiconductor type. This article also lists where UF *shall not be permitted.* Several examples of those areas are as service-entrance cable; in commercial garages; where exposed to direct rays of the sun, unless identified as sunlight resistant; or where subject to physical damage. **Figure 12-3** shows underground feeder cable, Type UF.

Underground feeder cable, Type UF is used for underground installations of branch circuits and feeder circuits and must be marked "Underground Feeder" cable. A UF cable is a listed factory assembly of one or more insulated conductors with an integral or an overall covering of nonmetallic material suitable for direct burial in the earth. It can be used in interior installations but must be installed by the same methods as nonmetallic-sheathed cable. Type UF cable is available in sizes from 14 AWG through 4/0 AWG (copper conductors) and from 12 AWG through 4/0 AWG (aluminum conductors). The current-carrying capacity (ampacity) of Type UF cable is found in *Table 310.16*, using the *60° Celsius* column. This cable must be buried according to *NEC Table 300.5*. Type UF cable may be used in direct exposure to the sun if the cable is a type *listed for sunlight resistance* or is *listed and marked for sunlight resistance.*

LIFE SKILLS

With a partner, discuss the requirements for the installation of UF cable. How do they compare to the requirements for MC cable?

Self-Check 3

1. Describe UF cable.
2. What *NEC* table covers UF burial requirements?

FIGURE 12-3 Underground feeder cable.

Courtesy of Southwire Company

12-4 NEC Requirements for the Installation of EMT

Before electrical metallic tubing (EMT) and related *NEC* requirements are described, the following section briefly describes several raceways.

Raceways

The *NEC* defines a raceway as an enclosed channel of metal or nonmetallic materials designed expressly for holding wires or cables. The following raceways are commonly used in residential wiring.

- *Rigid metal conduit (RMC)* is typically made of steel and is galvanized to enable it to resist rusting. It is sometimes referred to as *heavywall* conduit. RMC is often used as a **service mast** for a service entrance. A service mast is a piece of rigid metal conduit or intermediate metal conduit, usually 2 or 2½ inches (50.8 or 63.5 mm) in diameter, that provides service conductor protection and the proper height requirements for service drops. **Figure 12-4** shows an example of RMC with some associated fittings.

- *Intermediate metal conduit (IMC)* is a lighter version of RMC but can be used in all the locations that the heavier RMC may be used. Because it is lighter in weight, IMC is easier to handle during installation than RMC. **Figure 12-5** shows intermediate metal conduit (IMC) and associated fittings.

- *Rigid nonmetallic polyvinyl chloride conduit (PVC)* is an inexpensive conduit that can be used in a variety of residential wiring applications. There are two types of PVC conduit that electricians generally use: Schedule 40 has a heavy-duty wall thickness, and Schedule 80 has an extra-heavy-duty wall thickness. **Figure 12-6** shows rigid nonmetallic polyvinyl chloride conduit (PVC).

Service mast a piece of rigid metal conduit or intermediate metal conduit, usually 2 or 2½ inches in diameter, that provides service conductor protection and the proper height requirements for service drops.

FIGURE 12-4   Rigid metal conduit (RMC) and associated fittings: (A) one-hole strap, (B) conduit hanger, (C) Erickson coupling, (D) insulated bonding bushing, and (E) coupling.

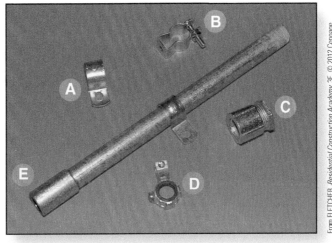

From FLETCHER. *Residential Construction Academy*, 3E. © 2012 Cengage Learning

FIGURE 12-5 Intermediate metal conduit (IMC) and associated fittings: (A) one-hole strap and (B) coupling.

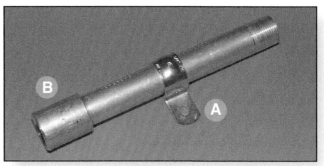

FIGURE 12-6 PVC and associated fittings: (A) integral coupling, (B) separate coupling, (C) connector, and (D) two-hole PVC strap.

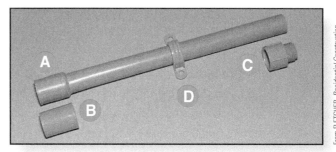

- *Flexible metal conduit (FMC)* is a raceway of circular cross section made of helically wound, formed, interlocked metal strips. The trade name for this raceway type is *Greenfield*. It looks very similar to BX cable but does not have the conductors already installed. It is up to the electrician to install the conductors in FMC. It is very flexible and is often used to connect appliances and other equipment in residential applications. **Figure 12-7** shows an example of FMC and some associated fittings.

FIGURE 12-7 Flexible metal conduit (FMC) and associated fittings: (A) 90° connector, (B) straight connector, and (C) one-hole strap.

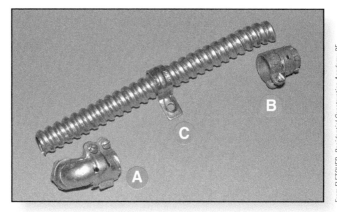

- *Electrical nonmetallic tubing (ENT)* is a flexible nonmetallic raceway that is being used more and more in residential wiring. Electricians often refer to it as *Smurf Tube* because its blue color reminded some electricians of the blue Smurf cartoon characters. **Figure 12-8** shows electrical nonnmetallic tubing (ENT) and associated fittings.

- *Liquidtight flexible metal conduit (LFMC) and liquidtight flexible nonmetallic conduit (LFNC)* are raceway types that are used where flexibility is desired in wet locations, such as outdoors. These raceways are not often used in residential wiring, but as an electrician, you should be able to recognize these raceway types for those times where they are used. A typical application for this wiring method is the connection to a central air-conditioning unit located outside a dwelling. **Figure 12-9** shows a liquidtight flexible metal conduit (LFMC) and associated fittings, and **Figure 12-10** shows a liquidtight flexible nonmetallic conduit (LFNC) and associated fittings.

FIGURE 12-8 ENT and associated fittings: (A) snap-on connector, (B) snap-on coupling, and (C) two-hole strap.

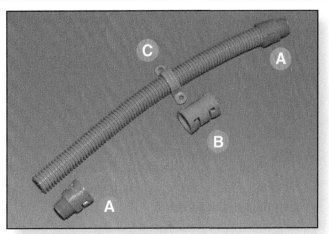

From FLETCHER. *Residential Construction Academy, 3E.* © 2012 Cengage Learning

FIGURE 12-9 LFMC and associated fittings: (A) 90° connector and (B) straight connector.

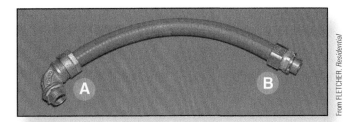

From FLETCHER. *Residential Construction Academy, 3E.* © 2012 Cengage Learning

FIGURE 12-10 LFNC and associated fittings: (A) straight connector and (B) 90° connector.

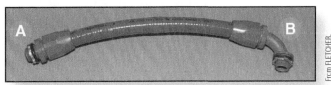

From FLETCHER. *Residential Construction Academy, 3E.* © 2012 Cengage Learning

Electrical Metallic Tubing (EMT)

Electrical metallic tubing (EMT) is another type of raceway, often referred to as *thinwall* conduit because of its very thin walls. **Figure 12-11** shows electrical metallic tubing (EMT).

Installation requirements for EMT are covered in *Article 358* of the *NEC*, which states where EMT *shall be permitted* to be used. These are some examples: both exposed and concealed; installed in concrete, in direct contact with the earth, or in areas subject to severe corrosive influences where protected by corrosive protection; or wet locations. This article also lists where EMT *shall not be permitted.* The following are several examples of those areas: where during installation or afterward it is subject to severe physical damage; where protected from corrosion solely by enamel; or for the support of luminaires or other equipment except conduit bodies no larger than the largest trade size of the tubing. EMT is easier to install than both rigid metal conduit (RMC) and intermediate metal conduit (IMC) because of its light weight. As an example, for size ½ inch (12.7 mm), 100 feet (30.4 m), the approximate weight is 30 pounds (13.6 kg). Electrical metallic tubing cannot be threaded. Fittings for EMT are set screw type or threadless compression. This type conduit is considered to be an equipment grounding conductor. EMT is available in sizes of ½ inch through 4 inches (12.7 mm through 101.6 mm) (galvanized steel) and 2 inches through 4 inches (50.8 mm through 101.6 mm) (aluminum) and comes in standard 10-foot (3 m) lengths.

LIFE SKILLS

Before you can understand the requirements for installing EMT, you must know about raceways. Find a partner and quiz each other on the specifics of the raceways that are identified in this section.

FIGURE 12-11 EMT and associated fittings: (A) compression connector, (B) compression coupling, (C) set-screw connector, (D) set-screw coupling, and (E) one-hole strap.

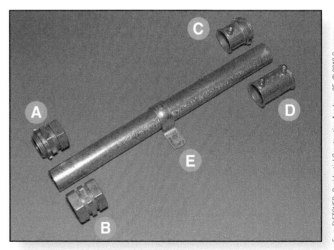

Self-Check 4

1. What is a raceway?
2. List three raceways commonly used in residential wiring.
3. Describe electrical metallic tubing and list its advantages as a raceway.

12-5 ■ Switches—Wiring Methods

Before describing how single-pole, three-way, and four-way switches are wired, this section provides a review of switch types.

Types of Switches

- Single-pole is used in 120-volt residential circuits to control a lighting outlet load from one specific location. This switch is the most commonly used type in residential wiring applications. **Figure 12-12** shows a single-pole switch.

- Double-pole is used to control a 240-volt load, such as a water heater, in a residential wiring system. Like the single-pole switch, the double-pole switch can control the load from only one specific location. **Figure 12-13** shows a double-pole switch. Remember, it is constructed much like a single-pole switch but has *four* terminal screws rather than two.

- Three-way switches are used to control lighting outlet loads from two separate locations, for example, at the top and bottom of a stairway. **Figure 12-14** shows a three-way switch. Recall that the three-way switch has *three* screw terminals and a grounding terminal.

FIGURE 12-12	Single-pole switch.

Courtesy of Hubbell Incorporated

FIGURE 12-13   Double-pole switch.

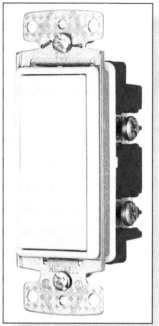

Courtesy of Hubbell Incorporated

FIGURE 12-14   Three-way switch.

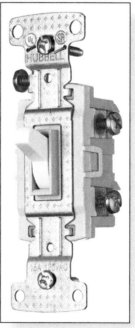

Courtesy of Hubbell Incorporated

- Four-way switches are used in combination with three-way switches; they allow for control of lighting outlet loads from three or more locations, such as in a room with three doorways, where the wiring plan calls for a switch controlling the room lighting to be located at each doorway. **Figure 12-15** shows a four-way switch. Note that the four-way switch looks almost identical to the double-pole switch.

FIGURE 12-15 Four-way switch.

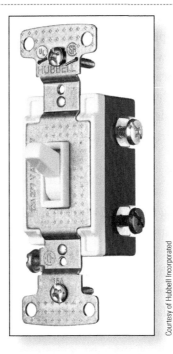

Courtesy of Hubbell Incorporated

However, like three-way switches, there is *no* ON or OFF designation listed on the toggle.

- A dimmer can be found in both single-pole and three-way configurations and is used to brighten or dim a lamp or lamps in a luminaire. **Figure 12-16** shows examples of dimmer switches. In this figure, note that dimmer switches have insulated pigtail wires rather than terminal screws.

FIGURE 12-16 Dimmer switches.

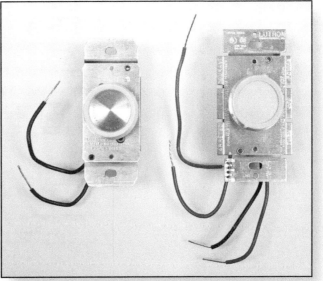

From FLETCHER, *Residential Construction Academy*, 3E. © 2012 Cengage Learning

Combination
switch a device
that consists of
more than one
switch type on the
same strap or yoke.

• A combination switch is a device that consists of more than one switch type on the same strap or yoke. **Figure 12-17** shows a combination switch. This allows the placement of multiple switches in a single-gang electrical box. For example, the wiring plans may call for two switches, a single-pole and a three-way switch, at a location next to a doorway. If it is found that there is not enough space between the building framing members to attach an electrical box, an alternative is to use a single-gang box and install a combination switch that has a single-pole switch and a three-way switch on the same strap.

NEC Requirements for Three- and Four-Way Switches

According to *404.15(A)* of the *NEC*, each switch must be marked with the current and voltage rating for the switch. The switch must be matched to the voltage and current for the circuit you are using the switch on. For example, a 20-amp-rated, 120-volt circuit wired with 12 AWG conductors having 16 amps of lighting load must have a switch with a voltage rating of at least 120 volts and a current rating of 20 amps (a standard switch rating). Many residential lighting circuits are wired with 14 AWG conductors protected with a 15-amp circuit breaker, and require switches with a 15-amp, 120-volt rating. This switch rating is the most common found in residential wiring.

NEC 404.2(A) requires three- and four-way switches to be wired so that all switching is done in the *ungrounded* circuit conductor. Single-pole switches are also wired to switch the ungrounded circuit conductor. *NEC 404.2(B)* states that switches *must not* disconnect the grounded conductor of a circuit.

FIGURE 12-17 Combination switch.

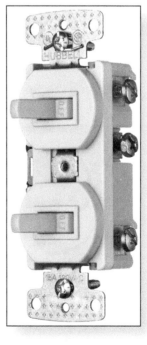

Courtesy of Hubbell Incorporated

TRADE TIP There is no need to ever connect a white insulated grounded conductor to any switch in a residential switching circuit.

Figure 12-18 illustrates how switching must be done in the ungrounded circuit connector.

NEC 404.2(B) actually says that switches must not disconnect the grounded conductor of a circuit. There is no need to ever connect a white insulated ground conductor to any switch in a residential switching circuit.

The *NEC* and Underwriters Laboratories refer to the switches used to control lighting outlets in residential applications as snap switches. A snap switch is a manually operated switch used in interior electric wiring and is usually used for the control of lighting or small motors. However, most electricians refer to them as *toggle switches* or simply as *switches*. Switches that are used to control lighting circuits are classified as *general-use snap switches*, and the requirements for these switches are the same for both the *NEC* and Underwriters Laboratories. The requirements are found in *Article 404* of the *NEC*.

Snap switch a manually operated switch used in interior electric wiring; usually used for the control of lighting or small motors.

Installing Single-Pole Switches

The most common switching used in residential wiring is single-pole switching. On a single-pole switch, two wires are connected to the two terminal screws on the switch. Both of these wires are considered "hot" ungrounded conductors. However, *only* the

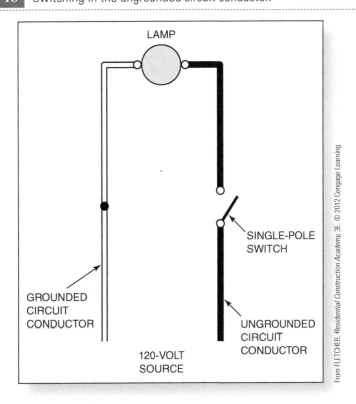

FIGURE 12-18 Switching in the ungrounded circuit conductor.

From FLETCHER. *Residential Construction Academy*, 3E. © 2012 Cengage Learning

conductor that feeds the switch is considered "hot." The conductor that goes from the switch to the lighting outlet is referred to as the *switch loop* or *switch leg*. The operation of a single-pole switch is shown in **Figure 12-19**.

In Figure 12-19(A), the toggle is in the OFF (down) position. Current cannot flow from one terminal to the other. Note that the *load* is *de-energized*. In (B), the toggle is in the ON (up) position, and now the internal contacts are closed. Current can flow from one terminal to the other. Now, the *load* is *energized*.

The following single-pole switching application uses nonmetallic-sheathed cable with nonmetallic boxes. Type NM and nonmetallic boxes are described here because they are most commonly used in residential wiring applications.

The simplest and most often used single-pole switching wiring configuration is where the electrical power feed is brought to the switch and then the cable continues on to the lighting load.

The following is the recommended procedure where installing a single-pole switch for a lighting load with the power source feeding the switch. **Figure 12-20(A)** illustrates a cabling diagram, and **(B)** shows a wiring diagram for this switching circuit.

Always wear safety glasses and other required PPE and observe all applicable safety rules.

FIGURE 12-19 Single-pole switch operation.

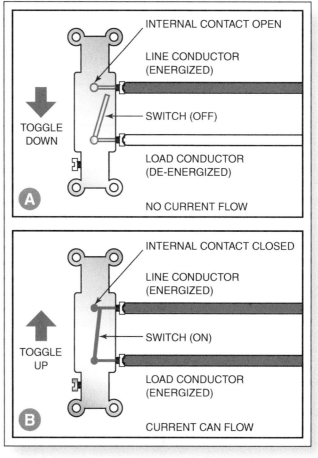

From FLETCHER. *Residential Construction Academy*, 3E. © 2012 Cengage Learning

Single-pole switching circuit controlling a lighting load with power source feeding the switch.

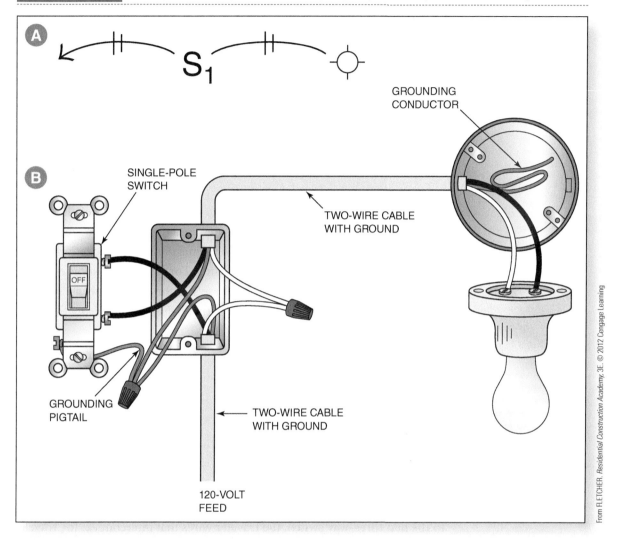

from FLETCHER. *Residential Construction Academy*, 3E. © 2012 Cengage Learning

At the <u>luminaire</u> box:

1. Connect the bare grounding conductors to the grounding connection of the luminaire. If the luminaire does not have any metal parts that must be grounded, simply fold the grounding conductor into the back of the box. Recall that it is *not* necessary to connect the circuit grounding wire to a nonmetallic box.

2. Connect the white insulated grounded conductor to the silver screw terminal or wire identified as the grounded wire on the luminaire.

3. Connect the black "hot" ungrounded conductor to the brass screw or a wire identified as the ungrounded conductor on the luminaire.

   At the switch box:

4. Connect the bare grounding conductors together, and using a bare pigtail, connect all the grounding conductors to the green grounding screw on the switch. Recall that a nonmetallic box does not require the circuit-grounding conductor to be connected to it.

5. Connect the white insulated grounded conductors together using a wirenut.

6. Connect the black "hot" conductor in the incoming power source cable to one of the terminal screws on the single-pole switch. It doesn't matter which terminal screw.

7. Connect the black "hot" conductor in the 2-wire cable (switch leg) going to the lighting outlet to the other terminal screw on the single-pole switch.

Installing Three-Way Switches

Three-way switches are used to control a lighting load from two different locations. You should recall that three-way switches have three terminal screws on them. One of the screws is called the *common* terminal and is colored black. The other two terminal screws are *both* the same color and are typically brass or bronze. These two terminals are called the *traveler* terminals. Some inexperienced electricians may find the connections for three-way switches confusing. For that matter, some *experienced* electricians may get confused when wiring these switches if it has been a while since they last wired one. There are some basic rules that help to make wiring three-way switches much easier:

- Install three-way switches in pairs. It is impossible to have a three-way switching circuit with only *one* three-way switch!

- Install a three-way *cable* between the two 3-way switches. Where using conduit, pull the three separate wires through the conduit between the two 3-way switches.

- Attach the black wire to the *common* terminals. Each three-way switch has a black "hot" feed conductor attached to it. The black insulated conductor goes to the lighting load attached to it.

- Where nonmetallic cable is used *and* where the power source feed is brought to the first three-way switch, the traveler wires that interconnect the traveler terminals of both switches are black and red in color.

- Where nonmetallic cable is used *and* where the power source feed is brought to the lighting outlet first, the traveler wires are red and white in color. The white traveler conductors have to be re-identified with black tape at each switch location.

- There is no marking on the ON or OFF position of the toggle on a three-way switch, so it does not make any difference which way it is positioned in the electrical device box.

The operation of three-way switches differs somewhat from that of single-pole switches. Single-pole switches either allow current flow through the switch and to the load or they do not. The internal design of a three-way switch always allows current to flow from the common terminal to either of the two traveler terminals. **Figure 12-21** shows how a three-way switch operates.

Refer to Figure 12-21(A). Note that a three-way switch has *two* traveler terminals and *one* common terminal. Figure 12-21(B) shows the switch toggle making a connection from the common terminal to one of the traveler terminals. Figure 12-21(C) shows the switch toggle in the opposite direction making a connection from the common terminal

FIGURE 12-21 The operation of a three-way switch.

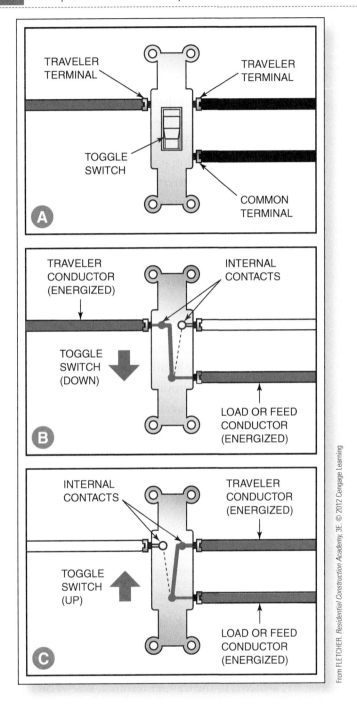

From FLETCHER. *Residential Construction Academy*, 3E. © 2012 Cengage Learning

to the other traveler terminal. In a three-way switch, there is always one *set* of contacts that are closed.

There are several three-way switch configurations, and where the power source is fed determines which one is used. The following application is probably the simplest of the three-way switching circuits. In this application, assume the use of nonmetallic-sheathed cable with nonmetallic boxes. Type NM and nonmetallic boxes are used in our application because they are most commonly used in residential wiring.

The following is the recommended procedure where installing a three-way switching circuit where the source of power comes into the first switch and is routed through it to the second switch (for the dual control) and then to the light. This configuration is sometimes called the *switch-switch-light* configuration. **Figure 12-22(A)** illustrates a cabling diagram and **(B)** a wiring diagram for a switching circuit that has two 3-way switches controlling a lighting load. The power source is feeding the first three-way switch location.

Note that as in *all* three-way switching circuits, the Neutral goes straight from the light to ground, although it must travel through several splices.

Always wear safety glasses and other required PPE, and observe all applicable safety rules.

At the <u>luminaire</u> box:

1. Connect the bare grounding conductors to the grounding connection of the luminaire. If the luminaire does not have any metal parts that must be grounded, simply fold the grounding conductor into the back of the box. Recall that it is *not* necessary to connect the circuit grounding wire to a nonmetallic box.

2. Connect the white insulated grounded conductor to the silver screw terminal or wire identified as the grounded wire on the luminaire.

3. Connect the black ungrounded conductor to the brass screw or a wire identified as the ungrounded conductor on the luminaire.

At the first three-way switch box:

4. Connect the bare grounding conductors together, and using a bare pigtail, connect all the grounding conductors to the green grounding screw on the switch.

5. Connect the white insulated grounded conductors together using a wirenut.

6. Connect the black "hot" conductor in the incoming power source cable to the common (black or odd color) terminal screw on three-way switch 1.

7. Connect the black and red traveler wires in the 3-wire cable going to three-way switch 2 to the traveler terminal screws on the three-way switch. Red and black traveler conductors can be connected to either traveler terminal.

At the second three-way switch box:

8. Connect the bare grounding conductors together, and using a bare pigtail, connect all the grounding conductors to the green grounding screw on the switch.

9. Connect the white insulated ground conductors together using a wirenut.

10. Connect the black conductor in the 2-wire cable going to the lighting outlet to the common (black or odd color) terminal screw on three-way switch 2.

11. Connect the black and red traveler wires in the three-way cable coming from three-way switch 1 to the traveler terminal screws on three-way switch 2. Again, red and black traveler conductors can be connected to either traveler terminal.

FIGURE 12-22 A switching circuit: (A) cabling diagram and (B) wiring diagram.

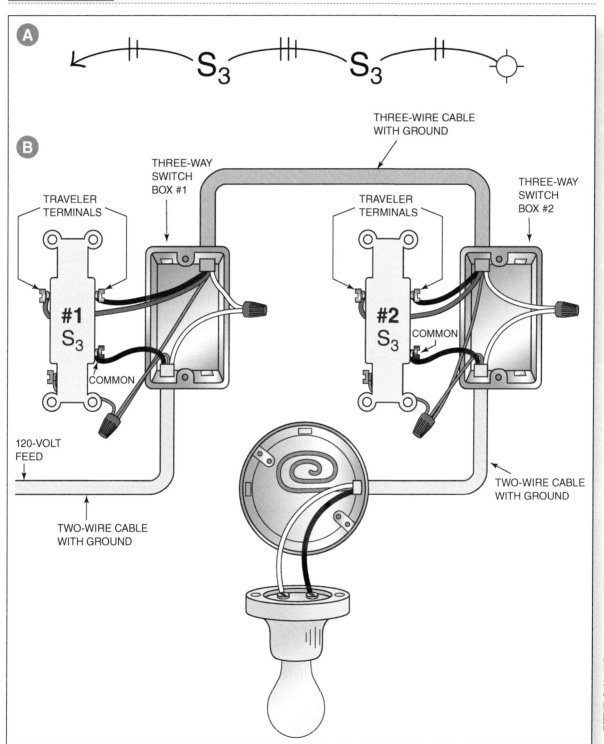

A

S_3 S_3

B

THREE-WIRE CABLE
WITH GROUND

THREE-WAY
SWITCH
BOX #1

THREE-WAY
SWITCH
BOX #2

TRAVELER
TERMINALS

TRAVELER
TERMINALS

#1
S_3

#2
S_3

COMMON

COMMON

120-VOLT
FEED

TWO-WIRE CABLE
WITH GROUND

TWO-WIRE CABLE
WITH GROUND

TRADE TIP *Before* you cut the three-way switch wire, typically called *three-way cable with ground*, measure carefully and cut accurately. It is better to have *more* than you need. You can always shorten the wires. Splicing another wire to make it longer is *not* recommended as it requires another accessible electrical box.

Installing Four-Way Switches

In order to control a lighting load from *more than two* locations, *four-way* switches are used along with three-way switches. Recall that four-way switches have four terminal screws on them. These four terminals are called *traveler* terminals. The four traveler screws are divided into two pairs. Each pair is the same color. One pair is usually brass or bronze, and the other pair is some other color like black. Before trying to understand the operation of a four-way connection, you must first master the common three-way switching circuit. Why? Because four-way switches are simply inserted into the wiring between three-way switches. Traveler wires from each three-way switch used are connected to the four-way switch screw terminals.

The following guidelines should be followed where wiring switching circuits with four-way switches:

- Four-way switches must always be installed between two 3-way switches. You do not have a properly operating circuit with only *one* four-way switch or a switching circuit with only four-way switches. As an example, if you wire a switching application that requires three switching locations for the same lighting load, you need two 3-way switches and one 4-way switch.

- Install a 3-wire cable between *all* four-way and three-way switches. If you are wiring with conduit, three separate wires must be pulled into the conduit between all the four-ways and three-ways.

- If you use nonmetallic cable, and the power source feed is brought to the first three-way switch in the circuit, the traveler wires that interconnect the traveler terminals of all four-way and three-way switches are black and red in color.

- If you use nonmetallic cable, and the power source feed is brought to the lighting outlet first, the traveler wires are red and white in color. It is important to re-identify the white traveler conductors at each switch location with black tape.

- Most four-way switch traveler terminals are vertically configured. This means that where the four-way switch is positioned in a vertical position, the top two screws have the same color and are a traveler terminal pair. The bottom screws are the other traveler terminal pair and have the same color. It is important to know that each traveler pair has a different color.

- There is no listing on the toggle for ON or OFF. So, like the three-way switch, it does not make any difference which way this switch is positioned in the electrical device box.

Four-way switches operate by having their internal contacts open or closed. This internal opening and closing is similar to single-pole and three-way switches. However, four-way switches have *four* sets of internal contacts. The contacts are opened or closed in a sequence determined by the position of the toggle. **Figure 12-23** shows the operation of a four-way switch.

Refer to Figure 12-23(A). In this circuit, where the toggle switch is placed in one position, connections are made vertically from the traveler terminals on the bottom of the switch to the traveler terminals on the top of the switch. In the circuit shown in Figure 12-23(B), where the switch toggle is placed in the opposite position, connections are made diagonally from the traveler terminals on the bottom to the traveler terminals on the top of the switch.

The following is the recommended procedure where installing a four-way switching circuit where the source of power comes into the first three-way switch. **Figure 12-24(A)** illustrates a cabling diagram, and **(B)** shows a wiring diagram for a switching circuit that has a four-way switch and two 3-way switches controlling a lighting load.

Always wear safety glasses and other required PPE and observe all applicable safety rules.

At the <u>luminaire</u> box:

1. Connect the bare grounding conductors to the grounding connection of the luminaire. If the luminaire does not have any metal parts that must be grounded, simply fold the grounding conductor into the back of the box. Recall that it is *not* necessary to connect the circuit grounding wire to a nonmetallic box.

2. Connect the white insulated grounded conductor to the silver screw terminal or wire identified as the grounded wire on the luminaire.

3. Connect the black ungrounded conductor to the brass screw or a wire identified as the ungrounded conductor on the luminaire.

FIGURE 12-23 Four-way switch operation.

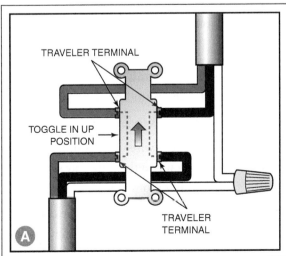
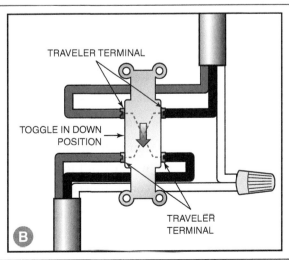

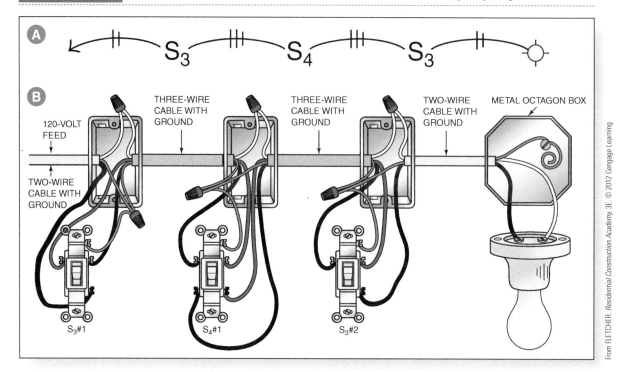

FIGURE 12-24 Switching circuit: one 4-way and two 3-way switches controlling a lighting load.

At three-way switch 1:

4. Connect the bare grounding conductors together, and using a bare pigtail, connect all grounding conductors to the green grounding screw on the switch.

5. Connect the white insulated grounded conductors together using a wirenut.

6. Connect the black "hot" conductor in the incoming power source cable to the common (black) terminal screw on the three-way switch.

7. Connect the black and red traveler wires in the 3-wire cable going to four-way switch 1 to the traveler terminal screws on three-way switch 1.

At four-way switch 1:

8. Connect the bare grounding conductors together, and using a bare pigtail, connect all the grounding conductors to the green grounding screw on the switch.

9. Connect the white insulated grounded conductors together using a wirenut.

10. Connect the black and red traveler wires in the 3-wire cable coming from three-way switch 1 to a pair of traveler terminal screws on the four-way switch. This connection can be made to the top pair *or* the bottom pair.

11. Connect the black and red traveler wires from the three-way cable coming from three-way switch 2 to the other pair of traveler terminal screws on four-way switch 1.

At three-way switch 2:

12. Connect the bare grounding conductors together, and using a bare pigtail, connect all the grounding conductors to the green grounding screw on the switch.

13. Connect the white insulated grounded conductors together using a wirenut.

14. Connect the black conductor in the 2-wire cable going to the lighting outlet to the common (black) terminal screw on three-way switch 2.

15. Connect the black and red traveler wires in the cable coming from four-way switch 1 to the traveler terminal screws on three-way switch 2.

Imagine you are an electrician and your apprentice asks about the differences between wiring single-pole, three-way and four-way switches. Write a note highlighting a few of the major points of differentiation.

Self-Check 5

1. List four types of switches commonly used in residential wiring.
2. Which *NEC* section covers switch current and voltage markings?
3. What does *404.2(A)* of the *NEC* state regarding the wiring of switches?
4. What is the purpose of a three-way switch?
5. Describe the operation of a three-way switch.
6. Briefly describe the procedure for wiring a three-way switch.
7. Briefly describe the procedure for wiring a four-way switch.

12-6 ▪ *NEC* Requirements for Replacing Existing Grounded and Ungrounded Receptacles

The *NEC* is very clear regarding the type of receptacle permitted to be used as a replacement for an existing receptacle. *NEC 406.3(D)* covers replacement rules.

The *NEC* details a retroactive ruling that requires a replacement receptacle in *any* location where the present code requires GFCI protection (example areas include bathrooms, kitchens, garages, etc.), and the replacement receptacle must be GFCI protected. A GFCI-type receptacle could provide this protection, or a branch circuit could be protected by a GFCI-type circuit breaker.

Existing wiring systems such as knob-and-tube, or older nonmetallic-sheathed cable wiring installed without an equipment grounding conductor, can still have a properly operating GFCI receptacle. Knob-and-tube wiring (K&T) was an early standardized method of electrical wiring that used knobs as insulators and ceramic tubes to isolate the wiring from neighboring wood. The GFCI receptacle does *not* need an equipment grounding conductor to protect against a deadly shock.

Knob-and-tube wiring an early standardized method of electrical wiring in structures that used knobs as insulators and ceramic tubes to isolate the wiring from neighboring wood.

Replacing Existing 2-Wire Receptacles Where a Grounding Means Does Exist

Replacing a 2-wire receptacle can be done easily if the wall box is properly grounded or where the branch-circuit wiring contains an equipment grounding conductor. Refer to **Figure 12-25.**

Figure 12-25(E) illustrates a grounded wall box. If this box is grounded by any method specified in *250.118* or if the branch-circuit wiring contains an equipment grounding conductor, as shown in (D), then an old style, 2-wire receptacle, as shown in (A), *must* be replaced with a grounding-type receptacle or a GFCI receptacle, as shown in Figure 12-25(B) and (C), respectively.

If a wall box is *not* grounded, *250.130(C)* permits an equipment grounding conductor to be run from the enclosure to be grounded, or from the green hexagon grounding screw of a grounding-type receptacle, to any accessible point on the grounding electrode system *or* to any accessible point on the grounding electrode conductor.

....

TRADE TIP The permission described in *250.130(C)* is *only for existing* installations. The labor involved in running an equipment grounding conductor to an accessible point on the grounding electrode system can be very expensive. A better choice is to make the replacement with a **GFCI** receptacle. Label it "No Equipment Ground."

....

FIGURE 12-25 Replacing a receptacle where a grounding means exists.

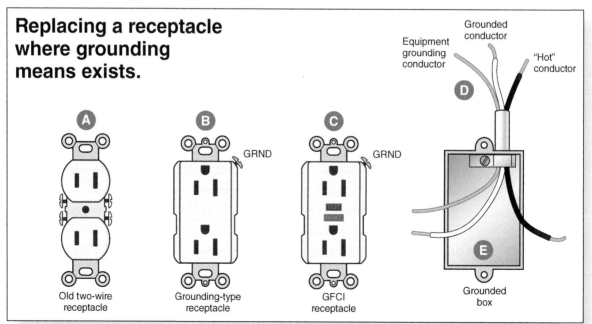

From MULLIN/SIMMONS. *Electrical Wiring Residential*, 17E. © 2012 Cengage Learning

Replacing Existing 2-Wire Receptacles Where a Grounding Means Does Not Exist

Figure 12-26 helps to illustrate choices to make where replacing a receptacle where grounding does *not* exist.

In Figure 12-26(E), note that this is a *nongrounded* box. Also note that (D) illustrates that there is *no* equipment grounding conductor run with circuit conductors. In this situation, an existing nongrounding-type receptacle, as shown in (A), may be replaced with a nongrounding-type receptacle that is the same type as shown in (A). It may be replaced with a GFCI type, as shown in (C); or a grounding-type receptacle, as shown in (B), *if* supplied through a GFCI receptacle (C); or a grounding-type receptacle (B) *if* a separate equipment grounding conductor is run from the receptacle to any accessible point on the grounding electrode system. This permission is covered in *250.130(C)*. Recall that installing an equipment conductor is labor intensive and costly. A better method is replacing the nongrounding-type receptacle with a GFCI type. However, where the replacement is with a GFCI-type receptacle, it *must* be labeled "No Equipment Ground," using the pressure-sensitive label furnished with the GFCI receptacle. *NEC 406.3(D)(3)* covers *marking* requirements regarding connections between receptacles.

LIFE SKILLS

Read through *Article 406 Receptacles, Cord Connectors, and Attachment Plugs (Caps)*, and then find a partner. Write three questions each about *NEC* requirements for replacing existing grounded and ungrounded receptacles. Then, quiz each other.

FIGURE 12-26   Replacing a receptacle where a grounding means does not exist.

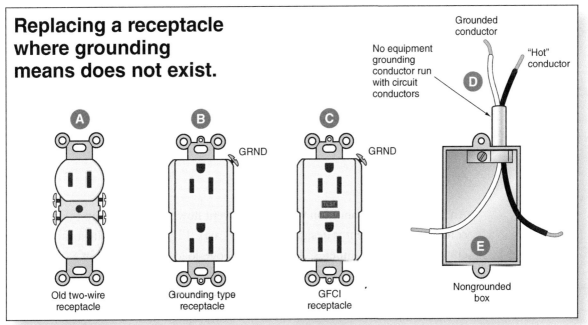

Self-Check 6

1. Which section of the *NEC* covers receptacle replacement requirements?
2. Can a GFCI receptacle be effective in protecting against shock in a wiring system without an equipment grounding conductor?
3. Why is the use of a GFCI receptacle a good choice for replacing a receptacle where a grounding means does not exist?

12-7 Five Types of Circuit Conditions

Recall that fuses and circuit breakers are sized by *matching* their ampere ratings to conductor ampacities and connected load currents. These protective devices sense overloads, short circuits, and ground faults, and protect the wiring and equipment from reaching dangerous temperatures.

As an electrician, most of the time you will (hopefully) encounter *normal* circuit conditions. However, overloads, shorts, and so on, occur. Even though fuses and circuit breakers do their job and protect wiring and equipment, you should be very familiar with the five possible circuit conditions:

- Normal
- Overload
- Short circuit
- Ground fault
- Open

Normal

> **Normal loading** where the current flowing is within the capability of the circuit and/ or the connected equipment.

Normal loading of a circuit is where the current flowing is within the capability of the circuit and/or the connected equipment. **Figure 12-27** illustrates a 15-ampere circuit carrying 10 amperes.

Refer to Figure 12-27. Note that there is a 240-volt source and a 24-ohm load. Ohm's law is used to determine the current through this circuit, which is 9.987 amperes. Source and conductor resistances are very small but *must* be used in calculating circuit current. In this circuit, the conductors safely carry the current. They do *not* get "hot," and the 15-ampere fuses do *not* open.

Overload

> **Overload** a condition where the current flowing is more than the circuit and/ or connected equipment is designed to safely carry.

An **overload** is a condition where the current flowing is more than the circuit and/or connected equipment is designed to safely carry. **Figure 12-28** illustrates a 15-ampere circuit carrying 20 amperes. A momentary overload is harmless, but a continuous overload condition causes the conductors and/or equipment to overheat, which is a potential cause of fire. The current flows through the *intended path*, the conductors and/or equipment.

FIGURE 12-27 Normally loaded circuit.

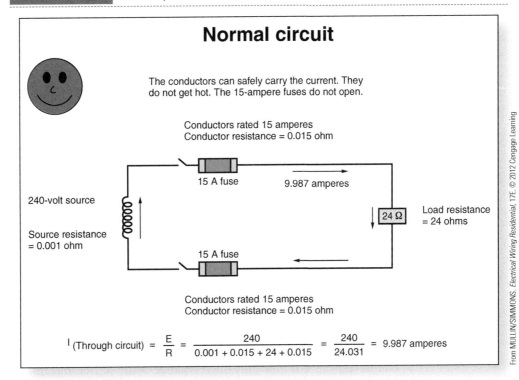

Normal circuit

The conductors can safely carry the current. They do not get hot. The 15-ampere fuses do not open.

Conductors rated 15 amperes
Conductor resistance = 0.015 ohm

15 A fuse   9.987 amperes

240-volt source

Source resistance = 0.001 ohm

24 Ω   Load resistance = 24 ohms

15 A fuse

Conductors rated 15 amperes
Conductor resistance = 0.015 ohm

$$I \text{ (Through circuit)} = \frac{E}{R} = \frac{240}{0.001 + 0.015 + 24 + 0.015} = \frac{240}{24.031} = 9.987 \text{ amperes}$$

FIGURE 12-28 Overloaded circuit.

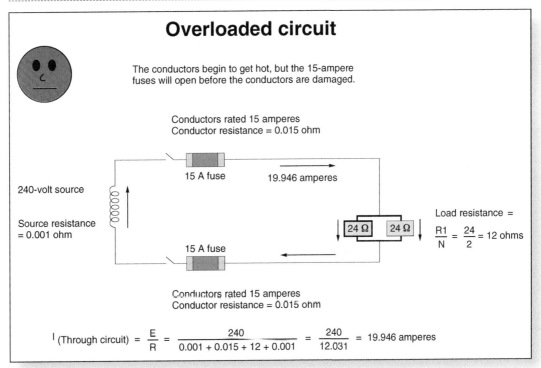

Overloaded circuit

The conductors begin to get hot, but the 15-ampere fuses will open before the conductors are damaged.

Conductors rated 15 amperes
Conductor resistance = 0.015 ohm

15 A fuse   19.946 amperes

240-volt source

Source resistance = 0.001 ohm

24 Ω   24 Ω

Load resistance = $\frac{R1}{N} = \frac{24}{2} = 12$ ohms

15 A fuse

Conductors rated 15 amperes
Conductor resistance = 0.015 ohm

$$I \text{ (Through circuit)} = \frac{E}{R} = \frac{240}{0.001 + 0.015 + 12 + 0.001} = \frac{240}{12.031} = 19.946 \text{ amperes}$$

From MULLIN/SIMMONS. *Electrical Wiring Residential*, 17E. © 2012 Cengage Learning

From MULLIN/SIMMONS. *Electrical Wiring Residential*, 17E. © 2012 Cengage Learning

Refer to Figure 12-28. In this circuit, an additional load is placed across the circuit load of 24 ohms. This *overloaded* condition allows *more* current to flow. Using Ohms law, the total circuit current is now 19.946 amperes. This level of current is above the current rating of the conductors. The conductors *begin to get hot*, but the 15-ampere fuses open *before* the conductors are damaged.

Short Circuit

Short circuit a condition where two or more conductors come in contact with one another, resulting in a current flow that bypasses the connected load.

A short circuit is a condition where two or more conductors come in contact with one another, resulting in a current flow that bypasses the connected load. A short circuit might be two "hot" conductors coming together, or it might be a "hot" conductor and a *grounded* conductor coming together. In either case, the current flows outside of the *intended path*. The only resistance is that of the conductors, the source, and the arc. This low resistance results in high levels of short-circuit current. The heat generated at the point of the arc *can* result in a fire. A short circuit is illustrated in **Figure 12-29**.

In the circuit shown in Figure 12-29, a wire across the 24-ohm load simulates a short circuit. Because the resistance of a short circuit is zero, Ohms law is used to determine the circuit current. The source voltage is still 240 volts, but the *total* circuit resistance is now only the resistance of the source and conductor resistances. The calculated current is 7742 amperes. This condition causes the conductors to get

FIGURE 12-29 Short circuit.

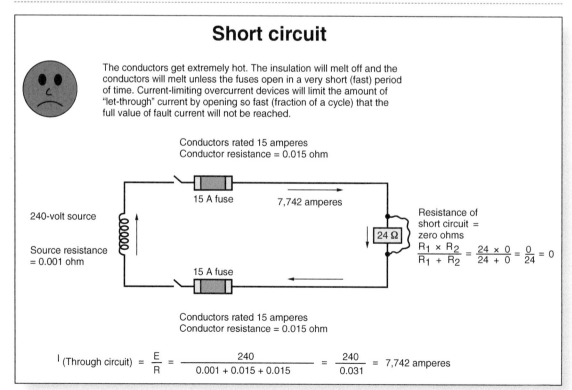

Short circuit

The conductors get extremely hot. The insulation will melt off and the conductors will melt unless the fuses open in a very short (fast) period of time. Current-limiting overcurrent devices will limit the amount of "let-through" current by opening so fast (fraction of a cycle) that the full value of fault current will not be reached.

Conductors rated 15 amperes
Conductor resistance = 0.015 ohm

15 A fuse   7,742 amperes

240-volt source

Source resistance = 0.001 ohm

15 A fuse

Conductors rated 15 amperes
Conductor resistance = 0.015 ohm

$24\ \Omega$

Resistance of short circuit = zero ohms

$$\frac{R_1 \times R_2}{R_1 + R_2} = \frac{24 \times 0}{24 + 0} = \frac{0}{24} = 0$$

$$I \text{ (Through circuit)} = \frac{E}{R} = \frac{240}{0.001 + 0.015 + 0.015} = \frac{240}{0.031} = 7{,}742 \text{ amperes}$$

extremely hot. Unless the fuses open rapidly, the insulation on the conductors melts. Because the fuses open rapidly, the full value of *fault* current is not reached.

Ground Fault

A ground fault is a condition where a "hot" conductor comes in contact with a grounded surface, such as a grounded metal raceway, metal water pipe, sheet metal, and so on, as shown in **Figure 12-30**.

A ground fault occurs where the current flows outside of its intended circuit. A ground fault can result in a flow of current greater than the circuit rating, in which case the overcurrent device opens. Ground faults can cause a current flow less than the circuit rating, in which case the overcurrent device does not open.

The calculation procedure for ground faults and short circuits is the same. However, the values of R (resistance) can vary greatly with unknown impedance of the ground return. Impedance is the total opposition to current flow. Many factors such as loose locknuts, bushings, or set screws on connectors and couplings can contribute to the resistance of the return ground path, making it extremely difficult to determine the actual ground-fault current values.

Impedance the total opposition to current flow.

Open

An **open circuit** is a condition where the circuit is *not closed* somewhere in the circuit. This open condition is illustrated in **Figure 12-31**.

Because there is an *open* in this circuit, there is *not* a complete and continuous path. Therefore, no current flows. Using Ohms law, current is effectively zero amperes. It is important to realize that in this or any series circuit, the source voltage (in this case 240 volts) appears across the open.

Open circuit a condition where the circuit is *not closed* somewhere in the circuit.

FIGURE 12-30 Ground fault.

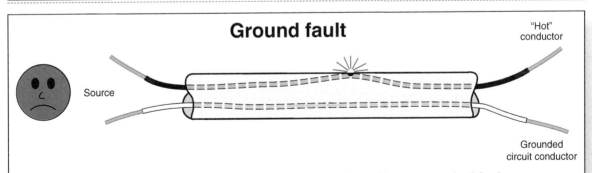

Ground fault

"Hot" conductor

Source

Grounded circuit conductor

"Hot" conductor comes in contact with metal raceway or other metal object. If the return ground path has low resistance (impedance), the overcurrent device protecting the circuit will clear the fault. If the return ground path has high resistance (impedance), the overcurrent device will not clear the fault. The metal object will then have a voltage to ground the same as the "hot" conductor has to ground. In house wiring, this voltage to ground is 120 volts. Proper grounding and ground-fault circuit-interrupter protection is discussed elsewhere in this text. The calculation procedure for a ground fault is the same as for a short circuit; however, the values of "R" can vary greatly because of the unknown impedance of the ground return path. Loose locknuts, bushings, set screws on connectors and couplings, poor terminations, rust, etc., all contribute to the resistance of the return ground path, making it extremely difficult to determine the actual ground-fault current values.

FIGURE 12-31 Open circuit.

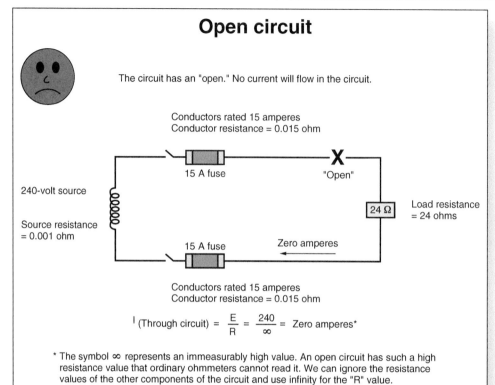

Open circuit

The circuit has an "open." No current will flow in the circuit.

Conductors rated 15 amperes
Conductor resistance = 0.015 ohm

15 A fuse   "Open"

240-volt source

Source resistance
= 0.001 ohm

Load resistance
= 24 ohms

24 Ω

15 A fuse   Zero amperes

Conductors rated 15 amperes
Conductor resistance = 0.015 ohm

$$\text{I (Through circuit)} = \frac{E}{R} = \frac{240}{\infty} = \text{Zero amperes*}$$

* The symbol ∞ represents an immeasurably high value. An open circuit has such a high resistance value that ordinary ohmmeters cannot read it. We can ignore the resistance values of the other components of the circuit and use infinity for the "R" value.

Caution: Line voltage (in this case 240 volts) appears across the "open."

LIFE SKILLS

With classmates, form a group of five. Then, have each group member give a 1- to 2-minute presentation discussing the key points of one of the five types of circuit conditions. After each presentation, the presenter takes questions from the rest of the group.

Self-Check 7

1. List the five possible circuit conditions.
2. Explain a short circuit.
3. Why is it difficult to determine an actual ground-fault current value?

Summary

- Nonmetallic-sheathed cable is a factory assembly of two or more insulated conductors having an outer sheathing of moisture-resistant, flame-retardant, nonmetallic material.

- Nonmetallic-sheathed cable, Type NMC is typically referred to as Romex and is used more than any other wiring methods in residential applications because it costs less to purchase and install.

- Armored cable, Type AC, usually referred to as BX cable, is a fabricated assembly of insulated conductors in a flexible metallic enclosure.

- Metal clad cable, Type MC is a factory assembly of one or more insulated circuit conductors enclosed in an armor of interlocking metal tape or a smooth or corrugated metallic sheath.

- Underground feeder cable, Type UF cable is a listed factory assembly of one or more insulated conductors with an integral or an overall covering of nonmetallic material suitable for direct burial in the earth.

- A raceway is an enclosed channel of metal or nonmetallic materials designed expressly for holding wires or cables.

- Electrical metallic tubing (EMT) is one type of raceway and is often referred to as thinwall conduit because of its very thin walls.

- Several of the most commonly used switches in residential applications include single-pole, double-pole, three-way, four-way, dimmer, and combination switches.

- According to *404.15(A)* of the *NEC*, each switch must be marked with the current and voltage rating for the switch, and the switch must be matched to the voltage and current of the circuit where it is being used.

- *NEC 404.2(A)* requires three- and four-way switches to be wired so that all switching is done in the ungrounded circuit conductor.

- The *NEC* and Underwriters Laboratories refer to the switches used to control lighting outlets in residential applications as snap switches; however, most electricians refer to them as toggle switches or simply as switches.

- Three-way switches have three terminal screws on them and are used to control a lighting load from two different locations.

- Three-way switches must always be installed in pairs.

- Four-way switches are used along with three-way switches to control a lighting load from more than two locations.

- The *NEC* requires a replacement receptacle in any location where the present *Code* requires GFCI protection, and the replacement receptacle must be GFCI protected.

- Replacing a 2-wire receptacle can be done easily if the wall box is properly grounded or where the branch-circuit wiring contains an equipment grounding conductor.

- If a wall box is not grounded, *250.130(C)* permits an equipment grounding conductor to be run from the enclosure to be grounded, or from the green hexagon grounding screw of a grounding-type receptacle, to any accessible point on the grounding electrode system *or* to any accessible point on the grounding electrode conductor.

- A popular method of replacing a nongrounding-type receptacle is to use a GFCI-type receptacle. However, where the replacement is with a GFCI-type receptacle, it must be labeled "No Equipment Ground."

- Five possible circuit conditions are normal, overload, short circuit, ground fault, and open.

Review Questions

True/False

1. Non-metallic sheathed cable is typically referred to as BX by most electricians. (True, False)

2. Metal clad cable is a factory assembly of one or more insulated circuit conductors enclosed in an armor of interlocking metal tape or a smooth or corrugated metallic sheath. (True, False)

3. Type UF cable may never be used in direct exposure to the sun under any circumstances. (True, False)

4. Metal clad cable contains no aluminum bonding wire. (True, False)

5. A raceway is an enclosed channel of metal or nonmetallic materials designed expressly for holding wires or cables. (True, False)

Multiple Choice

6. Which type raceway is often used as a service mast for a service entrance?
   A. RMC
   B. IMC
   C. PVC
   D. EMT

7. Which type raceway looks similar to BX cable, but does not have conductors installed and is often used to connect appliances and other equipment in residential applications?
   A. RMC
   B. IMC
   C. PVC
   D. FMC

8. Which type raceway is often referred to as thinwall and is easy to install because of its light weight?
   A. RMC
   B. PVC
   C. ENT
   D. EMT

9. Which type switch allows for control of lighting outlet loads from three or more locations?
   A. Combination
   B. Double-pole
   C. Three-way
   D. Four-way

10. Which type switch consists of more than one switch type on the same strap or yoke?
    A. Combination
    B. Double-pole
    C. Three-way
    D. Four-way

Fill in the Blank

11. *NEC* _____ states that each switch must be marked with the current and voltage rating for the switch.

12. *NEC 404.2(A)* requires three- and four-way switches to be wired so that all switching is done in the _____ circuit conductor.

13. The *NEC* and UL refer to the switches used to control lighting outlets in residential applications as _____ switches.

14. On a three-way switch, two terminal screws are the same color, usually brass or bronze, and called the _____ terminals.

15. The black-colored common terminal on a three-way switch should always have a _____ insulated wire attached to it.

16. _____ switches either allow current flow through the switch and to the load or they do not.

17. The three-way switch has _____ traveler terminals and one common terminal.

18. To control a lighting load from more than two locations, _____ switches are used.

19. Four-way switches must always be installed _____ two 3-way switches.

20. *NEC* _____ covers receptacle replacement requirements.

21. The recommended method for replacing a nongrounding-type receptacle is to use a _____ receptacle.

22. _____ loading of a circuit is where the current flowing is within the capability of the circuit and connected equipment.

23. A(n) _____ is a condition where the current is more than the circuit or the connected equipment is designed to safely carry.

24. A(n) _____ circuit is a condition where two or more conductors come in contact with one another, resulting in a current flow that bypasses the connected load.

25. A(n) _____ circuit is a condition where the circuit is not closed somewhere in the circuit.

Chapter 13

Wiring Calculations

Career Profile

Manufacturing Department Manager, **Jeremy Grosser, Rex Moore Electrical Contractors & Engineers**, strongly values the education process. "I would not have the success I have without education as a part of my journey," he says. "The more educated you are, the more valuable you are to an organization. Not because you can recite a code or formula but because you understand the cause and effect of a decision or action. This tends to lead you in a direction of correct choices and positive results."

Jeremy began his education very early in life: "I have always had a fascination with electricity. It started as a child. I would open up toys with motors and batteries to see what made them work," he notes. This interest led to "pursuing electronics courses in school. I also took courses in residential, commercial and industrial wiring and found great satisfaction in seeing the results of my hard work in the field. Understanding the theory of electricity gave me a great foundation for my current career. As a field installer I could understand why buildings had the systems they had and why *NEC* codes were written."

In his formal education, Grosser has earned an Associates of Science degree in Computer Integrated Electronics and completed a WECA electrical apprenticeship program. He has also had various opportunities to learn on the job:

In his early days in the field, Jeremy gained experience with "everything from jackhammering concrete in the rain to terminating conductors in large motor control centers." He then moved into a foreman position that required him to do work planning, problem resolution, and material management in all aspects of a project. In his current position at Rex Moore, Grosser has "earned the opportunity to develop a department that plays an integral part in ensuring projects can meet demanding budget and schedule requirements."

Grosser's advice to apprentice electricians reinforces his belief in the importance of education: "It takes time to become a confident electrician. Ask questions. Don't let pride get in your way—it's OK to say you don't know. People will have more respect for you if you say you don't know rather than giving them an incorrect answer. What you lack in knowledge, make up with extra effort. Don't ever stop learning and increasing your skills—the day you stop learning is the day the person behind you passes you by."

Chapter Outline

NEC REQUIREMENTS FOR CALCULATING BRANCH-CIRCUIT SIZING AND LOADING

CONDUIT FILL CALCULATIONS AS PER *NEC*

BOX FILL CALCULATIONS AND BOX SELECTION

PROPER CONDUCTOR SIZE AND OVERCURRENT DEVICE FOR A CIRCUIT

Key Terms

American Wire Gauge (AWG)

Ampacity

Appliance branch circuit

Bathroom branch circuit

Branch circuit

Conductivity

Device box

General-purpose branch circuit

Habitable floor area

Home runs

Individual branch circuit

Junction box

Laundry branch circuit

Chapter Objectives

After completing this chapter, you will be able to:

1. Determine the fundamental *NEC* requirements for calculating branch-circuit sizing and loading.
2. Perform conduit fill calculations as per *NEC Chapter 9, Tables 1–5*, and *Annex C*
3. Calculate box fill, and choose the correct size box.
4. Describe the proper size conductor and overcurrent device for a circuit, given a receptacle or switch.

Life Skills Covered

 Self-Advocacy

 Goal Setting

 Managing Stress

Life Skill Goals

The life skills covered in this chapter include Self-Advocacy, Goal Setting, and Managing Stress. As an electrician, you must often calculate one thing or another, and it won't always be easy. A solid understanding of how to perform wiring calculations prevents some stressful situations. So make sure you pay close attention as you go through the chapter. Everything that you learn will be used in the field. If you are able to fully understand and intelligently discuss this subject matter, it makes you a more attractive job candidate. Before you go any further, set a goal for yourself: Resolve to master the calculations contained within this chapter.

This chapter covers calculating branch-circuit sizing and load and determining the maximum size overcurrent devices for these circuits.

Electrical box selection is also discussed in this chapter, as well as box fill and conduit fill calculations as per *NEC* requirements. As an electrician, you need to know and follow the *NEC* requirements for calculating branch-circuit sizing and loading. Pay close attention to the information presented in this chapter, and learn these *NEC* requirements and the procedures necessary for using these important calculations!

Introduction

In a home's electrical system, feeder wires, also known as service-entrance conductors, supply power to the main panelboard and the branch circuits leaving the panelboard to power devices. The number and types of these branch circuits supplying power to various devices, referred to as the load, must be determined *before* an electrician can determine the size of the service-entrance ampacity required.

The *National Electrical Code (NEC)* details information and procedures that electricians must use to calculate the minimum size of branch circuits, feeders, and service-entrance conductors required for a residential electrical system. As an electrician, you must know and follow the *NEC* requirements and the procedures necessary for using these calculations.

13-1 *NEC* Requirements for Calculating Branch-Circuit Sizing and Loading

Branch circuit the circuit conductors between the final overcurrent device (fuse or circuit breaker) and the power and/or lighting outlets.

A **branch circuit** is defined by the *NEC* as the circuit conductors between the final overcurrent device (fuse or circuit breaker) and the power and/or lighting outlets. **Figure 13-1** shows a lighting branch circuit and a receptacle branch circuit. Note the connections from the final overcurrent protection devices to the lighting and receptacle outlets.

As an electrician, you must not only know what a branch circuit is but also have an understanding of the various types of branch circuits used in residential wiring. Types of branch circuits used in residential wiring applications include these:

- Lighting
- Receptacle
- General-purpose
- General lighting
- Small-appliance
- Laundry
- Bathroom
- Individual

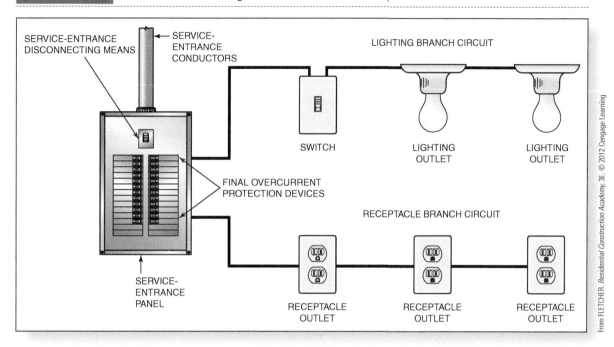

FIGURE 13-1 Branch circuits showing connections to overcurrent protection devices.

General Lighting Circuits

A branch circuit that has only lighting outlets is called a lighting branch circuit, and a branch circuit with only receptacles is referred to as a receptacle branch circuit. However, in residential wiring applications, these branch circuits are typically combined into what the *NEC* refers to as a general-purpose branch circuit. A general-purpose branch circuit is a branch circuit that supplies two or more receptacles or lighting outlets, or a combination of both. Televisions, computers, DVD players, and other household appliances can also be connected to this type of branch circuit using power cord-and-plug connections. An example of a general-purpose branch circuit would be a living room circuit that has a ceiling-mounted luminaire controlled by a wall switch and has several receptacles. The main objective of general-purpose branch circuits is the provision of lighting in rooms throughout the house. Lighting is provided by wall or ceiling luminaires controlled by switches or by simply plugging lamps into circuit receptacles. Because *lighting* is the main goal of such circuits, these general-*purpose* circuits are often referred to as general *lighting* circuits. **Figure 13-2** shows a *general lighting circuit*. Note that this circuit supplies electrical power to *both* lighting and receptacle outlets.

The outlets shown in the general lighting circuit of Figure 13-2 are installed with 14 AWG wire and an overcurrent protection device (fuse or breaker) rated at 15 amperes or by using 12 AWG wire protected by a circuit breaker or fuse rated at 20 amperes. Note and compare the differences between Figure 13-1, a lighting *and* receptacle branch circuit, and Figure 13-2, a general lighting circuit.

General-purpose branch circuit a branch circuit that supplies two or more receptacles or lighting outlets, or a combination of both.

FIGURE 13-2 A general lighting circuit.

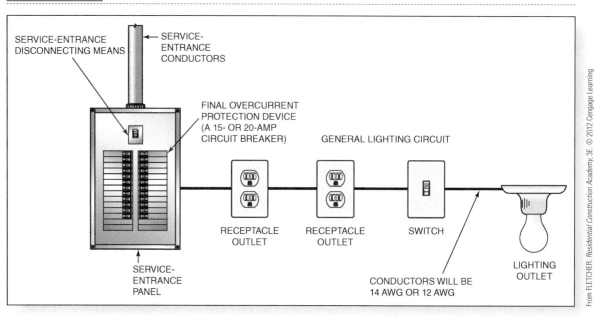

NEC Requirements and Calculations

Habitable floor area floor area or rooms in a residential occupancy used for living, sleeping, cooking, and eating, but excluding bath, storage and service areas, and corridors.

To calculate the *minimum* number of general lighting circuits required in a house, the **habitable floor area** must first be determined. Habitable floor area is floor area or rooms in a residential occupancy used for living, sleeping, cooking, and eating, but excluding bath, storage and service areas, and corridors. Once the floor area is determined, that area must be multiplied by the unit load per square foot for general lighting to get the *total* general lighting load in volt-amperes. The *NEC* states in *220.12* that a unit load of not less than that specified in *Table 220.12* for occupancies listed in the table is the minimum lighting load. *Table 220.12* is shown in **Figure 13-3**.

    Table 220.12 provides the minimum general lighting load per square foot for a number of building types. The unit load for a dwelling unit is listed in this table and is 3 volt-amperes per square foot. Outside dimensions of the dwelling unit are used to calculate floor area for each floor. This calculated floor area, however, does *not* include open porches, garages, or unused or unfinished spaces that cannot be adapted for future use. Attics, crawl spaces, and basements are examples of unused or unfinished spaces in dwelling units. However, a finished-off basement used as a playroom or family room *must* be included in the floor area calculation. The following example shows how to determine the minimum number of general lighting circuits:

    Example

    Given: A house with 2500 square feet of habitable living space.

    Solution:

    General lighting load in volt-amperes = 2500 square feet × 3 volt-amps per square foot[1]

---

[1]The unit load for a dwelling unit, according to *Table 220.12* is 3 volt-amperes.

FIGURE 13-3 *NEC Table 220.12. General Lighting Loads by Occupancy.*

TABLE 220.12 General Lighting Loads by Occupancy		
	Unit Load	
Type of Occupancy	**Volt-Amperes/ Square Meter**	**Volt-Amperes/ Square Foot**
Armories and auditoriums	11	1
Banks	39[b]	3½[b]
Barber shops and beauty parlors	33	3
Churches	11	1
Clubs	22	2
Court rooms	22	2
Dwelling units[a]	33	3
Garages—commercial (storage)	6	½
Hospitals	22	2
Hotels and motels, including apartment houses without provision for cooking by tenants[a]	22	2
Industrial commercial (loft) buildings	22	2
Lodge rooms	17	1½
Office buildings	39[b]	3½[b]
Restaurants	22	2
Schools	33	3
Stores	33	3
Warehouses (storage)	3	¼
In any of the preceding occupancies except one-family dwellings and individual dwelling units of two-family and multifamily dwellings:		
Assembly halls and auditoriums	11	1
Halls, corridors, closets, stairways	6	½
Storage spaces	3	¼

[a]See 220.14(J).
[b]See 220.14(K).

Therefore,

General lighting load in volt-amps = 2500 square feet $\times$ 3 volt-amps = 7500 volt-amperes

The general lighting load in amperes = general lighting load in volt-amps divided by 120 volts.[2] Therefore,

General lighting load in amps = 7500 volt-amps/120 volts = 62.5 amperes

To determine the minimum number of *15-amp* general lighting circuits, divide 62.5 amps by 15 amps. Therefore,

62.5 amps/15 amps = 4.16

We cannot use four circuits, as these circuits would be overloaded because 4.16 is greater than 4. So, we must increase the number of circuits to accommodate the calculated load. Therefore, a minimum of *five* 15-amp general lighting circuits must be installed.

To determine how many 20-amp-rated general lighting circuits should be installed, we simply divide the general lighting load in amps by 20 amps.

Small-Appliance Branch Circuits

Appliance branch circuit a branch circuit that supplies energy to one or more outlets to which appliances are to be connected and that has no permanently connected luminaires that are not part of an appliance.

The *NEC* defines appliance branch circuit *as a branch circuit that supplies energy to one or more outlets to which appliances are to be connected and that has no permanently connected luminaires that are not part of an appliance.** In residential wiring applications, a small-appliance branch circuit supplies electrical power to receptacles installed in kitchens, breakfast rooms, dining rooms, pantries, and other similar areas. A small-appliance branch circuit is illustrated in **Figure 13-4**.

In Figure 13-4, conductors are 12 AWG because the final overcurrent protection devices (circuit breakers) are rated at 20 amperes. *NEC 210.11(C)(1)* states that two or more 20-ampere-rated small-appliance branch circuits *must* be provided for all receptacle outlets specified by *210.52(B)*. And *210.52(B)* requires two or more 20-ampere circuits for all receptacle outlets for the small-appliance loads, including refrigeration equipment, in the kitchen, dining room, pantry, and breakfast room of a dwelling unit. No fewer than two small-appliance branch circuits must supply the countertop receptacle outlets in kitchens.

NEC 220.52(A) states that in each dwelling unit, the load must be computed at 1500 volt-amperes for each 2-wire small-appliance branch circuit. Therefore, a *minimum* load for small-appliance circuits in a house is calculated as follows:

1500 volt-amperes $\times$ 2 (minimum of 2 small-appliance branch circuits) = 3000 volt-amperes

If a dwelling unit is to have more than two small-appliance branch circuits installed, *each* circuit over two must also be calculated at 1500 volt-amperes.

---

[2]120 volts is the voltage of residential general lighting circuits.
*Reprinted with permission from NFPA 70-2011.

FIGURE 13-4 A small-appliance branch circuit.

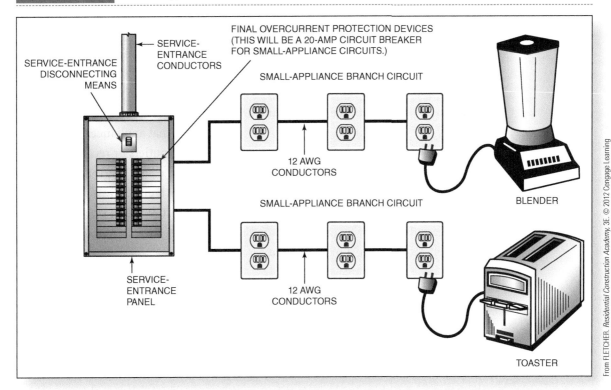

Laundry branch circuit the type of branch circuit used in residential wiring applications that supplies electrical power to laundry areas.

Laundry Branch Circuits

The type of branch circuit used in residential wiring applications that supplies electrical power to laundry areas is a **laundry branch circuit**. **Figure 13-5** shows a laundry branch circuit.

FIGURE 13-5 A laundry branch circuit.

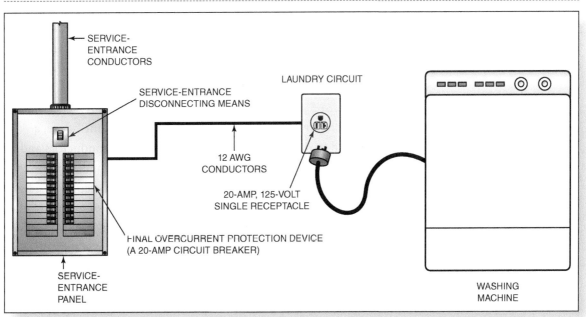

The conductors from the service panel to the receptacle are 12 AWG, and the receptacle is rated 20 amps at 125 volts. *NEC 210.11(C)(2)* requires the installation of at least one 20-ampere-rated laundry circuit in each dwelling.

A laundry area is considered to be a separate room or may be an area in a basement or garage. A laundry branch circuit is wired to provide electrical power to a washing machine and other items typically found in a laundry room. Examples are a clothes iron and portable dress steamer. If a gas-operated dryer is to be used, the laundry branch circuit supplies the 120-volt power for that appliance.

A laundry branch circuit must not have lighting outlets or receptacles in other rooms connected to it. *NEC 220.52(B)* states that a load of not less than 1500 volt-amperes is to be included for each 2-wire laundry branch circuit installed. The laundry circuit load is generally 1500 volt-amperes because most dwelling units have only one 2-wire laundry circuit installed. If more than one 2-wire laundry circuit is installed, 1500 volt-amperes must be calculated for each circuit.

Bathroom Branch Circuits

Bathroom branch circuit a circuit that supplies electrical power to a bathroom in a residential wiring application.

A circuit that supplies electrical power to a bathroom in a residential wiring application is a bathroom branch circuit. **Figure 13-6** shows a bathroom branch circuit.

A bathroom branch circuit supplies power for bathroom receptacles and bathroom luminaires as long as the circuit does *not* feed other bathrooms. If the circuit is wired so that it feeds more than one bathroom, only receptacle outlets can be fed by the bathroom circuit. A bathroom is an area including a basin that has one or more of the following: toilet, tub, or shower. The *NEC* states in *210.11(C)(3)* that in addition to the number of general lighting, small-appliance, and laundry branch circuits, at least *one* 20-ampere branch circuit must be provided to supply the bathroom receptacle outlet(s). There is, however, *no* volt-ampere value for a bathroom branch-circuit type in the calculation for total electrical load of a house.

Individual Branch Circuits

Individual branch circuit a circuit that provides electrical power to only one piece of electrical equipment.

An individual branch circuit is a circuit that provides electrical power to only *one* piece of electrical equipment. An individual branch circuit is shown in **Figure 13-7**.

One electric range *or* one dishwasher is an example of what an individual branch circuit supplies. In Figure 13-7, the range is wired using 8 AWG conductors, and the circuit is protected by a 40-amp circuit breaker. In the same figure, a dishwasher is connected to the service panel by using 12 AWG conductors and is protected by a 20-amp breaker. A circuit rating for such an appliance is to be calculated by using the ampere rating of the appliance. The *nameplate* on the appliance lists this information. An example is installing in a house a dishwasher having a nameplate rating of 10 amps at 120 volts. The total load for that appliance in volt-amperes can be calculated by multiplying the amperage of the appliance by the voltage: 10 amps $\times$ 120 volts = 1200 volt-amperes. The individual branch circuit for the dishwasher in the example needs a capacity of at least 1200 volt-amperes. The service-entrance calculation also includes this value. A volt-amp rating or wattage rating is listed on the

FIGURE 13-6 A bathroom branch circuit.

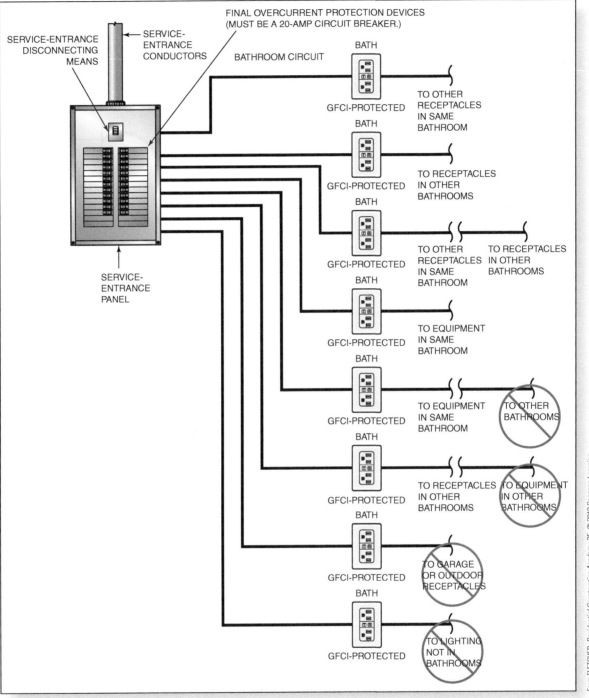

nameplate of some appliances. This information is used directly in the calculation for total electrical load of a house.

Individual branch-circuit loads for electric clothes dryers and household electric cooking appliances are based on the information found in the *NEC* in *220.54* for electric dryers and in *220.55* for electric ranges and other cooking appliances. The load for household electric clothes dryers in a dwelling unit *must* be 5000 watts (volt-amperes) or the nameplate rating, whichever is larger, for each dryer served.

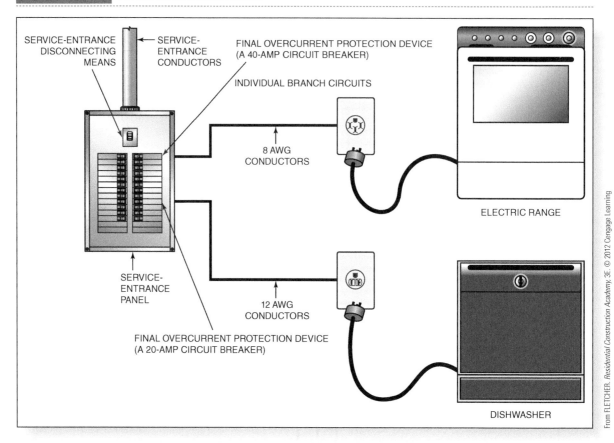

FIGURE 13-7 An individual branch circuit.

SERVICE-ENTRANCE DISCONNECTING MEANS

SERVICE-ENTRANCE CONDUCTORS

FINAL OVERCURRENT PROTECTION DEVICE (A 40-AMP CIRCUIT BREAKER)

INDIVIDUAL BRANCH CIRCUITS

8 AWG CONDUCTORS

ELECTRIC RANGE

SERVICE-ENTRANCE PANEL

12 AWG CONDUCTORS

FINAL OVERCURRENT PROTECTION DEVICE (A 20-AMP CIRCUIT BREAKER)

DISHWASHER

Ampacity of a Conductor

Ampacity
the current, in amperes, that a conductor can carry continuously under the conditions of use without exceeding its temperature rating.

The *NEC* defines ampacity as the current, in amperes, that a conductor can carry continuously under the conditions of use without exceeding its temperature rating. The *NEC* goes on to state in *210.19(A)(1)* that branch-circuit conductors are *required* to have an ampacity at least equal to the maximum electrical load to be served. An example of this rule is that if a branch circuit is to be wired that supplies electrical power to a load that draws 17 amperes, a wire size is needed that has an ampacity of at least 17 amperes. Anything less may cause the conductor to heat up and conductor insulation to melt.

The ampacity of a conductor to be used in residential wiring is found in *Table 310.15(B)(16)*, which was formerly *Table 310.16*. This table is shown in **Figure 13-8**.

Referring to *Table 310.16* in Figure 13-8, you see that the left side covers copper conductors, and the right side covers aluminum and copper-clad aluminum conductors. Most conductors used for branch-circuit wiring in residential applications are copper. So as an electrician, you will use the left side of this table more often. However, you should become familiar with the entire table. The following examples help you understand how ampacities are found using this table:

- A 14 AWG copper conductor with TW (60°C) insulation has an ampacity of *15* amps.

- A 10 AWG copper conductor with THW (75°C) insulation has an ampacity of *35* amps.

- A 6 AWG copper conductor with TW (60°C) insulation has an ampacity of *55* amps.

FIGURE 13-8 *NEC Table 310.15(B)(16)* is used to determine the minimum conductor size required for a specific ampacity.

TABLE 310.15(B)(16) (formerly TABLE 310.16) Allowable Ampacities of Insulated Conductors Rated Up to and Including 2000 Volts, 60°C Through 90°C (140°F Through 194°F), Not More Than Three Current-Carrying Conductors in Raceway, Cable, or Earth (Directly Buried), Based on Ambient Temperature of 30°C (86°F)*

	Temperature Rating of Conductor [See Table 310.104(A).]						
	60°C (140°F)	75°C (167°F)	90°C (194°F)	60°C (140°F)	75°C (167°F)	90°C (194°F)	
Size AWG or kcmil	Types TW, UF	Types RHW, THHW, THW, THWN, XHHW, USE, ZW	Types TBS, SA, SIS, FEP, FEPB, MI, RHH, RHW-2, THHN, THHW, THW-2, THWN-2, USE-2, XHH, XHHW, XHHW-2, ZW-2	Types TW, UF	Types RHW, THHW, THW, THWN, XHHW, USE	Types TBS, SA, SIS, THHN, THHW, THW-2, THWN-2, RHH, RHW-2, USE-2, XHH, XHHW, XHHW-2, ZW-2	Size AWG or kcmil
	COPPER			ALUMINUM OR COPPER-CLAD ALUMINUM			
18**	—	—	14	—	—	—	—
16**	—	—	18	—	—	—	—
14**	15	20	25	—	—	—	—
12**	20	25	30	15	20	25	12**
10**	30	35	40	25	30	35	10**
8	40	50	55	35	40	45	8
6	55	65	75	40	50	55	6
4	70	85	95	55	65	75	4
3	85	100	115	65	75	85	3
2	95	115	130	75	90	100	2
1	110	130	145	85	100	115	1
1/0	125	150	170	100	120	135	1/0
2/0	145	175	195	115	135	150	2/0
3/0	165	200	225	130	155	175	3/0
4/0	195	230	260	150	180	205	4/0
250	215	255	290	170	205	230	250
300	240	285	320	195	230	260	300
350	260	310	350	210	250	280	350
400	280	335	380	225	270	305	400
500	320	380	430	260	310	350	500
600	350	420	475	285	340	385	600
700	385	460	520	315	375	425	700
750	400	475	535	320	385	435	750
800	410	490	555	330	395	445	800
900	435	520	585	355	425	480	900
1000	455	545	615	375	445	500	1000
1250	495	590	665	405	485	545	1250
1500	525	625	705	435	520	585	1500
1750	545	650	735	455	545	615	1750
2000	555	665	750	470	560	630	2000

*Refer to 310.15(B)(2) for the ampacity correction factors where the ambient temperature is other than 30°C (86°F).
**Refer to 240.4(D) for conductor overcurrent protection limitations.

Reprinted with permission from NFPA 70®, *National Electrical Code®*, Copyright © 2010, National Fire Protection Association, Quincy, MA. This reprinted material is not the complete and official position of the NFPA on the re erenced subject, which is represented only by the standard in its entirety.

LIFE SKILLS

You will come across several calculations in this section and in this chapter as a whole. Each time you do, write the calculation down on a sheet of paper, and label it clearly. That way, when it is time to review, you will have all of the information readily available.

Self-Check 1

1. Define branch circuit.
2. List four types of branch circuits used in residential applications.
3. What might be connected to an individual branch circuit?
4. Define ampacity.

13-2 Conduit Fill Calculations as per *NEC*

In residential wiring applications, conduit of some type is very common for service-entrance installations. However, a raceway is almost never used for branch-circuit installation. It is easier and less expensive to use cable wiring methods for wiring houses. There are geographic areas that *require* that all wiring in a house be installed in a raceway wiring method.

The decision to use a certain type of raceway in a residential wiring installation is made based on the actual wiring application, an electrician's personal preference of raceway types, and what is required by any local electrical codes. Recall that rigid metal conduit (RMC) and intermediate metal conduit (IMC) are raceway types used for a mast-style service-entrance installation. Electrical metallic tubing (EMT) or rigid nonmetallic polyvinyl chloride conduit (PVC), as well as RMC and IMC, can be used for service-entrance installations that use a raceway for the installation of service conductors that *do not require* the raceway to extend above a roof (a mast-type service entrance). The *most common raceway* used for branch-circuit installation is *EMT*, mainly because of its relatively easy bending and connection techniques. It is also much less expensive than other metal raceways. Installations where flexibility is desired, such as connections to air-conditioning systems, often require a flexible raceway like flexible metal conduit (FMC) for use indoors or liquidtight flexible metal conduit (LFMC) for use outdoors.

The selection of the proper size raceway is determined by the number, size, and type of conductors to be installed in a particular type of raceway. To accurately make this determination, you must become familiar with certain tables located in *Chapter 9* and *Annex C* of the *NEC*.

The *NEC* has established *percent fill* values for conduit and tubing. *Table 1* in *Chapter 9* of the *NEC* specifies the *maximum* fill percentage of conduit or tubing. This code says that for an application where only one conductor is installed, the conduit cannot be filled to more than 53% of the conduit's cross-sectional area. For an

FIGURE 13-9   *Table 1* in *Chapter 9* of the *NEC* specifies the maximum fill percentage for conduit and tubing.

TABLE 1   Percent of Cross Section of Conduit and Tubing for Conductors

Number of Conductors	All Conductor Types
1	53
2	31
Over 2	40

application where only two conductors are installed, the conduit cannot be filled to more than 31% of the conductor's cross-sectional area. Three or more conductors are typically installed in conduit. In these applications, the conduit cannot be filled to more than 40% of its cross-sectional area. **Figure 13-9** shows *Table 1* in *Chapter 9* of the *NEC*.

TRADE TIP   *Annex C* of the *NEC* is used where conductors being installed are all of the same wire size and have the same insulation type. *Annex C* is found in the back of the *NEC* (beginning on page 70-744 in NFPA 70-2011) and lists the maximum number of conductors permitted in conduit or tubing. The maximum number of conductors permitted in a conduit according to *Annex C* takes into account the fill percentage requirements of *Table 1* in *Chapter 9* of the *NEC*. No additional calculations are needed.

For illustrative purposes only, **Figure 13-10** shows *Table 4* in *Chapter 9* of the *NEC*.

FIGURE 13-10   *Table 4* in *Chapter 9* of the *NEC*. Dimensional and percent fill data for EMT.

TABLE 4   Dimensions and Percent Area of Conduit and Tubing (Areas of Conduit or Tubing for the Combinations of Wires Permitted in Table 1, Chapter 9)

Article 358 — Electrical Metallic Tubing (EMT)

Metric Designator	Trade Size	Nominal Internal Diameter mm	Nominal Internal Diameter in.	Total Area 100% mm²	Total Area 100% in.²	60% mm²	60% in.²	1 Wire 53% mm²	1 Wire 53% in.²	2 Wires 31% mm²	2 Wires 31% in.²	Over 2 Wires 40% mm²	Over 2 Wires 40% in.²
16	½	15.8	0.622	196	0.304	118	0.182	104	0.161	61	0.094	78	0.122
21	¾	20.9	0.824	343	0.533	206	0.320	182	0.283	106	0.165	137	0.213
27	1	26.6	1.049	556	0.864	333	0.519	295	0.458	172	0.268	222	0.346
35	1¼	35.1	1.380	968	1.496	581	0.897	513	0.793	300	0.464	387	0.598
41	1½	40.9	1.610	1314	2.036	788	1.221	696	1.079	407	0.631	526	0.814
53	2	52.5	2.067	2165	3.356	1299	2.013	1147	1.770	671	1.040	866	1.342
63	2½	69.4	2.731	3783	5.858	2270	3.515	2005	3.105	1173	1.816	1513	2.343
78	3	85.2	3.356	5701	8.846	3421	5.307	3022	4.688	1767	2.742	2280	3.538
91	3½	97.4	3.834	7451	11.545	4471	6.927	3949	6.119	2310	3.579	2980	4.618
103	4	110.1	4.334	9521	14.753	5712	8.852	5046	7.819	2951	4.573	3808	5.901

This table lists dimensions and areas for electrical metallic tubing (EMT). This is only one of 12 charts located in *Chapter 9, Table 4* of the *NEC* for the various types of raceways.

LIFE SKILLS

How does the material covered in this section relate to what you are doing in the field? Write an example of an instance where you may need to perform conduit fill calculations during your apprenticeship. Being prepared for what you will encounter on the job helps you to build confidence.

Self-Check 2

1. Why are cable wiring methods used in residential wiring applications?
2. Where in the *NEC* are percent fill values covered for conduit and tubing?

13-3 ▪ Box Fill Calculations and Box Selection

In this section, you learn the difference between a device box and a junction box. You see how box fill is calculated and what determines the correct size box.

Box Types and Selection

Device box an electrical device that is designed to hold devices such as switches and receptacle.

There are different electrical box types. For residential wiring applications, electrical boxes come in metal and plastic nonmetallic types. Electrical boxes are generally classified as device boxes and junction boxes. A device box is an electrical device that is designed to hold devices such as switches and receptacles. Device boxes are designed as a 3 in. × 2 in. × 3½ in. metal device box or a 20-cubic-inch plastic nail-on device box. **Figure 13-11** shows a metal device box, sometimes called

FIGURE 13-11 A metal device box.

From ZACHARIASON. *Electrical Materials*, 1E. © 2008 Cengage Learning

FIGURE 13-12 A nonmetallic device box.

From ZACHARIASON. *Electrical Materials*, 1E. © 2008 Cengage Learning

a handy box, and **Figure 13-12** shows a nonmetallic (single-gang fiberglass) device box.

A junction box is a box whose purpose is to provide a protected place for splicing electrical conductors. Devices are *not* installed in junction boxes. **Figure 13-13** shows a junction box.

Junction box a box whose purpose is to provide a protected place for splicing electrical conductors.

In some cases, conductors are spliced together in a box that also has a device in it. This particular box serves as both a device box and a junction box.

Electricians use electrical plans to determine what needs to be installed and at what locations. If the plans show a duplex receptacle, the electrician will install a *device box*. If the symbol for a ceiling-mounted incandescent luminaire is shown on the plans, an *outlet box* designed to accommodate a luminaire must be installed. If a *junction box* symbol is indicated on the plans, a box type that is designed to accommodate the appropriate number of conductors coming into that box must be installed. Paying close attention to the electrical plans ensures the right boxes are installed at the proper locations!

The electrical contractor decides on whether to use metal or nonmetallic boxes. Some plans call for only metal boxes or only nonmetallic boxes. Most residential wiring applications for new houses use nonmetallic boxes at those locations in the house where they are appropriate. Typically, houses are wired using a combination of metal and nonmetallic boxes.

FIGURE 13-13 A junction box.

© Bass/www.Shutterstock.com

FIGURE 13-14   Metal device boxes: (A) single gang, (B) two gang, and (C) three gang.

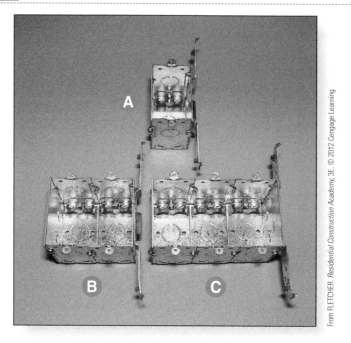

From FLETCHER. *Residential Construction Academy*, 3E. © 2012 Cengage Learning

Electrical plans also indicate whether a device box must be single gang, two gang, three gang, or more. **Figure 13-14** shows metal device boxes.

If the symbols on the electrical plan indicate that two single-pole switches are to be located at a specific spot, a two-gang box has to be installed to accommodate the two switches.

TRADE TIP   Where using metal device boxes that are gangable, you may have to remove the sides of the boxes and configure them in a way that results in an electrical box that accommodates the required number of devices at that location.

Plastic nonmetallic electrical boxes are designed to accommodate a specific number of devices. An example is a location that requires three switches. A nonmetallic electrical box already configured as a three-gang box is used. Plastic nonmetallic device boxes are shown in **Figure 13-15**.

Another consideration when selecting an electrical box is whether it is to be used in *new* work or *old* work. If selecting a box for new construction, the *new work* boxes are used. These type boxes have *no ears* on them and are designed to be attached directly to framing members such as studs and joists. However, if wiring an existing house, often called *old* work or *remodel* work, this requires *old work* boxes that have ears on them. Securing electrical boxes directly in a wall or ceiling material such as sheetrock requires special mounting accessories to hold them in place.

FIGURE 13-15 Plastic nonmetallic device boxes: (A) single gang, (B) two gang, and (C) three gang.

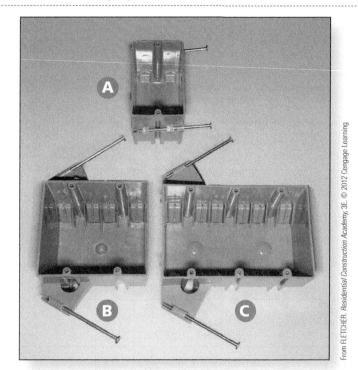

From FLETCHER. *Residential Construction Academy*, 3E. © 2012 Cengage Learning

Box Fill and Sizing Electrical Boxes

After an electrician selects the proper box type for a wiring application, the size of the box must be determined. The electrical box must be large enough to accommodate the required number of conductors and other electrical devices, conductors, and fittings that will be placed in the box. Sizing an electrical box is an important part of wiring. An electrical box that is sized too small makes it difficult to install electrical devices and other items, such as pigtails and wirenuts, in the box. An improperly sized box can become a safety hazard. You should *never* jam wires into a box. This practice may not only damage the insulation on the wires but may also result in heat buildup within the box that can further damage the insulation on the wire. In *Article 314*, the *NEC* covers proper procedures for determining a minimum size box for the number of electrical conductors and other items that will be installed in the box. As an electrician, you need to become familiar with *Tables 314.16(A)* and *314.16(B)*. Where conductors are the same size, the proper metal box size can be selected by referring to *Table 314.16(A)*. Where conductors are of different sizes and nonmetallic boxes are installed, refer to *Table 314.16(B)* and use the cubic-inch volume for the particular size of wire being used. **Figures 13-16(A)** and **(B)** show these tables.

It should be noted that *Tables 314.16(A)* and *314.16(B)* do *not* take into consideration the space taken by luminaire studs, cable clamps, hickeys, switches, pilot lights, or receptacles that may be installed in the electrical box. These devices need additional space. Table 13-1 is a listing derived from *314.16(B)(1)* through *314.16(B)(5)*. This information should be considered where sizing electrical boxes.

FIGURE 13-16(A) *NEC Table 314.16(A).*

TABLE 314.16(A) Metal Boxes

Box Trade Size			Minimum Volume		Maximum Number of Conductors* (arranged by AWG size)						
mm	in.		cm³	in.³	18	16	14	12	10	8	6
100 × 32	(4 × 1¼)	round/octagonal	205	12.5	8	7	6	5	5	2	2
100 × 38	(4 × 1½)	round/octagonal	254	15.5	10	8	7	6	6	5	3
100 × 54	(4 × 2⅛)	round/octagonal	353	21.5	14	12	10	9	8	7	4
100 × 32	(4 × 1¼)	square	295	18.0	12	10	9	8	7	6	3
100 × 38	(4 × 1½)	square	344	21.0	14	12	10	9	8	7	4
100 × 54	(4 × 2⅛)	square	497	30.3	20	17	15	13	12	10	6
120 × 32	(4¹¹⁄₁₆ × 1¼)	square	418	25.5	17	14	12	11	10	8	5
120 × 38	(4¹¹⁄₁₆ × 1½)	square	484	29.5	19	16	14	13	11	9	5
120 × 54	(4¹¹⁄₁₆ × 2⅛)	square	689	42.0	28	24	21	18	16	14	8
75 × 50 × 38	(3 × 2 × 1½)	device	123	7.5	5	4	3	3	3	2	1
75 × 50 × 50	(3 × 2 × 2)	device	164	10.0	6	5	5	4	4	3	2
75× 50 × 57	(3 × 2 × 2¼)	device	172	10.5	7	6	5	4	4	3	2
75 × 50 × 65	(3 × 2 × 2½)	device	205	12.5	8	7	6	5	5	4	2
75 × 50 × 70	(3 × 2 × 2¾)	device	230	14.0	9	8	7	6	5	4	2
75 × 50 × 90	(3 × 2 × 3½)	device	295	18.0	12	10	9	8	7	6	3
100 × 54 × 38	(4 × 2⅛ × 1½)	device	169	10.3	6	5	5	4	4	3	2
100 × 54 × 48	(4 × 2⅛ × 1⅞)	device	213	13.0	8	7	6	5	5	4	2
100 × 54 × 54	(4 × 2⅛ × 2⅛)	device	238	14.5	9	8	7	6	5	4	2
95 × 50 × 65	(3¾ × 2 × 2½)	masonry box/gang	230	14.0	9	8	7	6	5	4	2
95 × 50 × 90	(3¾ × 2 × 3½)	masonry box/gang	344	21.0	14	12	10	9	8	7	4
min. 44.5 depth	FS — single cover/gang (1¾)		221	13.5	9	7	6	6	5	4	2
min. 60.3 depth	FD — single cover/gang (2⅜)		295	18.0	12	10	9	8	7	6	3
min. 44.5 depth	FS — multiple cover/gang (1¾)		295	18.0	12	10	9	8	7	6	3
min. 60.3 depth	FD — multiple cover/gang (2⅜)		395	24.0	16	13	12	10	9	8	4

Where no volume allowances are required by 314.16(B)(2) through (B)(5).

FIGURE 13-16(B) *NEC Table 314.16(B).*

TABLE 314.16(B) Volume Allowance Required per Conductor

Size of Conductor (AWG)	Free Space Within Box for Each Conductor	
	cm³	in.³
18	24.6	1.50
16	28.7	1.75
14	32.8	2.00
12	36.9	2.25
10	41.0	2.50
8	49.2	3.00
6	81.9	5.00

Some examples of using *Tables 314.16(A)* and *314.16(B)* follow:

Example 1

A *nonmetallic* box is used and all conductors are the *same* size. The nonmetallic box is marked as having a volume of 21 cubic inches. The box contains no luminaire stud or cable clamps. How many 12 AWG conductors are permitted in this box?

Using *Table 314.16(B)*, shown in Figure 13-16(B), the volume requirement for a 12 AWG conductor is 2.25 cubic inches. The maximum number of 12 AWG conductors is calculated as follows:

$$21/2.25 = 9.3$$

TABLE 13-1 Additional Allowances to Be Considered for Electrical Box Sizing

Quick checklist for possibilities to be considered when determining proper size boxes. See *314.16(B)(1)* through *314.16(B)(5)*.	
• If box contains no fittings, devices, fixture studs, cable clamps, hickeys, switches, receptacles, or equipment grounding conductors:	• Refer directly to *Table 314.16(A)* or *Table 314.16(B)*.
• **Clamps**. If box contains one or more internal cable clamps:	• Add a single-volume based on the largest conductor in the box.
• **Support Fittings**. If box contains one or more luminaire studs or hickeys:	• Add a single-volume for each type based on the largest conductor in the box.
• **Device or Equipment**. If box contains one or more wiring devices on a yoke:	• Add a double-volume for each yoke based on the largest conductor connected to a device on that yoke. Some large wiring devices, such as a 30-ampere dryer receptacle, require a 2-gang box.
• **Equipment Grounding Conductors**. If a box contains one or more equipment grounding conductors:	• Add a single-volume based on the largest equipment grounding conductor in the box.
• **Isolated Equipment Grounding Conductor**. If a box contains one or more additional "isolated" (insulated) equipment grounding conductors as permitted by 250.146(D) for "noise" reduction:	• Add a single-volume based on the largest equipment grounding conductor in the box.
• For conductors less than 12 in. (300 mm) long that are looped or coiled in the box without being spliced:	• Add a single-volume for each conductor that is looped or coiled through the box.
• For conductors 12 in. (300 mm) or longer that are looped or coiled in the box without being spliced:	• Add a double-volume for each conductor that is looped or coiled through the box.
• For conductors that originated outside of the box and terminate inside the box:	• Add a single-volume for each conductor that originates outside the box and terminates inside the box.
• If no part of the conductor leaves the box, as with a "jumper" wire used to connect three wiring devices on one yoke, or pigtails:	• Don't count this (these). No additional volume required.
• For small equipment grounding conductors or not more than four conductors smaller than 14 AWG that originate from a luminaire canopy or similar canopy (like a fan) and terminate in the box:	• Don't count this (these). No additional volume required.
• For small fittings, such as locknuts, bushings, and wire connectors:	• Don't count this (these). No additional volume required.

In order to comply with the *NEC*, 9 conductors shall be installed. (Installing 10 conductors would be a *Code* violation.)

Example 2

A *metal* device box is used and all conductors are the *same* size. The box contains one luminaire stud and two internal cable clamps. The wiring method is two 14/2 NM with ground cables.

Four 14 AWG conductors	4
One luminaire stud	1
Two cable clamps (count only one)	1
Two 14 AWG equipment grounding conductors (count only one)	1
Total	7

Refer to *Tables 314.16(A)* and *314.16(B)* and note that a 3 in. × 2 in. × 2¾ in. device box can be used for this example.

NEC 314.16 states that boxes and conduit bodies shall be of sufficient size to provide free space for all enclosed conductors. This section also covers box volume and box fill calculations. The total box volume is determined by adding the individual volumes of the box components. Box components include the box itself plus any attachments to it, such as a plaster ring, extension ring, or luminaire dome cover. A 4 in. × 1½ in. square box and a ¾ in. raised plaster ring being used together are shown in **Figure 13-17**.

....

TRADE TIP When calculating total volume in an electrical box, the box itself and any raised cover attached must be included. The cubic-inch capacity of the cover must be marked on it in order for the volume of the cover to be included.

....

Total box fill is the sum of the volume allowances for all items that contribute to box fill. The volume allowance for each item is based on the volume listed in

FIGURE 13-17 A square box and plaster ring.

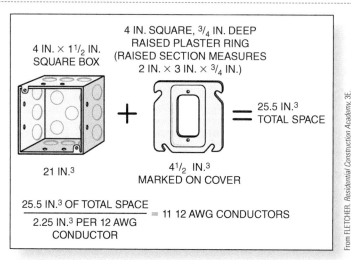

4 IN. × 1½ IN. SQUARE BOX

4 IN. SQUARE, ¾ IN. DEEP RAISED PLASTER RING (RAISED SECTION MEASURES 2 IN. × 3 IN. × ¾ IN.)

25.5 IN.³ TOTAL SPACE

21 IN.³

4½ IN.³ MARKED ON COVER

$$\frac{25.5\ \text{IN.}^3\ \text{OF TOTAL SPACE}}{2.25\ \text{IN.}^3\ \text{PER 12 AWG CONDUCTOR}} = 11\ 12\ \text{AWG CONDUCTORS}$$

Table 314.16(B) for the conductor size indicated. **Figure 13-18** is a box selection guide used to determine the maximum number of conductors permitted for the more common metal box styles and sizes used in residential wiring.

FIGURE 13-18 Metal electrical box selection guide.

BOX SELECTION GUIDE
FOR METAL BOXES GENERALLY USED FOR RESIDENTIAL WIRING

DEVICE BOXES

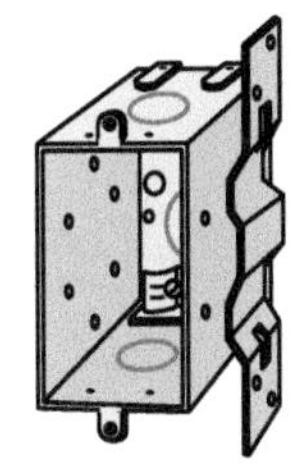

WIRE SIZE	3x2x2½ (12.5 IN.³)	3x2x2¾ (14 IN.³)	3x2x3½ (18 IN.³)
14 AWG	6	7	9
12 AWG	5	6	8
10 AWG	5	5	7

SQUARE BOXES

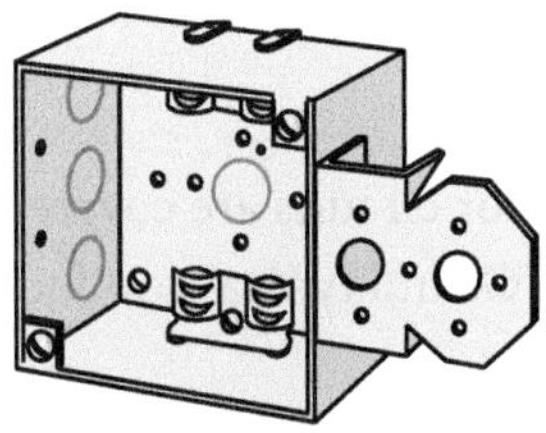

WIRE SIZE	4x4x1½ (21 IN.³)	4x4x2⅛ (30.3 IN.³)
14 AWG	10	15
12 AWG	9	13
10 AWG	8	12

OCTAGON BOXES

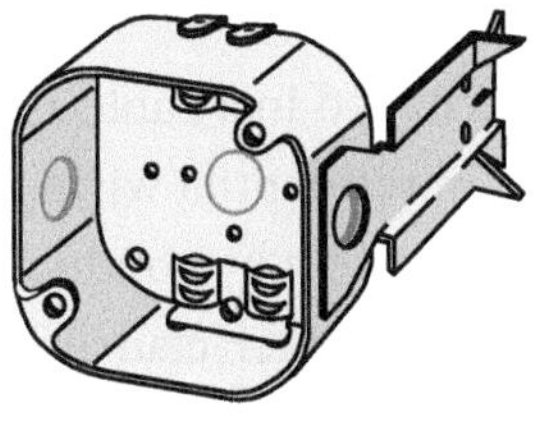

WIRE SIZE	4x1½ (15.5 IN.³)	4x2⅛ (21.5 IN.³)
14 AWG	7	10
12 AWG	6	9
10 AWG	6	8

HANDY BOXES

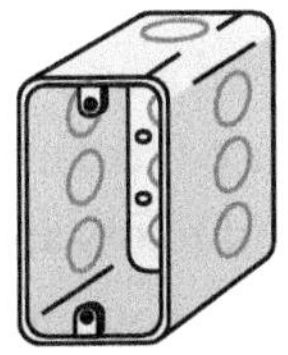

WIRE SIZE	4x2⅛x1½ (10.3 IN.³)	4x2⅛x1⅞ (13 IN.³)	4x2⅛x2⅛ (14.5 IN.³)
14 AWG	5	6	7
12 AWG	4	5	6
10 AWG	4	5	5

RAISED COVERS

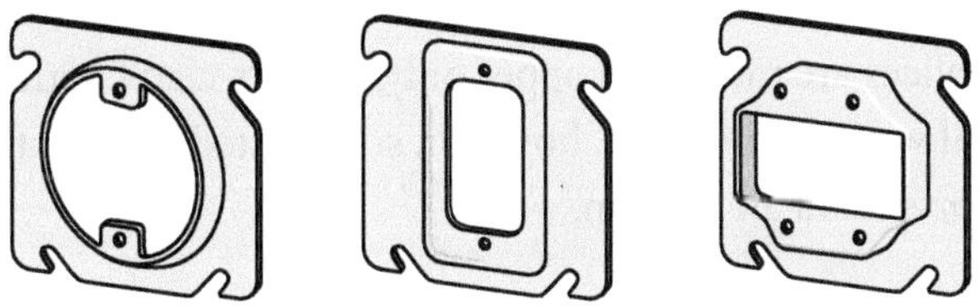

WHERE RAISED COVERS ARE MARKED WITH THEIR VOLUME IN CUBIC INCHES, THAT VOLUME MAY BE ADDED TO THE BOX VOLUME TO DETERMINE MAXIMUM NUMBER OF CONDUCTORS IN THE COMBINED BOX AND RAISED COVER.

NOTE: BE SURE TO MAKE DEDUCTIONS FROM THE ABOVE MAXIMUM NUMBER OF CONDUCTORS PERMITTED FOR WIRING DEVICES, CABLE CLAMPS, FIXTURE STUDS, AND GROUNDING CONDUCTORS. THE CUBIC INCH (IN.³) VOLUME IS TAKEN DIRECTLY FROM *TABLE 314.16(A)* OF THE *NEC.*® NONMETALLIC BOXES ARE MARKED WITH THEIR CUBIC INCH CAPACITY.

Set a goal to be able to understand and explain the various types of boxes listed in this chapter and their use. Try testing yourself by writing a short description of each without looking at the text or your notes.

Self-Check 3

1. List the two types of electrical boxes used in residential wiring applications.
2. List the two general classifications of electrical boxes.
3. Where are electrical boxes with *ears* used?
4. Which *NEC* section states that boxes and conduit bodies shall be of sufficient size to provide free space for all enclosed conductors?

13-4 Proper Conductor Size and Overcurrent Device for a Circuit

Before *NEC* requirements regarding overcurrent devices for circuits are covered, a review of conductor types, conductor sizes and the AWG, and typical conductor applications is provided in the following sections.

Conductors

Residential wiring applications typically use conductors installed in a cable assembly. Very seldom are raceways with installed conductors used in residential wiring. Current is delivered to various loads in a residential wiring system by conductors. The most commonly used conductors are copper, aluminum, and copper-clad aluminum. Copper is the most popular conductor because of its strength and its high level of conductivity. Conductivity is a measure of the ability of a substance to conduct electricity.

Conductivity a measure of the ability of a substance to conduct electricity.

Aluminum has a higher resistance value and is not as good a conductor as copper. Therefore, a larger-size aluminum conductor must be used to carry the same amount of current as a copper conductor. Aluminum conductors are generally used in larger sizes because of their lower cost and lighter weight. You must use antioxidant compound on exposed aluminum at all terminations, as per manufacturers' recommendations, because aluminum oxidizes more quickly than copper.

Copper-clad aluminum has an aluminum core and an outer coating of copper. Because most wiring devices are rated for copper only, copper-clad conductors are rarely used in residential wiring. There are, however, some listed devices that can be used with copper *or* copper-clad aluminum.

TRADE TIP   Always check the wiring device for a marking that indicates what conductor type is listed and should be used.

TABLE 13-2 Conductor and Terminal Identification Markings

Type of Device	Marking on Terminal or Conductor	Conductor Permitted
15- or 20-ampere receptacles and switches	CO/ALR	Aluminum, copper, copper-clad aluminum
15- and 20-ampere receptacles and switches	NONE	Copper, copper-clad aluminum
30-ampere and greater receptacles and switches	AL/CU	Aluminum, copper, copper-clad aluminum
30-ampere and greater receptacles and switches	NONE	Copper only
Screwless pressure terminal connectors of the push-in type	NONE	Copper or copper-clad aluminum
Wire connectors	AL	Aluminum
Wire connectors	AL/CU or CU/AL	Aluminum, copper, copper-clad aluminum
Wire connectors	CC	Copper-clad aluminum only
Wire connectors	CC/CU or CU/CC	Copper or copper-clad aluminum
Wire connectors	CU or CC/CU	Copper only
Any of the above devices	COPPER or CU ONLY	Copper only

From FLETCHER. *Residential Construction Academy*, 3E. © 2012 Cengage Learning

Table 13-2 shows typical markings used to determine the type of conductors permitted with devices and connectors.

American Wire Gauge (AWG)

American Wire Gauge (AWG) the system used to size the conductors used in residential wiring.

The system used to size the conductors used in residential wiring is referred to as the **American Wire Gauge (AWG)**. In the AWG system, the smaller the number, the larger the conductor, and the larger the number, the smaller the conductor. For example, a 6 AWG conductor is larger than a 12 AWG conductor, and a 14 AWG conductor is smaller than an 8 AWG conductor. The *NEC* recognizes building wire from 18 AWG up to 4/0.

TRADE TIP   1/0, 2/0, 3/0, and 4/0 are sometimes listed as 0, 00, 000, and 0000, respectively. These wire sizes are pronounced as 1 *aught*, 2 *aught*, 3 *aught*, and 4 *aught*.

Some common conductor sizes and their applications are shown in Table 13-3.

NEC Requirements

The following are *NEC* references (tables and requirements) as they pertain to conductors and overcurrent protection devices for circuits. As an electrician, you should be very familiar with these requirements.

TABLE 13-3 Residential Wiring—Conductor Applications

Copper Conductor Size	Overcurrent Protection (Fuse or Circuit Breaker)	Typical Applications
18 AWG	Circuit transformers provide overcurrent protection.	Low-voltage wiring for thermostats, chimes, security, remote control, home automation systems, and so on.
16 AWG	Circuit transformers provide overcurrent protection.	Same applications as above. Good for long runs to minimize voltage drop.
14 AWG	15 amperes	Typical lighting branch circuits.
12 AWG	20 amperes	Small-appliance branch circuits for the receptacles in kitchens and dining rooms. Also, laundry receptacles and workshop receptacles. Often used for lighting branch circuits, as well as some smaller electric water heaters.
10 AWG	30 amperes	Most electric clothes dryers, built-in ovens, cooktops, central air conditioners, some electric water heaters, heat pumps.
8 AWG	40 amperes	Electric ranges, ovens, heat pumps, some large electric clothes dryers, large central air conditioners, heat pumps.
6 AWG	50 amperes	Electric furnaces, heat pumps, feeders to subpanels.
4 AWG	70 amperes	Electric furnaces, feeders to subpanels.
3 AWG and larger	100 amperes	Main service entrance conductors, feeders to subpanels, electric furnaces.

NEC 210.19(A)(1) covers the minimum ampacity and size of branch-circuit conductors. This section states that branch-circuit conductors shall have an ampacity not less than the maximum load to be served. Where a branch circuit supplies continuous loads or any combination of continuous and noncontinuous loads, the minimum branch-circuit conductor size, before the application of any adjustment or correction factors, shall have an allowable ampacity not less than the noncontinuous load plus 125% of the continuous load.

The current-carrying capacity, or ampacity of conductor, must not be less than the rating of the overcurrent device protecting that conductor. This topic is covered in both *210.19* and *210.20.*

The ampere rating of the branch-circuit overcurrent protection device (circuit breaker or fuse) determines the rating of the branch circuit. As an example, if a 20-amp conductor is protected by a 15-amp circuit breaker, the circuit is considered to be a 15-amp branch circuit.

NEC 210.21(B) covers receptacles and branch-circuit ratings. **Figure 13-19** shows the requirements of *210.21(B)* and *Table 210.21(B)(3),* where the branch circuit supplies two or more receptacles or outlets in homes.

NEC 210.3 covers branch-circuit ratings. This section states that branch circuits are rated in accordance with the maximum permitted ampere rating or setting of the overcurrent device. This section further states that where conductors of

FIGURE 13-19 Receptacle ratings for various size circuits.

TABLE 210.21(B)(3)	Receptacle Ratings for Various Size Circuits
Circuit Rating (Amperes)	**Receptacle Rating (Amperes)**
15	Not over 15
20	15 or 20
30	30
40	40 or 50
50	50

higher ampacity are used for any reason, the ampere rating or setting of the specified overcurrent device shall determine the circuit rating. An example of where conductors of higher ampacity might be used is for long home runs. A home run is the branch-circuit wiring from the panel to the first outlet or junction box in that circuit. A branch circuit (home run) with an overcurrent device rated at 15 amps uses 14 AWG conductors. However, in an effort to reduce the voltage drop in this long home run, 12 AWG conductors may be used. As mentioned above, this is permitted by *210.3*.

Home runs the branch-circuit wiring from the panel to the first outlet or junction box in that circuit.

NEC 210.24 is listed as *Branch-Circuit Requirements—Summary*, and *Table 210.24* provides a summary of minimum requirements. **Figure 13-20** shows this table.

Table 210.24 provides *only* a summary of minimum requirements. For specific requirements applying to each branch, review *210.19, 210.20,* and *210.21*.

NEC 240.4 states that circuit conductors must be protected against overcurrent in accordance with their ampacities, as specified in *Table 310.15(B)(16)*. Refer back to Figure 13-8.

NEC 240.6(A) lists the standard sizes of fuses and fixed-trip circuit breakers.

FIGURE 13-20 Summary of branch-circuit requirements.

TABLE 210.24 Summary of Branch-Circuit Requirements	15 A	20 A	30 A	40 A	50 A
Circuit Rating	**15 A**	**20 A**	**30 A**	**40 A**	**50 A**
Conductors (min. size):					
Circuit wires[1]	14	12	10	8	6
Taps	14	14	14	12	12
Fixture wires and cords — see 240.5					
Overcurrent Protection	**15 A**	**20 A**	**30 A**	**40 A**	**50 A**
Outlet devices:					
Lampholders permitted	Any type	Any type	Heavy duty	Heavy duty	Heavy duty
Receptacle rating[2]	15 max. A	15 or 20 A	30 A	40 or 50 A	50 A
Maximum Load	**15 A**	**20 A**	**30 A**	**40 A**	**50 A**
Permissible load	See 210.23(A)	See 210.23(A)	See 210.23(B)	See 210.23(C)	See 210.23(C)

[1] These gauges are for copper conductors.
[2] For receptacle rating of cord-connected electric-discharge luminaires, see 410.62(C).

Understanding *NEC* requirements keeps you from making mistakes when selecting the proper conductor size and overcurrent devices for circuits being installed. You *must* know that where using 14 AWG wire, the overcurrent device is 15 amps and where running 12 AWG wire, the protection device is rated at 20 amps. You must also understand that device ratings connected to these circuits *must* match. As an example, a circuit using 14 AWG conductors requires switches and receptacles rated at 15 amps.

LIFE SKILLS

The introduction to this section suggests that you will be well served to review conductor types, conductor sizes, the AWG, and typical conductor applications before learning the *NEC* requirements regarding overcurrent devices for circuits. Can you think of any instances in the field where understanding background information before tackling a problem might make things easier?

Self-Check 4

1. List the three most commonly used conductors in residential wiring.
2. List three typical applications in residential wiring that use 12 AWG conductors.
3. What is a home run with reference to residential wiring?

Summary

- A branch circuit is defined by the *NEC* as the circuit conductors between the final overcurrent device (fuse or circuit breaker) and the power and/or lighting outlets.
- Types of branch circuits used in residential wiring applications include lighting, receptacle, general-purpose, general lighting, small-appliance, laundry, bathroom, and individual.
- To calculate the minimum number of general lighting circuits required in a house, the habitable floor area (rooms used for living) must first be determined.
- Ampacity is the current, in amperes, that a conductor can carry continuously under the conditions of use without exceeding its temperature rating. *NEC* 210.19(A)(1) states that branch-circuit conductors are required to have an ampacity at least equal to the maximum electrical load to be served.
- Most conductors used for branch-circuit wiring in residential applications are copper because of copper's strength and high level of conductivity.
- In residential wiring applications, conduit of some type is very common for service-entrance installations; however, a raceway for branch-circuit installation is almost never used.
- It is easier and less expensive to use cable wiring methods where wiring houses.

- The decision to use a certain type of raceway in a residential wiring installation is made based on the actual wiring application, an electrician's personal preference of raceway types, and what is required by any local electrical codes.

- *Table 1* in *Chapter 9* of the *NEC* specifies the *maximum* fill percentage of conduit or tubing.

- For residential wiring applications, electrical boxes come in metal and plastic nonmetallic types and are generally classified as device boxes, outlet boxes, and junction boxes.

- Where using metal device boxes that are gangable, you may have to remove the sides of the boxes and configure them in a way that results in an electrical box that accommodates the required number of devices at that location.

- Plastic nonmetallic electrical boxes are designed to accommodate a specific number of devices.

- An electrical box that is sized too small makes it difficult to install electrical devices and other items such as pigtails and wirenuts in the box. An improperly sized box can become a safety hazard.

- *Article 314* covers proper procedures for determining a minimum size box for the number of electrical conductors and other items that are installed in the box.

- The most commonly used conductors in residential wiring are copper, aluminum, and copper-clad aluminum.

- The system used to size the conductors used in residential wiring is referred to as the American Wire Gauge (AWG).

- In the AWG system, the smaller the number, the larger the conductor, and the larger the number, the smaller the conductor.

Review Questions

True/False

1. A branch circuit is defined as the circuit conductors between the final overcurrent device and the power and/or lighting outlets. (True, False)

2. A branch circuit that has only lighting outlets is called a general-purpose branch circuit. (True, False)

3. To calculate the minimum number of general lighting circuits required in a house, the habitable floor area must first be determined. (True, False)

4. An appliance branch circuit is a branch circuit that supplies power to one or more outlets to which appliances are to be connected and that has no permanently connected luminaires that are not part of an appliance. (True, False)

5. *NEC 220.52(A)* states that in each dwelling unit, the load must be computed at 500 volt-amperes for each 2-wire small-appliance branch circuit. (True, False)

Multiple Choice

6. *NEC 210.11(C)(3)* states that in addition to the number of general lighting, small-appliance, and laundry branch circuits, at least how many 20-amp branch circuits must be provided to supply the bathroom receptacle outlets?
   A. One
   B. Two
   C. Three
   D. Four

7. The current that a conductor can carry continuously under the conditions of use without exceeding its temperature rating is _____.
   A. Amperage
   B. Ampacity
   C. Conductivity
   D. Volt-amperes

8. Most conductors used for branch-circuit wiring in residential applications are _____.
   A. Copper-clad aluminum
   B. Iron-clad aluminum
   C. Copper
   D. Aluminum

9. The most common raceway used for branch-circuit installation is _____.
   A. RMC
   B. IMC
   C. PVC
   D. EMT

10. *Table 1* in *Chapter 9* of the *NEC* specifies that the maximum fill percentage of conduit or tubing with three or more conductors installed is _____.
    A. 31
    B. 41
    C. 40
    D. 53

Fill in the Blank

11. *Annex* _____ of the *NEC* is used where conductors being installed are all of the same wire size and have the same insulation type.

12. A(n) _____ box is an electrical device that is designed to hold devices such as switches and receptacles.

13. A(n) _____ box is a box that is designed for the mounting of a receptacle or a luminaire.

14. A(n) _____ box is a box whose purpose is to provide a protected place for splicing electrical conductors.

15. _____ electrical boxes are designed to accommodate a specific number of devices.

16. Where wiring an existing house, electrical boxes with _____ are required.

17. *Article* _____ covers proper procedures for determining a minimum size box for the number of electrical conductors and other items installed in the box.

18. *NEC* _____ states that boxes and conduit bodies shall be of sufficient size to provide free space for all enclosed conductors.

19. Total box fill is the _____ of the volume allowances for all items that contribute to box fill.

20. Copper is the most popular conductor because of its strength and its high level of

_____.

21. Always check the wiring device for a(n) _____ that indicates what conductor type is listed and should be used.

22. In the AWG system, the _____ the number, the larger the conductor.

23. The _____ of a conductor must not be less than the rating of the overcurrent device protecting that conductor.

24. A home run is the branch-circuit wiring from the _____ to the first outlet or junction box in that circuit.

25. *NEC* _____ lists the standard sizes of fuses and fixed-trip circuit breakers.

ACTIVITY 13–1   Calculating Box Fill

The standard method for determining adequate box size calculates the total box volume, and from that subtracts the total box fill, to ensure compliance. Using this method, refer to **Figure 13-21**, which is a standard 3 in. $\times$ 2 in. $\times$ 3½ in. device box (18 in.^{3}), and determine whether the box is adequately sized.

Items Contained within Box	Volume Allowance	Unit Volume Based on *Table 314.16(B)* (in.^{3})	Total Box Fill (in.^{3})
Total			

FIGURE 13-21   Device box.

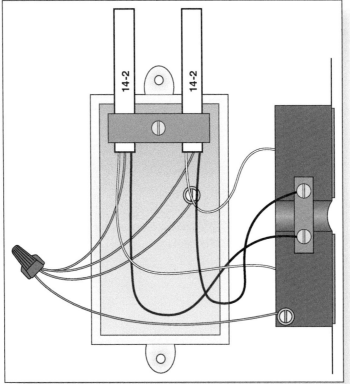

© Cengage Learning 2013

ACTIVITY 13–2   Calculating Box Fill

The standard method for determining adequate box size calculates the total box volume, and from that subtracts the total box fill, to ensure compliance. Using this method, refer to **Figure 13-22**, which shows two standard gangable device boxes containing conductors of different sizes, and determine whether the box is adequately sized.

Items Contained within Box	Volume Allowance	Unit Volume Based on *Table 314.16(B)* (in.³)	Total Box Fill (in.³)
Total			

FIGURE 13-22   Two standard device boxes.

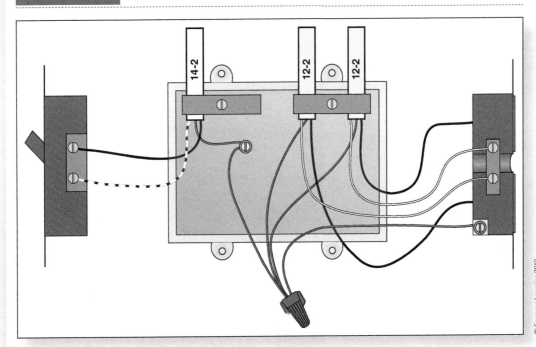

© Cengage Learning 2013

Chapter 14

Wiring Requirements

Career Profile

Larry Carlyle, a retired electrical contractor and current apprentice instructor, is a lifelong member of the electrical field. Of entering the field he says, "My father started an electrical contracting business when I was 13 years old. I would help him during school vacations, which gave me an understanding as to what the trade was about. Over the years, his business grew, and once I finished college, I joined him in the business. He eventually retired, and I took over the business and ran it for the next 20 years." He describes the responsibilities he held as the owner of a small contracting business: "I was responsible for finding the work, estimating, managing the employees, purchasing materials required for the jobs, attending jobsite meetings, preparing billings, and monitoring the financial health of the business." He explains that running a business also involves "finding and keeping good employees, finding and securing profitable jobs, then managing those jobs so they are [cost-effective]." Supervising a contracting business means "being constantly alert to potential problems and recognizing them before they occur," he concludes. Like his fellow instructors in apprenticeship programs, Carlyle knows that the electrical industry offers many job opportunities. He notes, "The construction end of it allows you to be exposed to the other opportunities such as [jobs for] estimators, project managers, purchasing agents, and owners." Therefore, he tells students, you should "learn all you can and aspire to higher levels."

Chapter Objectives

After completing this chapter, you will be able to:

1. Determine locations of receptacles, switches, and luminaires for a residential dwelling as per *NEC*.

2. List locations where GFCI protection is required in a residential dwelling unit.

3. Draw a cable layout for a given master bedroom.

Life Skills Covered

 Communication Skills

 Cooperation and Teamwork

Life Skill Goals

The life skills that we're covering in this chapter are Communication Skills and Cooperation and Teamwork. As an electrician, you're going to be dealing with wiring requirements all the time, so pay extra attention to the content. Take time to talk with your classmates about what you're learning. You may be surprised; sometimes the simple act of explaining something out loud can help you understand the subject better.

Where You Are Headed

This chapter covers *NEC* requirements for locating receptacles, switches, and luminaires in residential wiring applications. Ground-fault protection and *Code* requirements are emphasized, as GFCI protection locations are also detailed

in this chapter. As an electrician, on the jobsite you will review floor and electrical plans. You may then be called upon to sketch cable diagrams and make the wiring work according to the *NEC* in the most efficient, cost-effective manner. In this chapter, a cable layout diagram is developed for a master bedroom in a residential dwelling. Make sure you are very familiar with the *NEC* requirements described in this chapter, and become proficient in drawing cable layout diagrams.

Introduction

You have learned the importance of the *NEC Code* requirements in previous chapters. This chapter covers *NEC* requirements for locating receptacles, switches, and luminaires in residential wiring applications. *NEC* requirements are further covered as they pertain to ground-fault protection, and *Code* requirements are emphasized as GFCI protection locations are detailed. The last section in this chapter describes how a cable layout diagram is developed and covers how to draw a cable layout diagram for a master bedroom in a residential dwelling.

14-1 *NEC* Requirements for Locating Receptacles, Switches, and Luminaires for a Residential Dwelling

Before *NEC* requirements are described, it is important to have a basic understanding of how a circuit is laid out and who generally lays these circuits out. Although *NEC* requirements are always followed, there are many possible circuit arrangements.

Circuit Layout

In every installation, *NEC* requirements *must* be followed. With residential wiring applications, the job of laying out the circuits typically belongs to the electrician. The electrician, always mindful of *NEC* requirements, has to figure out how to lay out electrical circuits in an efficient and cost-effective way.

In some installations, generally the larger and more costly residences, an architect may be responsible for the circuit design. However, sometimes certain procedures or methods must be done to meet *Code* requirements that the architect may *not* have included on the prints.

Following are some possible circuit arrangements:

- Run more than one circuit when wiring a room. Refer to **Figure 14-1**.

If, for example, circuit "A" develops a problem, circuit "B" continues to supply power to the light and the other outlets in this room.

- Be cost-effective when wiring receptacle outlets. **Figure 14-2** shows how receptacle outlets can be connected *back to back*.

FIGURE 14-1   Two circuits feeding one room.

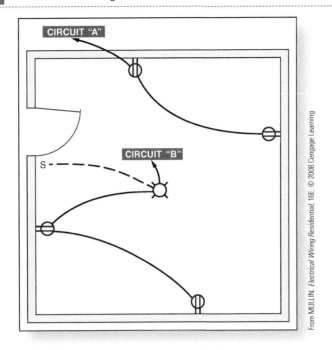

From MULLIN. *Electrical Wiring Residential*, 16E. © 2008 Cengage Learning

FIGURE 14-2   Receptacle outlets connected back to back.

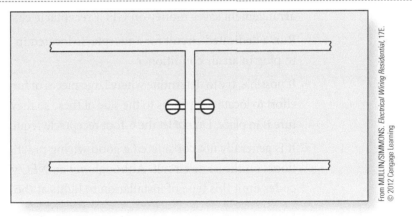

From MULLIN/SIMMONS. *Electrical Wiring Residential*, 17E.
© 2012 Cengage Learning

Because the distance between outlets is so small, the cost of materials can be reduced. However, *building codes* must be consulted before installing back-to-back outlets in **fire-rated walls**. A fire-rated wall, or fire wall, is made of fireproof material to prevent the spread of a fire from one part of a dwelling to another.

Fire-rated wall made of fireproof material to prevent the spread of a fire from one part of a dwelling to another; also called a fire wall.

- A way to keep the outlet box at the lighting outlet or the wiring compartment on recessed cans less crowded is to run the branch circuit *first* to the switch and *then* to the controlled lighting outlet (see **Figure 14-3**). When the grounded circuit conductor is present at the switch location, it is easier to install switching devices that require a connection to the grounded circuit conductor.

- Use a 2-wire NMB cable (2 "hot" conductors, neutral, and ground) to supply two circuits for a room. One circuit is used for lighting and one for receptacles (sharing the neutral).

Lighting outlet wiring.

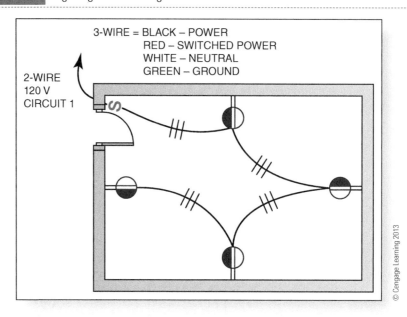

3-WIRE = BLACK – POWER
RED – SWITCHED POWER
WHITE – NEUTRAL
GREEN – GROUND

2-WIRE
120 V
CIRCUIT 1

© Cengage Learning 2013

- When wiring countertop receptacles, use *one* GFCI receptacle and off the *load* side of the GFCI receptacle, connect remaining countertop (standard) receptacles. If a ground fault occurs in one of the standard receptacles, the GFCI trips. This arrangement saves money on GFCI receptacle costs.

- Run a dedicated circuit for a receptacle located in a room where you might need to plug in an air conditioner.

- If possible, try to determine where large pieces of furniture will be placed. Make an effort to locate receptacles to the side of these so they will be accessible once the furniture is in place. Do *not* let the 6-foot receptacle requirement prevent this procedure.

- It is generally not considered a good wiring practice to include outlets on different floors on the same circuit. Although *not* an *NEC* violation, some local building codes limit this type of installation to lights at the head and foot of a stairway.

Arc-Fault Circuit Interrupter (AFCI) a device intended to provide protection from the effects of arc faults by recognizing characteristics unique to arcing and by functioning to de-energize the circuit when an arc fault is detected.

Because arc-fault circuit interrupters (AFCIs) are now part of *NEC* requirements as covered in *210.12*, wiring general-purpose branch circuits has become more complicated. An arc-fault circuit interrupter is a device intended to provide protection from the effects of arc faults by recognizing characteristics unique to arcing and by functioning to de-energize the circuit when an arc fault is detected. A decision must be made as to whether to put all bedroom receptacles, for example, and lighting outlets on one AFCI-protected branch circuit or whether to connect the receptacle and lighting outlets on different branch circuits, which means that *more* than one AFCI circuit breaker is required. This increases cost.

TRADE TIP When you are unsure whether a luminaire is suitable for running wires to and beyond it to another luminaire or part of the circuit, the installation should be made so that *only* the two wires connect to that luminaire.

NEC Requirements on Receptacle Locations

NEC 210.52 states where receptacle outlets must be installed in a dwelling unit. In order to meet or exceed the *Code* requirements, an electrician must be familiar with this information. These requirements provide permitted locations for electrical boxes that electricians install during the rough-in stage of wiring. The rough-in stage in an electrical installation is the time when the raceways, cable, boxes, and other electrical equipment are installed; this electrical work must be completed before any construction work can be done that covers wall and ceiling surfaces. *NEC 406* covers requirements that apply to dwelling unit receptacles that are rated 125 volts and 15 or 20 amperes *and* that are not part of a luminaire or an appliance. These receptacles are generally used to supply lighting and general-purpose electrical equipment and are in addition to the ones that are more than 5½ ft (1.7 m) above the floor, located within cupboards and cabinets, or controlled by a wall switch.

NEC 210.52(A) states that in every kitchen, family room, dining room, living room, parlor, library, den, sunroom, bedroom, recreation room, or similar room or area of dwelling units, receptacle outlets *must* be installed in accordance with the general provisions as follows: Receptacles must be installed so that no point measured horizontally along the floor line in any wall space is more than 6 ft (1.8 m) from a receptacle outlet. This installation provision requires electricians to install electrical boxes for receptacles during rough-in of wiring. These boxes are installed so that no point in any wall space is more than 6 ft (1.8 m) from a receptacle. This requirement allows for the placement of an appliance or lamp anywhere in a room near a wall and be within 6 ft (1.8 m) of a receptacle. This accommodates table lamps with cords that are typically 6 ft long. **Figure 14-4** illustrates the not more than 6-ft placement of receptacles. Installing receptacles in this not only follows *Code* requirements but also eliminates the need for long extension cords running down and around the walls of a room.

NEC 210.52(C) states the required location of receptacles at countertop locations in a dwelling unit kitchen or dining room. This section provides important requirements that state that electrical boxes for the receptacles must be installed and wired *before* kitchen cabinets and countertops are installed. Electrical boxes must be accurately located so they are not hidden behind cabinets or other kitchen equipment.

Rough-in stage in an electrical installation, the time when the raceways, cable, boxes, and other electrical equipment are installed; this electrical work must be completed before any construction work can be done that covers wall and ceiling surfaces.

FIGURE 14-4 Receptacle placement.

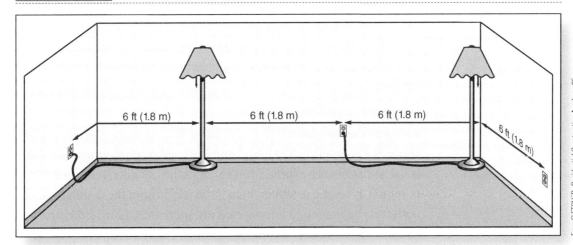

6 ft (1.8 m)   6 ft (1.8 m)   6 ft (1.8 m)   6 ft (1.8 m)

From FLETCHER. *Residential Construction Academy*, 3E.
© 2012 Cengage Learning

FIGURE 14-5 Dwelling unit receptacles service countertop spaces in a kitchen.

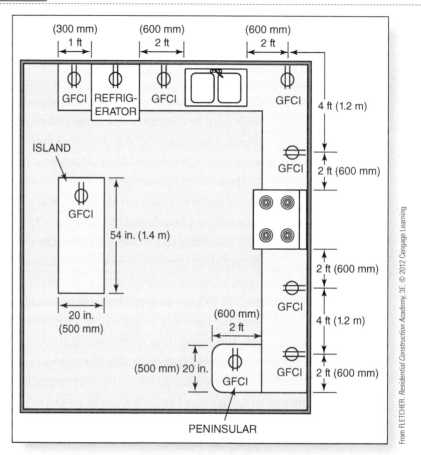

The following are requirements related to receptacles that serve countertops in a kitchen or dining room: **Figure 14-5** illustrates kitchen receptacles serving countertop spaces.

- A receptacle outlet must be installed at each wall counter space that is 12 in. (300 mm) or wider. Receptacle outlets must be installed so that no point along the wall line is more than 24 in. (600 mm), measured horizontally, from a receptacle outlet in that space.

- At least one receptacle outlet must be installed at each island counter space, with a long dimension of 24 in. (600 mm) or greater and a short dimension of 12 in. (300 mm) or greater.

- At least one receptacle outlet must be installed at each peninsular counter space, with a long dimension of 24 in. (600 mm) or greater and a short dimension of 12 in. (300 mm) or greater. A peninsular countertop is measured from the connecting edge.

- Countertop spaces separated by range tops, refrigerators, or sinks are considered as separate countertop spaces.

- Receptacle outlets must be located above, but not more than 20 in. (500 mm) above, the countertop. On island and peninsular countertops where the countertop is flat across its entire surface (no backsplashes, dividers, etc.) and there is no way to mount a receptacle within 20 in. (500 mm) above the countertop, receptacle outlets are permitted to be mounted not more than 12 in. (300 mm) below the countertop. However, receptacles mounted below a countertop cannot be located where the countertop extends more than 6 in. (150 mm) beyond its support base, such as at bar-type eating areas in a kitchen. **Figure 14-6** shows how receptacles must be mounted as per *210.52(C)*.

NEC 210.52(D) requires one wall receptacle in each bathroom of a dwelling unit to be installed adjacent and within 36 in. (900 mm) of the outside edge of the basin. **Figure 14-7** is an illustration of a receptacle installed within 36 in. (900 mm) of a bathroom basin.

This receptacle is required in addition to any receptacle that may be part of any luminaire. A receptacle outlet is required adjacent to each additional basin unless the basins are in close proximity.

NEC 210.52(E) states that for a one-family dwelling and each unit of a two-family dwelling that is at grade level, at least one receptacle outlet accessible at grade level and not more than 6.5 ft (2.0 m) above grade must be installed at the front and back of the dwelling. This requirement helps to eliminate running extension cords through doors and windows to provide power to radios, bug zappers, appliances, or decorations located on a porch, deck, or balcony.

NEC 210.52(F) states the requirement of having at least *one* receptacle outlet installed for the laundry. Recall that a 20-amp branch circuit supplies the laundry receptacle outlets, and other outlets are not permitted on this circuit.

FIGURE 14-6 Mounting receptacles in a countertop.

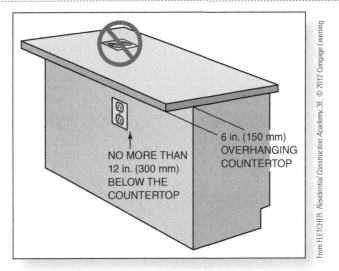

6 in. (150 mm) OVERHANGING COUNTERTOP

NO MORE THAN 12 in. (300 mm) BELOW THE COUNTERTOP

From FLETCHER. *Residential Construction Academy*, 3E. © 2012 Cengage Learning

FIGURE 14-7 A receptacle installed within 36 in. (900 mm) of a bathroom basin.

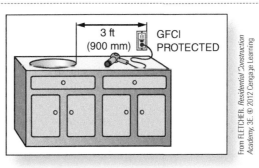

3 ft (900 mm)   GFCI PROTECTED

From FLETCHER. *Residential Construction Academy*, 3E. © 2012 Cengage Learning

NEC 210.52(G) states that for a one-family dwelling, at least one receptacle outlet, in addition to any provided for laundry equipment, must be installed in each basement, each attached garage, and each detached garage with electric power. Finished basements with one or more habitable rooms must have a receptacle outlet installed as per this *Code* requirement.

NEC 210.52 requires that in dwelling units, hallways of 10 ft (3 m) or more in length must have at least one receptacle outlet.

NEC Requirements for Locating Lighting Outlets

The *NEC* covers requirements for providing lighting in *210.70*. **Figure 14-8** is a comprehensive illustration showing how lighting outlets in dwellings *must* be installed as per *210.70*.

Habitable Rooms

NEC 210.70(A)(1) states that at least one wall switch–controlled lighting outlet be installed in every habitable room and bathroom. Switched receptacles meet the intent

FIGURE 14-8 Dwelling unit lighting outlets as required by *210.70*.

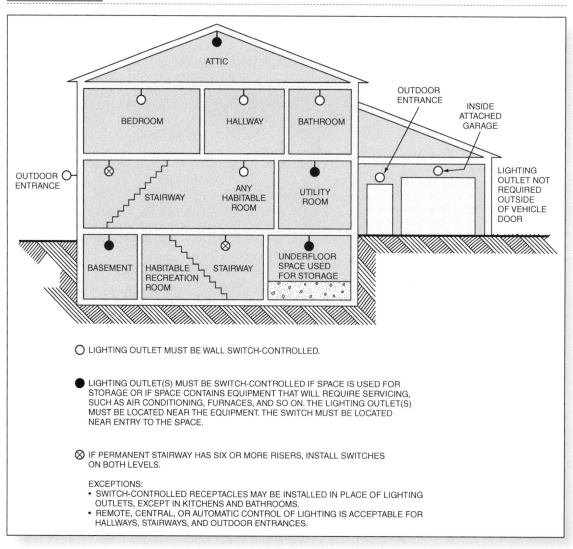

○ LIGHTING OUTLET MUST BE WALL SWITCH-CONTROLLED.

● LIGHTING OUTLET(S) MUST BE SWITCH-CONTROLLED IF SPACE IS USED FOR STORAGE OR IF SPACE CONTAINS EQUIPMENT THAT WILL REQUIRE SERVICING, SUCH AS AIR CONDITIONING, FURNACES, AND SO ON. THE LIGHTING OUTLET(S) MUST BE LOCATED NEAR THE EQUIPMENT. THE SWITCH MUST BE LOCATED NEAR ENTRY TO THE SPACE.

⊗ IF PERMANENT STAIRWAY HAS SIX OR MORE RISERS, INSTALL SWITCHES ON BOTH LEVELS.

EXCEPTIONS:
• SWITCH-CONTROLLED RECEPTACLES MAY BE INSTALLED IN PLACE OF LIGHTING OUTLETS, EXCEPT IN KITCHENS AND BATHROOMS.
• REMOTE, CENTRAL, OR AUTOMATIC CONTROL OF LIGHTING IS ACCEPTABLE FOR HALLWAYS, STAIRWAYS, AND OUTDOOR ENTRANCES.

From MULLIN/SIMMONS. *Electrical Wiring Residential*, 17E. © 2012 Cengage Learning

of this requirement except in kitchens and bathrooms. Bedrooms, living rooms, and similar rooms very often do *not* have ceiling luminaires. Lighting is provided by cord-and-plug-connected table, floor, or swag lamps.

> TRADE TIP Providing convenient control of cord-and-plug-connected lighting requires the wiring of split-circuit receptacles. During the rough-in stage, remember that the *always "hot"* portion of a split-circuit receptacle qualifies for the required receptacle. Also recall that the *switched* portion of a split-circuit receptacle is *not* considered to be one of the required receptacles, but rather is considered to be in addition.

Additional Locations

NEC 210.70(A)(2)(a) states that at least one wall switch–controlled lighting outlet must be installed in hallways, stairways, attached garages, and detached garages with power. There is *no Code* requirement that says a detached garage must have electric power. However, if it does, then it must have at least one GFCI receptacle, and it must have wall switch–controlled lighting.

Tool sheds or other similar accessory units are sometimes considered storage or utility rooms that come under the requirements of *210.70(A)(3)* and therefore require a wall switch–controlled lighting outlet.

NEC 210.70(A)(2)(b) states the requirement to have at least one wall switch–controlled lighting outlet installed to provide illumination on the exterior side of outdoor entrances or exits that have grade-level access to dwelling units, attached garages, and detached garages if they have electric power. However, a vehicle door in a garage is *not* considered to be an outside entrance or exit.

Stairways

NEC 210.70(A)(2)(c) requires that where there are six or more stair risers on an interior stairway, wall switch–controlled stairway lighting must be provided on each floor level and any landing level that has an entryway.

There is an exception to *(A)(2)(a)*, *(A)(2)(b)*, and *(A)(2)(c)*. It states that in hallways, stairways, and outdoor entrances, remote, central, or automatic control of lighting shall be permitted. An example is an outdoor floodlight controlled by a motion sensor or photocell. This light does *not* require a separate switch. However, *(A)(2)(c)* also *requires* that this automatic sensor have an "override" switch as *part of the unit.*

Basements, Attics, Storage, and Other Equipment Spaces

NEC 210.70(A)(3) covers storage or equipment spaces. It states that spaces that are used for storage or contain equipment that requires servicing, such as attics, underfloor spaces, utility rooms, and basements, have installed at least one lighting outlet containing a switch or controlled by a wall switch. Some equipment types that may require servicing are furnaces, dehumidifiers, submersible pumps, air conditioners, and so on. The lighting outlet must be at or near the equipment that may require servicing, and the control is permitted to be a wall switch or a lighting outlet that has an integral switch such as a pullchain switch.

LIFE SKILLS

This unit begins by stating that before *NEC* requirements for wiring can be understood, it is important to have a basic understanding of how a circuit is laid out and who generally lays these circuits out. Together with a few classmates, discuss three things: How is a circuit laid out? Who generally lays circuits out? Why is it important to know the answer before learning about the *NEC* requirements?

Self-Check 1

1. List two possible circuit arrangements found in residential wiring applications.
2. What is an arc-fault circuit interrupter?
3. What *NEC* section covers where receptacle outlets are to be located in a dwelling unit?

14-2 ■ *NEC* Requirements for GFCI Protection Locations for a Residential Dwelling

The *NEC* is very clear about *what locations* require ground-fault protection. The following section details the *NEC* requirement, lists those locations, and describes GFCI circuit breakers and GFCI receptacles.

NEC Requirements

In an effort to protect against electric shocks, *NEC* requires that ground-fault circuit interrupters (GFCI) be provided for all single-phase, 125-volt, 15- and 20-amp receptacles in dwelling unit locations that are listed in *210.8(A)(1)* through *(8)*. Ground-fault circuit protection can be provided by using GFCI circuit breakers or with GFCI receptacles. Ground-fault protection is required in the following areas of a residential dwelling:

- Bathrooms
- Garages
- Outdoors
- Crawl spaces
- Unfinished basements
- Kitchens
- Sinks in locations other than kitchens where receptacles are installed within 6 ft (1.8 m) of the outside edge of the sink.

GFCI protection is also required for 15- and 20-amp, 125-volt receptacles in boathouses and for boat hoists, outdoors, and for receptacles serving the countertop in kitchens.

GFCI Receptacles and Circuit Breakers

Ground-fault circuit protection can be provided with different wiring arrangements. **Figure 14-9** illustrates how a GFCI is installed that interrupts or shuts off the *entire* circuit if a ground fault in the range of 4 to 6 milliamps develops.

Figure 14-10 shows how a GFCI receptacle is installed. Notice that *only* the receptacle is shut off when a ground fault greater than 6 milliamps occurs.

Both GFCI methods described provide ground-fault protection. GFCI receptacles break *both* the ungrounded ("hot") and grounded conductors. However, GFCI circuit breakers interrupt, or break *only* the ungrounded ("hot") conductor.

LIFE SKILLS

Many older houses don't have GCFI protection in all of the locations that the *NEC* requires. How would you explain to a customer the need to upgrade outlets in an older home?

Self-Check 2

1. What *NEC* section covers GFCI requirements for residential dwellings?
2. List dwelling locations listed in the *NEC* that require GFCI protection.

FIGURE 14-9 GFCI as a part of the branch-circuit overcurrent device.

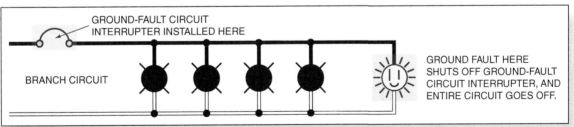

FIGURE 14-10 GFCI as an integral part of a receptacle outlet.

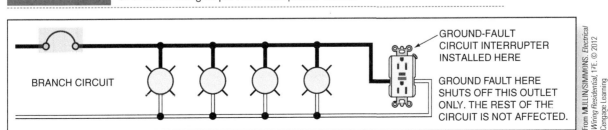

14-3. Cable Layout for a Master Bedroom in a Residential Dwelling

Before describing the procedure for drawing out a cable diagram for a residential dwelling, it is a good time to review symbols. Symbols were covered in Chapter 10 and can be seen in Figures 10-14 through 10-17. Keep in mind that these symbols are the very same ones that you see on construction floor plans and electrical prints.

For residential dwellings, generally symbols indicating the location of receptacles, lighting, and appliances are shown on the construction *floor* plans. However, this information (outlet count, estimated load, etc.) may be found in the electrical plans.

In residential wiring, it is the electrician's responsibility, once on site, to acquire the plans and then determine how to make it work. He or she needs to follow all *Code* requirements and make sure the job is accomplished in the most cost-effective manner. The electrician determines how to lay out the branch circuits. He or she needs to decide the number of outlets on a circuit, how to run the cables, and from where the branch-circuit home runs will be taken.

Electricians generally just sketch out a cable layout rather than a detailed wiring diagram. A cable layout, or cabling diagram, is a drawing that shows the number of conductors in each cable and how cables are run and what devices are connected in a specific area or room. After referring to the plans, an electrician makes a cable layout of all lighting and receptacle outlets. An example of a cable layout is shown in **Figure 14-11**.

If needed, a wiring diagram is drawn up next. A wiring diagram is a schematic drawing of an electrical system. However, most journeyman electricians do *not* prepare detailed wiring diagrams, because from experience, they know what connections must be made. These electricians know the size and type of boxes and cables that are needed.

The cable layout in Figure 14-10 provides an example. An electrician knows that two 3-way switches, three receptacles, and a lighting outlet should be used. The electrician knows what type boxes are required and where they should be located. He or she knows what cables are necessary, how to lay out the wiring, and what connections must be made.

Cable layout
a drawing that shows the number of conductors in each cable, how cables are run, and what devices are connected together in a specific area or room.

FIGURE 14-11 A cable layout.

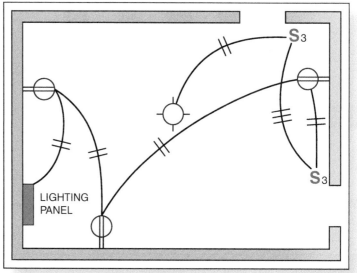

© Cengage Learning 2013

Cable Layout for a Master Bedroom

To sketch out a cable layout for any room, the electrician must first consult the residential construction floor plan. This plan provides the physical dimensions, including closet dimensions, doorways, and windows. **Figure 14-12** shows a typical residential floor plan.

Recall that for residential dwellings, symbols indicating the location of receptacles, lighting, and appliances are typically shown on the construction floor plan and not on a separate *electrical* plan.

Electrical plans and tables, when included, provide an outlet count and estimated load. Such a table, which is used for our master bedroom example, is shown in Table 14-1.

Table 14-1 and the construction floor (electrical) plan provide the electrician with the necessary information to make a cable layout. In our master bedroom example, the circuit has four split-circuit receptacle outlets, one outdoor weatherproof GFCI receptacle outlet, two closet luminaires, each on a separate switch, plus a ceiling fan or luminaire, one telephone outlet, and one television outlet.

In addition, an outdoor bracket luminaire is located adjacent to the sliding door and is controlled by a single-pole switch just inside the sliding door.

The split-circuit receptacle outlets are controlled by two 3-way switches. One is located next to the sliding door. The use of split-circuit receptacles offers the advantage of having switch control of one of the receptacles at a given outlet, whereas the other receptacle remains *live* at all times.

FIGURE 14-12 Typical residential floor plan.

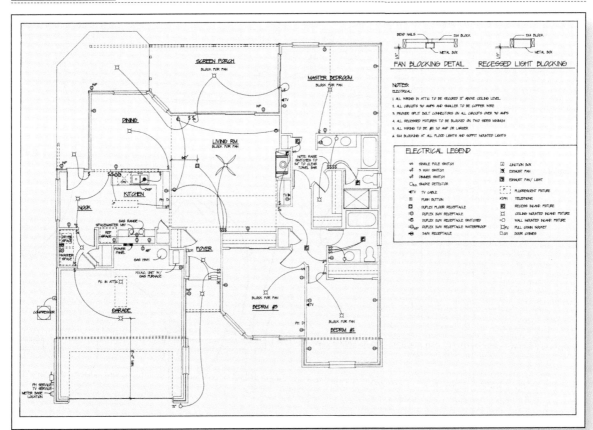

TABLE 14-1 Master Bedroom Outlet Count and Estimated Load

Description	Quantity	Watts	Volt-Amperes
Receptacles @ 120 watts each	4	480	480
Weatherproof receptacle	1	120	120
Outdoor bracket luminaire	1	150	150
Clothes closet luminaires One 75-W lamp each	2	150	150
Ceiling fan/light Three 50-W lamps	1	150	150
Fan motor (0.75 @ 120 V)		80	90
TOTALS	9	1130	1140

From MULLIN/SIMMONS. *Electrical Wiring Residential*, 17E. © 2012 Cengage Learning

Next to the bedroom door are a three-way switch and a ceiling fan/light control, which is installed in a separate two-gang box.

Therefore, armed with the above listed information, the plan that shows what is to be installed, and designated locations, the electrician can make a cable layout of all lighting and receptacle outlets. **Figure 14-13** shows each of the devices just described and how cables will be run.

FIGURE 14-13 Cable layout for a master bedroom.

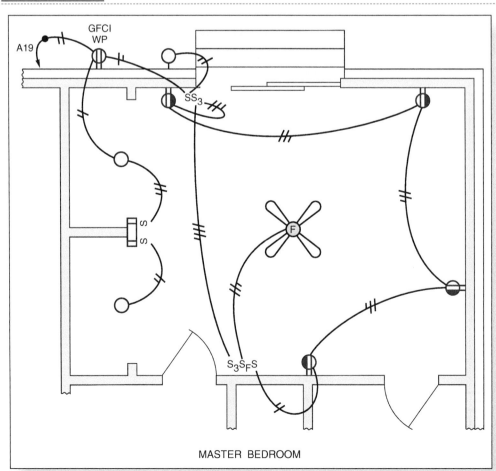

From MULLIN/SIMMONS. *Electrical Wiring Residential*, 17E. © 2012 Cengage Learning

Let's follow a section of what the electrician has laid out in this diagram. Refer to Figure 14-13, and notice the home run near the circuit designation A19 in the upper left of this layout. A 2-wire cable is to be run to the outdoor weatherproof GFCI receptacle outlet labeled GFCI WP. From this outdoor receptacle, the electrician has decided to run a 2-wire cable to one of the closet luminaires.

Keep in mind that the electrician first consulted the plans and then determined what was to be installed and where. Then a cable layout was sketched. This cable layout is just one way of laying out the circuits; certainly, other layouts are possible. The most important underlying factors are that it be wired according to the *NEC* and as efficiently and cost effectively as possible.

LIFE SKILLS

You can't understand cable layouts if you don't understand the symbols covered in Chapter 10. Make flashcards featuring these symbols, and take turns with a classmate, quizzing each other.

Self-Check 3

1. For residential wiring applications, who determines how to lay out the branch circuits?
2. What is a cable layout diagram?

Summary

- In every installation, *NEC* requirements must be followed and with residential wiring applications, the job of laying out the circuits typically belongs to the electrician.
- Arc-fault circuit interrupters (AFCIs), covered in *210.12*, are devices intended to provide protection from the effects of arc faults by recognizing characteristics unique to arcing and by functioning to de-energize the circuit when an arc fault is detected. Arc-fault circuits are required everywhere in a dwelling other than the kitchen and areas that require GFI.
- *NEC 210.52* states where receptacle outlets must be installed in a dwelling unit.
- *NEC* requirements provide permitted locations for electrical boxes that electricians install during the rough-in stage of wiring.
- The *NEC* covers requirements for providing lighting in *210.70*.
- *NEC 210.70(A)(1)* states that at least one wall switch–controlled lighting outlet be installed in every habitable room and bathroom.
- *NEC 210.70(A)(2)(a)* states that at least one wall switch–controlled lighting outlet must be installed in hallways, stairways, attached garages, and detached garages with power.

- *NEC 210.70(A)(2)(c)* requires that where there are six or more stair risers on an interior stairway, wall switch–controlled stairway lighting must be provided on each floor level and any landing level that has an entryway.
- *NEC 210.70(A)(3)* states that spaces that are used for storage or contain equipment that requires servicing, such as attics, underfloor spaces, utility rooms, and basements, have at least one lighting outlet installed containing a switch or controlled by a wall switch.
- In an effort to protect against electric shocks, *NEC* requires that ground-fault circuit interrupters (GFCI) be provided for all single-phase, 125-volt, 15- and 20-amp receptacles in dwelling unit locations that are listed in *210.8(A)(1)* through *(8)*.
- When a ground-fault circuit breaker is installed, it interrupts or shuts off the entire circuit if a ground fault in the range of 4 to 6 milliamps develops.
- Dwelling unit locations listed in *210.8(A)(1)* through *(8)* that require GFCI protection are as follows: bathrooms, garages, outdoors, crawl spaces, unfinished basements, kitchens, sinks—in locations other than kitchens where receptacles are installed within 6 ft (1.8 m) of the outside edge of the sink.
- Both GFCI methods described provide ground-fault protection. GFCI receptacles break both the ungrounded ("hot") and grounded conductors. However, GFCI circuit breakers interrupt, or break only the ungrounded ("hot") conductor.
- For residential dwellings, generally symbols indicating the location of receptacles, lighting, and appliances are shown on the construction floor plans.
- Electricians generally sketch out a cable layout diagram, which is a drawing that shows the number of conductors in each cable, how cables are run, and what devices are connected in a specific area or room.
- To sketch out a cable layout for any room, the electrician must first consult the residential construction floor plan, which provides the physical dimensions, including closet dimensions, doorways, and windows.

Review Questions

True/False

1. In almost every installation, *NEC* requirements must be followed. (True, False)

2. A way to keep an outlet box at the lighting outlet or the wiring compartment on recessed cans less crowded is to run the branch circuit first to the switch and then to the controlled lighting outlet. (True, False)

3. Arc-fault circuit interrupters are not yet part of *NEC* requirements. (True, False)

4. The rough-in stage of an electrical installation is when the raceways, cable, boxes, and other electrical equipment are installed. (True, False)

5. Receptacles must be installed so that no point measured horizontally along the floor line in any wall space is more than 8 ft from a receptacle outlet. (True, False)

Multiple Choice

6. Which *NEC* section covers location of receptacles at countertop locations in a dwelling unit kitchen or dining room?
   A. *210.52(A)*
   B. *210.52(B)*
   C. *210.52(C)*
   D. *210.52(D)*

7. The *NEC* requires one wall receptacle in each bathroom of a dwelling unit to be installed adjacent and within how many inches of the outside edge of the basin?
   A. 16
   B. 26
   C. 36
   D. 46

8. *NEC 210.52* requires that in dwelling units, hallways of _____ feet or more in length must have at least one receptacle outlet.
   A. 4
   B. 6
   C. 8
   D. 10

9. Providing convenient control of cord-and-plug-connected lighting requires the wiring of what type of receptacles?
   A. GFCI
   B. Single
   C. Split-circuit
   D. Multiple

10. *NEC 210.70(A)(2)(c)* requires that where there are _____ or more risers on an interior stairway, wall switch–controlled stairway lighting must be provided.
   A. 3
   B. 4
   C. 5
   D. 6

Fill in the Blank

11. Ground-fault circuit protection can be provided using GFCI circuit breakers or with GFCI _____.

12. When using a GFCI receptacle, only the receptacle is shut off when a ground fault greater than _____ milliamps occurs.

13. For residential dwellings, typically symbols showing the location of receptacles, lighting, and appliances are shown on the _____ floor plans.

14. A(n) _____ diagram is a drawing that shows the number of conductors in each cable and how the cables are run.

15. A(n) _____ is a schematic drawing of an electrical system.

Chapter 15

Green Technology and the Electrical Industry

Career Profile

As the Director of Education at Independent Electrical Contractors (IEC) Chesapeake, **Jim Deal** works daily with students and instructors in electrical apprenticeship programs. And Jim wants students in such programs to know that "if they apply themselves, there is no limit to what they can accomplish," because the electrical field is "very broad ... with great opportunities for advancement." Jim's own career path shows the wisdom of his statements. After serving in the military, he became a field electrician, completed a 4-year apprenticeship program, became a journeyman electrician, and ultimately attained a Master Electrician license. Early on in his training, Jim took a job at the University of Maryland Hospital; he then developed his skills and gained a range of experiences that qualified him for the position of plant operations manager for all mechanical and electrical systems at the facility. Jim's day-to-day responsibilities in his current position include making sure students are adhering to the program and that instructors are teaching material in an understandable, implementable way. He wants pre-apprentices to remember that above all else, apprenticeships offer a knowledge of "how to use resources to solve problems." Thus, he advises, "Make sure that you're not just memorizing material. [Work to] understand the material and be able to apply what you learn in the field."

Chapter Outline

GREEN TECHNOLOGY

SOLAR AND WIND TECHNOLOGIES AND OTHER GREEN ENERGY
SOURCES

U.S. GREEN BUILDING COUNCIL (USGBC) AND LEADERSHIP
IN ENERGY AND ENVIRONMENTAL DESIGN (LEED)

GREEN TECHNOLOGY EMPLOYMENT OPPORTUNITIES
FOR ELECTRICIANS

Key Terms

Biomass

Distributed generation

Drag

Energy audit

Energy carrier

Energy conservation

Geothermal energy

Geothermal heat pump

Greenhouse gas

Green technology

Hydrocarbons

Hydroelectric power

Hydrogen fuel cell

Lift

Photovoltaic system

Renewable energy

Retrofitting

Rotor

Solar energy

Technology

Wind turbine

Chapter Objectives

After completing this chapter, you will be able to:

1. Define green technology, and list four major goals of this developing technology.

2. Describe the importance of solar voltaic panels and wind turbine products and the need to integrate these technologies seamlessly into the nation's electric power grid.

3. Explain the impact of the U.S. Green Building Council's Leadership in Energy and Environmental Design (LEED) Green Building Rating System in measuring a building's environmental impact.

4. List several green technology employment opportunities for electricians.

Life Skills Covered

 Critical Thinking   Self-Advocacy

Life Skill Goals

Green is the new buzzword in the electrical field and the construction industry in general. Environmental friendliness and energy efficiency are at the top of most new homebuyers' list. The life skills in the lesson are designed to help you position yourself to take advantage of the trend. One of the life skills covered is Critical Thinking. By thinking critically about green technology issues, you will be able to understand why certain options are more useful or practical than others. Also covered in this chapter is Self-Advocacy. If green energy truly is the wave of the future, then understanding its usefulness and applications will position you well as you begin your career. As you go through this chapter, ask yourself, "What can I learn from this chapter that will give me an advantage when I'm beginning my career?"

The field of green technology encompasses a continuously evolving group of methods and materials, from techniques for generating energy to nontoxic cleaning products. This technology is beneficial for the environment. Energies such as solar, wind, and water are examples of green technologies. It is expected that this field will bring innovation and changes in the way we live and the way we treat our planet. For you, as an electrician, green technology brings job opportunities. This chapter introduces developing green technologies, such as solar, wind, and others, and their goals. The U.S. Green Building Council is introduced, and Leadership in Energy and Environmental Design (LEED) is described, emphasizing applications in electrical construction and industrial electrical industries. The chapter concludes by discussing green technology employment opportunities. Learn about this evolving technology so you can position yourself to take advantage of the "green-collar" jobs for electricians!

Introduction

You may have heard the term *green technology* or, as it's sometimes referred to, environmental technology, or clean technology. So, what is green technology, and what impact will it have on us, especially those of us in the electrical industry? Green technology describes the activities people are pursuing and the steps being taken to promote a cleaner environment and preserve natural resources. Recycling and reducing waste that pollutes water and air are ways you can help the green technology movement. Research and development in methods to integrate renewable energy technologies, such as solar and wind power, into the electric power system will also be instrumental in the evolving green technology movement.

15-1 Green Technology

Technology
the application of knowledge for practical purposes.

Green technology the application of environmental science and green chemistry to conserve the natural environment and resources, and to curb the negative impacts of human involvement.

Technology refers to the application of knowledge for practical purposes.
Green technology is the application of environmental science and green chemistry to conserve the natural environment and resources, and to curb the negative impacts of human activity. Sustainable development is at the core of green technology, and the idea that we all must take an interest in the well-being of our planet and support those involved with developing this evolving green technology.

Some of the main objectives of green technology are listed here:

- *Sustainability*—meeting the needs of our generation without damaging or depleting natural resources—in other words, meeting present needs without putting at risk the needs of future generations.

- *"Cradle to cradle" design*—ending the "cradle to grave" cycle of manufactured products by developing products that can be fully recycled, reclaimed, or re-used.

* *Source reduction*—reducing waste and pollution by changing methods of production and consumption.

* *Innovation*—developing alternatives to technologies such as fossil fuel or chemical-intensive agriculture, which have been shown to increase health risks and cause damage to the environment.

* *Viability*—supporting and using technologies and products that are environmentally friendly, moving forward with urgency to work within these technologies, and creating new employment opportunities that truly protect the planet.

We can all be a part of this exciting technology. If we are, we, along with generations that follow, will reap the benefits.

There are lots of reasons to "go green" when it comes to the *electrical* industry. Going green saves money and the environment. Currently, buildings account for approximately 50% of all greenhouse gas emissions, and the design and construction of *green* buildings will have a significant impact on the effort to reduce these emissions. A greenhouse gas is a gas that contributes to the greenhouse effect by absorbing infrared radiation, for example, carbon dioxide and chlorofluorocarbons. By joining forces with agencies and businesses that are developing green energy solutions, as an electrician you become part of an electrical construction industry that is transforming the design and construction of buildings so they are environmentally responsible, profitable, and healthy places to live and work.

Greenhouse gas a gas that contributes to the greenhouse effect by absorbing infrared radiation, for example, carbon dioxide and chlorofluorocarbons.

Green Energy Solutions

Energy solutions are specific to the individual assets and objectives of each customer. However, each typically includes energy conservation, energy efficiency, and responsible energy production. Combined, these objectives provide an integrated energy solution that powers buildings and communities with less waste and more efficiency. A closer look at these objectives follows.

Energy Conservation

Energy conservation achieved through efficient energy use, whereby energy use is decreased while achieving a similar outcome, or by reduced consumption of energy services.

Energy conservation is achieved through efficient energy use, whereby energy use is decreased while achieving a similar outcome, or by reduced consumption of energy services.

Electrical systems that automatically adjust lighting and environmental controls as daylight and occupancy change are called *energy management systems (EMSs)*. An EMS helps to reduce building and maintenance costs and thereby energy waste.

Energy Efficiency

Generally, the first step a building owner can take to cut operating costs is to reduce the power it takes to operate the building, a step that also helps the environment. However, significant energy efficiency requires more efficient materials and equipment to generate, transfer, and use energy. You can only save so much by *conserving* energy.

Energy audit
an assessment of
the energy needs
and efficiency
of a building or
buildings.

Energy audits are becoming more commonplace. An energy audit conducted by contractors is an assessment of the energy needs and efficiency of a building or buildings. Once an audit is conducted, the contractor provides a list of methods that help to reduce power use, as well as suggestions to improve energy efficiency.

Responsible Energy Production

Alternative energy sources like solar photovoltaic (PV) panels and distributed generation will prove to be valuable, but only if the technology and equipment for alternative power is *designed and installed* correctly. When you need shingles replaced on your roof, you call a roofer. When you want PV panels installed, you call an electrician! The same safety and performance considerations apply for powering a building using alternative energy sources as they do for traditional electrical transmission and distribution.

LIFE SKILLS

Choose a partner, and have a discussion about the trend toward green technology. Is it good or bad for the electrical field? Why or why not? (Critical Thinking)

Self-Check 1

1. What is green technology?
2. List three green energy solutions.

15-2 Solar and Wind Technologies and Other Green Energy Sources

**Renewable
energy** generated
from natural
resources, such
as sunlight,
wind, rain, tides,
and geothermal
heat, which are
regenerative
(i.e., naturally
replenished).

Before we take a closer look at alternative energy sources, let's understand why these renewable energy sources are important. Renewable energy is generated from natural resources such as sunlight, wind, rain, tides, and geothermal heat, which are regenerative (i.e., naturally replenished). Following are several benefits of renewable energy sources:

- Environmental benefits—Renewable energy technologies are clean sources of energy that do not have a negative impact on the environment that conventional energy technologies do.

- Conservation of natural resources—Renewable energy does not "use up" our natural resources, and it will not run out. Other sources of energy such as oil and gas may someday be depleted.

- Increased economic development and creation of new jobs—Most renewable energy investments are spent on materials and workmanship to build and maintain the facilities that generate them, rather than on costly energy imports. Renewable energy investments are generally spent within the United States, often in the same state and even the same city. What does this mean to you? Energy dollars stay home to create jobs and fuel local economies and do not go overseas. Moreover, renewable energy technologies developed and built in the United States are being *sold* overseas, increasing *our* trade revenues.

- A reduced reliance on imported fuel and electricity—In recent years, the United States has increased its dependence on foreign oil supplies. This increased dependence impacts more than just our national energy policy.

U.S. national laboratories and other agencies are researching and developing methods to integrate renewable energy technologies into the electric power system. The following section provides information on solar and wind energy sources as well as several other evolving energy technologies.

Solar Energy

Solar energy
radiant energy
from the sun
that is converted
into thermal or
electrical energy.

Solar energy, or power from the sun, is a vast and inexhaustible resource. Radiant energy from the sun is converted into thermal or electrical energy. Solar energy technologies use the sun's energy and light to provide heat, light, hot water, electricity, and even cooling for homes, businesses, and industry.

Solar energy technologies include the following:

- Photovoltaic systems that produce electricity directly from sunlight
- Solar hot water that is water heated with solar energy
- Solar electricity, which uses the sun's heat to produce electricity
- Passive solar heating and daylighting that use solar energy to heat and light buildings
- Solar process space heating and cooling, involving industrial and commercial uses of the sun's heat.

Photovoltaic
system a system
that uses solar cells
to convert light into
electricity.

Our focus in this section on solar energy technologies is to describe **photovoltaic (PV) systems**, sometimes referred to as solar cell systems. A photovoltaic system uses solar cells to convert sunlight directly into electricity. You may be familiar with solar cells used to power calculators and watches. These are made with semiconductor materials similar to those used in computers or other solid-state devices. When these materials absorb sunlight, the solar energy knocks electrons loose from their atoms, allowing the electrons to flow through the material to produce electricity. This process of converting light (photons) to electricity (voltage) is referred to as the photovoltaic (PV) effect. A single solar cell is shown in **Figure 15-1**.

Solar cells are usually combined into modules that hold about 40 cells. A number of these modules are mounted in PV arrays that can measure up to several yards on a side. A solar cell array mounted on a roof is shown in **Figure 15-2**.

FIGURE 15-1 Single solar cell.

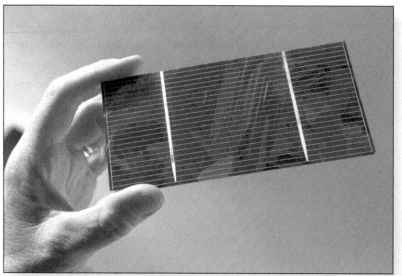

© ason/www.Shutterstock.com

FIGURE 15-2 Solar cell array mounted on a roof.

© ssuaphotos/www.Shutterstock.com

These flat-plate PV arrays can be mounted at a fixed angle facing south, or they can be mounted on a tracking device that follows the sun, allowing them to capture the greatest amount of sunlight throughout the day. **Figure 15-3** shows solar cell panels mounted on a tracking device.

FIGURE 15-3 Solar cell panels mounted on a tracking device.

Several connected PV arrays can provide enough power for a household, as shown in **Figure 15-4**.

To be used for a large electric utility or industrial applications, hundreds of arrays can be interconnected to form a single, large PV system.

The performance of a solar cell is measured in terms of its efficiency at turning sunlight into electricity. Only sunlight of certain energies is efficient enough to create electricity. Unfortunately, much of the sunlight is reflected or absorbed by the material

FIGURE 15-4 PV arrays mounted on a house.

that makes up the solar cell. Therefore, a typical commercial solar cell yields only about a 15% efficiency. This means that only about one-sixth of the sun's energy striking the cell generates electricity. With such low efficiencies, larger arrays are required, and this means higher cost. Research labs and the U.S. Department of Energy are working toward improving solar cell efficiencies while holding down the cost per cell.

During the last few years, thanks to lower costs and strong incentives, the PV industry has placed more focus on home, business, and utility-scale systems that are attached to the power grid. In some locations, it is less expensive for utilities to install solar panels than to upgrade the transmission and distribution system to meet new electricity demand.

This **distributed-generation** approach provides a new model for the utilities of the future. Distributed generation refers to a variety of small, modular, power-generating technologies that can be combined to improve the operation of the electricity delivery system. Small generators, spread throughout a city and controlled by computers, could replace the large coal and nuclear plants that dominate the landscape now.

Wind Energy

For hundreds of years, we have been harnessing the wind's energy. Harnessing the wind is one of the cleanest, most sustainable ways to generate electricity. We have come a long way since windmills were used for pumping water or grinding grain. **Figure 15-5** shows a traditional windmill.

Today, the windmill's modern equivalent, a **wind turbine**, uses the wind's energy to generate electricity. A wind turbine is a machine that captures the energy of wind and transfers the motion to a generator shaft to produce electricity. Wind turbines are shown in **Figure 15-6**.

Distributed generation a variety of small, modular, power-generating technologies that can be combined to improve the operation of the electricity delivery system.

Wind turbine a machine that captures the energy of wind and transfers the motion to a generator shaft to produce electricity.

FIGURE 15-5 Traditional windmill.

FIGURE 15-6   Wind turbines.

According to the U.S. Department of Energy, wind power contributed 42% of all new U.S. electricity-generating capacity in 2008. For the fourth consecutive year, wind power was the second-largest *new* resource added to the U.S. electrical grid, behind natural gas but ahead of new coal. The United States now claims the world's largest wind energy capacity.

Wind turbines, like windmills, are mounted on a tower to capture the most energy. At 100 ft (30.4 m) or more above ground, they take advantage of the faster and less turbulent wind, catching the wind's energy with their propeller-like blades. **Figure 15-7** shows a close-up of a wind turbine.

Typically, two or three blades are mounted on a shaft to form a rotor, the rotating armature of a motor or generator.

Rotor the rotating armature of a motor or generator.

FIGURE 15-7   A close-up of a wind turbine.

Lift an upward force that counteracts the force of gravity, produced by changing the direction and speed of a moving stream of air.

Drag the longitudinal retarding force exerted by air or other fluid surrounding a moving object.

A turbine blade acts much like an airplane wing. When the wind blows, a pocket of low-pressure air forms on the downwind side of the blade. The low-pressure air pocket then pulls the blade toward it, causing the rotor to turn. This is called lift, an upward force that counteracts the force of gravity and is produced by changing the direction and speed of a moving stream of air. The force of lift is actually much stronger than the wind's force against the front side of the blade, which is called drag. Drag is defined as the longitudinal retarding force exerted by air or other fluid surrounding a moving object. The combination of lift and drag causes the rotor to spin like a propeller, and the turning shaft spins a generator to make electricity.

Wind turbines can be used as stand-alone applications, or they can be connected to a utility power grid. Some wind turbine applications are combined with a photovoltaic (PV) solar cell system. **Figure 15-8** shows a roof with solar panels and a wind turbine next to the dwelling.

A large number of wind turbines can be built close together to form a wind plant. This type of arrangement is required for utility-scale sources of wind energy. There are several electricity providers today that use wind plants to supply power to their customers.

FIGURE 15-8 A roof with solar panels, and a wind turbine next to the dwelling.

Generally, stand-alone wind turbines are used for such applications as water pumping or communications. However, homeowners, farmers, and ranchers in windy areas can also use wind turbines as a way to reduce their electric bills. **Figure 15-9** shows how a farm pastureland sustains wind-powered turbines.

Small wind systems also have potential as distributed energy resources. Recall that distributed resources refer to a variety of small, modular, power-generating technologies that can be combined to improve the operation of the electricity delivery system.

It was determined by a comprehensive study conducted by the U.S. Department of Energy that expanding wind power from a little more than 1% in 2007 to 20% by 2030 is feasible, affordable, and will not affect the reliability of the nation's power supply. Aside from confirming that it could be done, the study estimated that achieving this goal would create over 500,000 new U.S. jobs, reduce global warming, and save 4 trillion gallons of water. Other benefits would include greatly improved air and water quality for future generations, and much less dependence on fossil fuels and their escalating costs.

Other Alternative Green Energy Sources

Solar and wind are currently considered the most popular *green* energy sources. The following section, however, describes several other evolving energy technologies.

Geothermal

Geothermal energy thermal energy generated and stored in the earth.

Geothermal energy is thermal energy generated and stored in the earth. The word *geothermal* originates from the Greek root *geo*, meaning "earth," and word part *thermos*, meaning "heat." Geothermal energy is clean and sustainable. Resources of geothermal energy range from shallow ground to hot water and hot rock found a few miles beneath

FIGURE 15-9   Pastureland and wind turbines.

the earth's surface. Drilling deeper brings you to the extremely high temperatures of the molten rock called magma.

In almost every geographic location, the upper 10 ft (3 m), or shallow ground, of the earth's surface maintains a nearly constant temperature between 50°F and 60°F (10°C to 15.5°C). Geothermal heat pumps can tap into this resource to heat and cool buildings. A geothermal pump, or ground source heat pump (GSHP), is a central heating and/or cooling system that pumps heat to or from the ground. It uses the earth as a heat source (in the winter) or a heat sink (in the summer). A geothermal heat pump system consists of a heat pump, an air delivery system (ductwork), and a heat exchanger, which is a system of pipes buried in the shallow ground near the building. **Figure 15-10** illustrates a geothermal heat pump diagram.

In the winter, the heat pump removes heat from the heat exchanger and pumps it into the indoor air delivery system. In the summer, the process is reversed, and the heat pump moves heat from the indoor air into the heat exchanger. The heat removed from the indoor air during the summer can also be used to provide a free source of hot water.

Geothermal heat pump a central heating and/or cooling system that pumps heat to or from the ground. It uses the earth as a heat source (in the winter) or a heat sink (in the summer); also called ground source heat pump (GSHP).

FIGURE 15-10 A geothermal heat pump diagram.

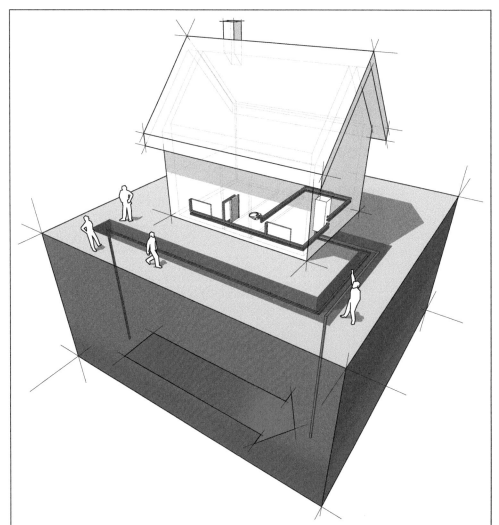

National laboratories continue to research and develop geothermal technologies that include geothermal electricity production, geothermal direct use, and geothermal heat pumps. The geothermal resources are theoretically more than adequate to supply humanity's energy needs, but only a very small fraction may be profitably exploited. Drilling and exploration for deep resources is very expensive.

Hydroelectric

Hydroelectric power the force, or energy, of moving water that is captured and turned into electricity.

The force, or energy, of moving water can be captured and turned into electricity. This is called **hydroelectric power**, hydropower, or simply waterpower. Years ago and long before the development of electric power, hydropower was used for irrigation and the operation of various machines, such as watermills, textile machines, sawmills, dock cranes, and domestic lifts.

Today, exploiting the movement of water to generate electricity, known as hydroelectric power, is the largest source of renewable power in the United States and worldwide.

The most common type of hydroelectric power plant uses a dam on a river to store water in a reservoir. **Figure 15-11** shows the Hoover Dam located in the Black Canyon of the Colorado River on the border of Arizona and Nevada. This dam's generators provide power for public and private utilities in Nevada, Arizona, and California.

Water released from the reservoir flows through a turbine, spinning it, which in turn activates a generator to produce electricity. But hydroelectric power doesn't necessarily require a large dam. Some hydroelectric power plants just use a small canal to channel the river water through a turbine. A hydroelectric power station installed on a river is shown in **Figure 15-12**.

FIGURE 15-11 The Hoover Dam.

FIGURE 15-12 A hydroelectric power station on a river.

© Aleksandr Kurganov/www.Shutterstock.com

Another type of hydroelectric plant, called a *pump storage plant*, can even store power. The power is sent from a power grid into the electric generators. The generators then spin the turbines backward, which causes the turbines to pump water from a river or lower reservoir to an upper reservoir, where power is stored. To *use* the power, the water is released from the upper reservoir back down into the river or lower reservoir. This spins the turbines forward, activating the generators to produce electricity.

A small hydroelectric power system can produce enough electricity for a home, farm, or ranch.

Biomass

Biomass organic matter used as a fuel, especially in a power station for the generation of electricity.

Biomass is a renewable energy source. Biomass is organic matter used as a fuel, especially in a power station for the generation of electricity. It is biological material from living or recently living organisms. The use of biomass or bioenergy is not new. As a matter of fact, we started using biomass energy thousands of years ago, ever since people started burning wood to cook food or to keep warm.

Wood remains our largest biomass energy resource. **Figure 15-13** shows wood chips used for biomass combustion.

However, many other sources of biomass can now be used. Some other sources include plants, residues from agriculture or forestry, and the organic component of municipal and industrial wastes. Even fumes from landfills can be used as a biomass energy source.

Greenhouse gas emissions can be drastically reduced by using biomass energy. Biomass generates approximately the same amount of carbon dioxide as fossil fuels. So, why use biomass instead of fossil fuels? Every time a *new* plant grows, carbon dioxide is actually *removed* from the atmosphere. The *net* emission of carbon dioxide is zero as

FIGURE 15-13 Wood chips for biomass combustion.

© Ulrich Mueller/www.Shutterstock.com

long as plants continue to be replenished for biomass energy purposes. Also consider the fact that although fossil fuels have their origin in ancient biomass, they are *not* considered biomass by the generally accepted definition because they contain carbon that has been *out* of the carbon cycle for a very long time. Their combustion therefore disturbs the carbon dioxide content in the atmosphere.

The three major biomass energy technology applications are these:

- Biofuels—Converting biomass into liquid fuels for transportation.
- Biopower—Burning biomass directly, or converting it into a gaseous fuel or oil, to generate electricity.
- Bioproducts—Converting biomass into chemicals for making products that typically are made from petroleum.

Hydrogen Energy and the Fuel Cell

The simplest known element is hydrogen. You may recall that the hydrogen atom has only one proton and one electron. This element is the most abundant element in the universe. Although hydrogen is plentiful and its atomic structure is simple, it does not occur naturally as a gas on the earth. Hydrogen is always combined with other elements. The most familiar example is water, which is a combination of hydrogen and oxygen (H_2O).

Many organic compounds, such as **hydrocarbons**, are comprised of high levels of carbon. A hydrocarbon is a compound of hydrogen and carbon, such as any of those that are the chief components of petroleum and natural gas. Hydrocarbons make up many of our currently used fuels, such as gasoline, natural gas, methanol, and propane. The separation of hydrogen from hydrocarbons is accomplished by heating in a process called reforming. Natural gas is used most often in the reforming process to make hydrogen.

Because hydrogen is high energy and because an engine can burn pure hydrogen that produces no pollution, NASA has used liquid hydrogen since the 1970s to propel

Hydrocarbon
a compound of hydrogen and carbon, such as any of those that are the chief components of petroleum and natural gas.

Hydrogen fuel cell an electrochemical cell in which the energy of a reaction between a fuel (liquid hydrogen) and an oxidant (liquid oxygen) is converted directly and continuously into electrical energy.

the space shuttle and other rockets into space. As a matter of fact, hydrogen fuel cells power the shuttle's electrical systems, and the by-product produced is water, which the crew drinks. A hydrogen fuel cell is an electrochemical cell in which the energy of a reaction between a fuel (liquid hydrogen) and an oxidant (liquid oxygen) is converted directly and continuously into electrical energy.

When hydrogen and oxygen are combined in a fuel cell, electricity, heat, and water are produced. Fuel cells can be compared to batteries, which produce usable electric power by way of a chemical reaction. The main difference *and* advantage is the fuel cell produces electricity as long as fuel (hydrogen) is supplied, and it never loses its charge.

Energy carrier moves and delivers energy in a usable form to consumers.

Fuel cells promise to be a technology that can be a source of heat and electricity for buildings, and an electrical power source for electric motors propelling vehicles. Hydrogen, much like electricity, can become a viable energy carrier. An energy carrier moves and delivers energy in a usable form to consumers. Remember, energy sources like sun and wind cannot produce energy all the time. However, they can produce electric energy and hydrogen, which can be stored until it is needed. Hydrogen can then be transported to locations where it is needed.

LIFE SKILLS

Which of the green energy sources listed in this unit do you think will have the largest impact on your career? Why do you feel that way? Form an opinion, and then discuss it with some of your classmates.

Self-Check 2

1. What is meant by renewable energy?
2. What is solar energy?
3. What is a photovoltaic system?
4. What is a wind turbine?
5. What is geothermal energy, and what is a geothermal heat pump?
6. What is hydroelectric power, and how is "hydro" electricity typically generated?
7. What is biomass and what is our largest biomass energy resource?

15-3 U.S. Green Building Council (USGBC) and Leadership in Energy and Environmental Design (LEED)

The U.S. Green Building Council (USGBC) is an industry organization made up of many parts of the construction industry, including owners, designers, and contractors. The USGBC promotes the construction of environmentally friendly, high-performance buildings through its sponsorship of the Leadership in Energy and Environmental Design (LEED) Green Building Rating System. The purpose of this rating system is to provide an objective standard for certifying that a building is environmentally friendly or green. As a result of the public's rising concern about the environment and rising energy costs, there is a growing movement among public and private building owners to have their buildings LEED certified. Concern for the environment has prompted the creation of many initiatives. LEED is one of these initiatives. LEED is an internationally recognized environmental program and is currently deployed in more than 30 countries around the world. It provides a means of verifying that a building or group of buildings were designed and built in a way that improves energy savings, water efficiency, indoor environmental quality, and carbon dioxide emissions reduction.

Although the foundation for LEED certification is laid during the design process, the design intent must be implemented through the construction process. The electrical contracting firm needs to be aware of LEED requirements because they can impact material; and equipment procurement as well as construction requirements and costs.

LEED recognizes achievements and promotes expertise in *green* building through a comprehensive system, offering project certification based on points, professional accreditation, training, and practical resources.

Over 40% of what LEED certification covers is the work that electrical contractors perform, including atmosphere and lighting control, on-site renewable energy generation and management, construction materials and lighting component selection, and light-pollution reduction. To be clear, when a building owner decides to seek LEED certification, the electrical contractor likely has the most influence on the project's success.

Electrical contractors need to become adept in meeting LEED certification for new construction, energy auditing and **retrofitting** existing buildings to improve efficiency, and installing renewable energy technologies safely and efficiently. Retrofitting is the fitting in or adding of new technology or features to an older structure or building.

Buildings that have gone through energy audits and were retrofitted with electrical efficiencies are the *greenest* buildings. This practice keeps construction-related waste out of our landfills.

Building owners have come to realize that operating expenses can be cut by *reducing* the power it takes to operate their building. This practice also reduces a building's environmental impact. The greatest savings come when changes are made to *lighting* and *HVAC*.

Energy audits provide *efficiency* opportunities of a structure and its day-to-day operation. As an electrician, you should become familiar with *total energy management (TEM)*. Most TEM programs work with a checklist that helps with analyzing the best energy options, materials, fixtures, and environmental controls to *minimize* waste and *maximize* efficiency. A credentialed auditing process can help to reduce operating expenses almost immediately.

According to the U.S. Energy Information Agency, homes and commercial buildings use 71% of the electricity in the United States, and this number will rise to 75% by 2025. Opportunities abound for reducing the enormous amount of energy consumed by buildings. Green technology applications in electrical construction and

Retrofitting the fitting in or adding of new technology or features to an older structure or building.

industrial electrical industries include lighting and design class, efficiency of motors, and controlling of loads (lighting, HVAC, etc.) and building operation through programmable logic controllers (PLCs).

Projects applying for Leadership in Energy and Environmental Design (LEED) status are growing just as fast as the rest of the green construction market. With many LEED credits being electric and energy related, you, as an electrician, have an opportunity to work on these projects. However, you need to become familiar with credit requirements, and you should look into becoming a LEED accredited professional (AP).

LIFE SKILLS

How does having a firm understanding of LEED certification make you a more attractive job candidate? What sort of advantages does a job candidate with LEED expertise have over a candidate that lacks an exhaustive understanding?

Self-Check 3

1. What is the U.S. Green Building Council, and why is this group important in the areas of green technology?
2. What is Leadership in Energy and Environmental Design, and what is the purpose of its Green Building Rating System?
3. Why are total energy management programs important?

15-4 Green Technology Employment Opportunities for Electricians

The number of jobs in America's emerging clean energy economy grew nearly two and a half times faster than overall jobs between 1998 and 2007, according to a June 2009 report released by The Pew Charitable Trusts. Pew developed a clear, data-driven definition of the clean energy economy and conducted the first-ever hard count across all 50 states of the actual jobs, companies, and venture capital investments that supply the growing demand for environmentally friendly products and services. With the continuing growth of green jobs, it may be a great time to begin a career in what looks to be a driving industry of the future.

Skills and sustainability are interconnected issues, and nowhere is this more relevant than in the electrical sector. The electrical industry plays a front line role in the drive for energy efficiency. Electricians not only influence both residential and commercial industries, but they are on the frontline to advise and install the technology that will take the green agenda forward.

From retrofitting energy-efficient lighting to installing photovoltaic panels, the people best placed to effectively advise both residential and commercial customers on the most effective measures they can take, as well as to actually install the technology, are *electricians*.

The following section lists several *green* areas of opportunities.

Green Areas of Opportunities

- Solar photovoltaic (PV) installations
- Wind turbine installations
- Mass transit and light rail projects
- *Smart* grid transmission systems
- Commercial and residential structure retrofitting
  - Re-lamping (lighting systems upgrades)
  - Daylight harvesting (photo sensors for detecting light levels, dimmers, motion sensors, timers, etc.)
  - HVAC (specific areas such as installing electrical consumption economizers or programmable thermostats that optimize efficiency of HVAC equipment).
  - Energy management systems and monitoring devices (installation of devices).

Green Jobs

As listed in the previous section, there are several green areas of opportunity for electricians. Green jobs are available. You may be interested in some of the more familiar jobs such as installing or working with solar panels or wind turbines or retrofitting buildings. Following is a listing of just a few of the more *unique* green jobs that are becoming available as the green technology industry continues to evolve.

- Parking-lot outlets *installation* electrician (outlets used to charge electric vehicles)
- *Energy* contractor
- Electric vehicle electrician
- Wind turbine manufacturing electrician

Your search for traditional electrician jobs can be supplemented by using Internet Web sites. One site that should be mentioned here is *greenjobsearch.org*. This is a popular Web site used by employers listing green jobs and by job seekers looking specifically for green technology jobs.

The number of jobs in the clean energy economy grew nearly two and a half times faster than overall jobs between 1998 and 2007. What can you do now, during your education and later during your apprenticeship, to qualify yourself for a job in the green energy field?

Self-Check 4

1. List four green areas of opportunities.

2. List three unique green jobs.

Summary

- Green technology is the application of environmental science and green chemistry to conserve the natural environment and resources, and to curb the negative impacts of human involvement.

- Electricians should become part of an electrical construction industry that is transforming the design and construction of buildings so they are environmentally responsible, profitable, and healthy places to live and work.

- Energy solutions typically include energy conservation, energy efficiency, and responsible energy production.

- Renewable energy is energy generated from natural resources such as sunlight, wind, rain, tides, and geothermal heat, which are regenerative (i.e., naturally replenished).

- U.S. national laboratories and other agencies are researching and developing methods to integrate renewable energy technologies, such as solar and wind power, into the electric power system.

- Solar energy is radiant energy from the sun that is converted into thermal or electrical energy.

- A photovoltaic system is a system that uses solar cells to convert light into electricity.

- The performance of a solar cell is measured in terms of its efficiency at turning sunlight into electricity; a typical commercial solar cell yields only about a 15% efficiency.

- Distributed generation, a new model for the utilities of the future, refers to a variety of small, modular, power-generating technologies that can be combined to improve the operation of the electricity delivery system.

- A wind turbine is a machine that captures the energy of wind and transfers the motion to a generator shaft to produce electricity.

- According to the U.S. Department of Energy, wind power contributed 42% of all new U.S. electricity-generating capacity in 2008.

- Wind turbines, like windmills, are mounted on a tower to capture the most energy. At 100 ft (30.4 m) or more above ground, they take advantage of the faster and less turbulent wind.

- Geothermal energy is thermal energy generated and stored in the earth. Geothermal heat pumps can tap into this resource to heat and cool buildings.

- A geothermal pump, or ground source heat pump (GSHP), is a central heating and/or cooling system that pumps heat to or from the ground. It uses the earth as a heat source (in the winter) or a heat sink (in the summer).

- Hydroelectric power, also called hydropower or simply waterpower, the energy of moving water, can be captured and turned into electricity.

- Biomass, a renewable energy source, is organic matter used as a fuel, especially in a power station for the generation of electricity. Wood remains our largest biomass energy resource.

- When hydrogen and oxygen are combined in a fuel cell, electricity, heat, and water are produced. The fuel cell produces electricity as long as fuel (hydrogen) is supplied, and it never loses its charge.

- The U.S. Green Building Council (USGBC) is an industry organization made up of many parts of the construction industry, including owners, designers, and contractors.

- The USGBC promotes the construction of environmentally friendly, high-performance buildings through its sponsorship of the Leadership in Energy and Environmental Design (LEED) Green Building Rating System.

- The purpose of the LEED rating system is to provide an objective standard for certifying that a building is environmentally friendly or green.

- Over 40% of what LEED certification covers is the work that electrical contractors perform, including atmosphere and lighting control, on-site renewable energy generation and management, construction materials and lighting component selection, and light-pollution reduction.

- Electricians work on new construction projects, energy auditing, and retrofitting, to improve efficiency and install renewable energy technologies safely and efficiently.

- Total energy management (TEM) programs work with a checklist that helps with analyzing the best energy options, materials, fixtures, and environmental controls to *minimize* waste and *maximize* efficiency.

- Green technology applications in electrical construction and industrial electrical industries include lighting and design class, efficiency of motors, and controlling of loads (lighting, HVAC, etc.) and building operation through programmable logic controllers (PLCs).

- From retrofitting energy-efficient lighting to installing photovoltaic panels, the people best placed to effectively advise both residential and commercial customers on the most effective measures they can take, as well as to actually install the technology, are electricians.

- The Web site *greenjobsearch.org* is used by employers listing green jobs and by job seekers looking specifically for green technology jobs.

Review Questions

True/False

1. The term *technology* refers to the application of knowledge for practical purposes. (True, False)

2. Developing technologies such as fossil fuel or chemical-intensive agriculture benefits the environment. (True, False)

3. Design and construction of green buildings will have a significant impact on the effort to increase greenhouse gas emissions.   (True, False)

4. Energy solutions include energy conservation, energy efficiency, and responsible energy production.   (True, False)

5. Energy management systems help to reduce building and maintenance costs and energy waste.   (True, False)

Multiple Choice

6. An assessment of energy needs and the efficiency of a building is referred to as a(n) _____.
   A. Energy conservation
   B. Energy production
   C. Energy audit
   D. Renewable energy list

7. Energy generated from natural resources, such as sunlight, wind, or geothermal heat, is referred to as _____.
   A. Reciprocal energy
   B. Renewable energy
   C. Distributed energy
   D. Retrofitted energy

8. Energy technologies that use energy from the sun to provide heat, light, and electricity are _____.
   A. Solar technologies
   B. Geothermal technologies
   C. Hydroelectric technologies
   D. Biomass technologies

9. What is a system that uses solar cells to convert light into electricity?
   A. Turbine
   B. Hydroelectric
   C. Geothermal
   D. Photovoltaic

10. A typical commercial solar cell yields about _____.
   A. 5% efficiency
   B. 15% efficiency
   C. 25% efficiency
   D. 35% efficiency

Fill in the Blank

11. _____ generation refers to a variety of small, modular power-generating technologies that can be combined to improve the operation of the electricity delivery system.

12. A wind _____ is a machine that captures the energy of wind and produces electricity.

13. _____ energy is thermal energy generated and stored in the earth.

14. The force of moving water that is captured and turned into electricity is called _____ power.

15. _____ is organic matter used as a fuel, especially in a power station for the generation of electricity.

16. When hydrogen and oxygen are combined in a fuel _____, electricity, heat, and water are produced.

17. An energy _____ moves and delivers energy in a usable form to consumers.

18. The purpose of the Leadership in Energy and Environmental Design's _____ is to provide an objective standard for certifying that a building is environmentally friendly.

19. _____ is the adding of new technology or features to an older structure or building.

20. A popular area of opportunity in the evolving green technology workforce is commercial and residential structure _____.

ACTIVITY 15-1 Energy Source Evaluation

Complete the following chart, using the information gained in Chapter 15 and from your class discussions.

Energy Source	Benefits of the Source	Drawbacks of the Source	Examples of the Energy Source Being Used in Your Community	Job-Related Opportunities
Solar				

(Continues)

(CONTINUED)

Energy Source	Benefits of the Source	Drawbacks of the Source	Examples of the Energy Source Being Used in Your Community	Job-Related Opportunities
Wind				

Chapter 16

The Job Search

Career Profile

Keith Chitwood, Manager of Apprenticeship Education Programs for Western Electrical Contractors Association (WECA), became an instructor in the electrical field as a way of "giving back." Early on in his search for employment, he was hired by a contractor and soon began attending orientation for the electrical apprenticeship program with ABC. He spent the next four years going to school and working for that same contractor. Upon completing the apprenticeship, he was asked by ABC to impart to others the knowledge he had received in his apprenticeship by teaching other apprentices. Chitwood was hooked: He continued teaching apprenticeship for the next 18 years. He was then recruited from the teaching field first for a position with a large electrical contractor in San Diego and later for a position with a large electrical contractor in Sacramento. When a downturn in the economy eliminated his position, he was hired by WECA's education department, where he teaches the first year of the commercial electrical apprenticeship program. Keith says that "a typical class day will include lecture/discussion and group activity as well as lab activities which allow students time for hands-on instruction," and he identifies as his greatest challenge "communicating as much information as possible in the short time that students are in my class." Chitwood strives to provide his students with such a wealth of material because though "the construction industry is in the grips of a recession now," at some point "the economy will recover and construction will turn the corner—the future holds vast opportunities." He advises incoming apprenticeship students to "remember that everyone goes through rough patches, with work sometimes being cyclical. To minimize that impact, plan ahead, gather as much knowledge as possible in school, and demonstrate both your knowledge and your willingness to learn new tasks on the jobsite. Take every educational opportunity that is offered to you."

Chapter Outline

UNDERSTANDING THE HIRING PROCESS

PREPARING FOR THE JOB SEARCH

COMPLETING THE JOB APPLICATION

INTERVIEWING SUCCESSFULLY

Key Terms

Career objective

Cover letter

Form I-9

Indenture agreement

Professional reference

Résumé

Want ad

Chapter Objectives

After completing this chapter, you will be able to:

1. List the different ways of seeking employment in the electrical trades.
2. List the advantages of completing an apprenticeship program in order to enter the electrical trades.
3. Name the basic qualifications a job candidate must meet before seeking work in the electrical trades.
4. Explain how to compile an educational history, an occupational history, a professional references list, and a skills inventory.
5. Name the documents required by potential employers for the completion of the I-9 form.
6. Describe how to develop a list of questions to ask at a job interview.
7. Describe how to fill out a job application successfully.
8. Describe how to write an effective cover letter for a job application.
9. Describe how to construct an effective résumé that can be transmitted electronically to potential employers.
10. List behaviors that make positive impressions on potential employers at job interviews.

Life Skill Covered

Self-Advocacy

In this chapter, the life skill goals are all about extolling your own virtues. Never is Self-Advocacy more necessary than during the job search. Especially in today's market, there are plenty of job applicants, but not as many job openings. Now is the time to start separating yourself from the crowd. As you read through this chapter, there are several questions that you should be asking yourself. What sets you apart from the crowd? What can you offer an employer that is above and beyond the job description? What can you do during your apprenticeship to put yourself in a position to get hired upon its completion?

Where You Are Headed

To obtain work as an electrician's helper or to complete an apprenticeship in the electrical trades, you must, of course, secure a job that allows you to put into practice what you are learning in your courses; on-the-job training is a key component of your education. Beginning a job search can be overwhelming. But if you make yourself aware of what potential employers want to see in your application materials, you can create professional documents that should allow you to move smoothly through the process of landing the kind of position you seek.

Introduction

Finding an entry-level job in the electrical trades is a multistep process. To efficiently work through the job-search process, you must begin by defining for yourself the kind of position you want to obtain. Once you have defined your career goals, you must find listings for jobs of the sort you seek and then prepare application materials that show you to be directly qualified for the jobs outlined in the ads you decide to respond to. These materials comprise a résumé, cover letters, and/or application forms. Then, research the companies that invited you to interview, and prepare to demonstrate your knowledge, maturity, and reliability to your interviewers.

Want ad an advertisement printed in a newspaper or posted on job-search Web sites by an employer seeking to fill a job opening. A want ad lists the kinds of work required and the qualifications a potential employee must meet; sometimes called a classified ad or a help-wanted ad.

16-1 Understanding the Hiring Process

There are typically two routes to getting a job in the electrical trades: by traditional job hunting and through an apprenticeship program.

Traditional Job Hunting

You can look through the want ads to find a job in the electrical trades. The benefit of this approach is that you don't have to find funding for schooling, and you don't have to spend time taking classes.

Keep in mind that the benefits of enrolling in an apprenticeship program outweigh those of traditional job hunting, however. These benefits are discussed later in this chapter.

Search the Want Ads

In the past, job seekers turned to the classified ads in their local newspapers to find job openings. Although classified ads are still published in newspapers, most companies prefer to advertise open positions on the Internet. Many Web sites are devoted to listing help-wanted ads.

To begin a traditional job search, outside an apprenticeship program you might begin by using such terms as *electrician jobs, electrician helper jobs, journeyman electrician jobs*, and *industrial electrician jobs* to search one or more of these Web sites:

- google.com
- careerbuilder.com
- jobs.monster.com
- indeed.com
- linkedin.com
- tiptopjob.com

A search of each site should yield listings of positions that you can consider in building a list of jobs that you are qualified to apply for.

A posting for an entry-level position in the electrical trades specifies how much education and experience a job candidate must have to apply for the position. It also

FIGURE 16-1 Searching the want ads.

names additional requirements the candidate has to fulfill at the time of application. Perhaps most importantly, it lists the duties of the position and explains when and where those responsibilities are carried out.

A typical online job posting looks something like this:

Construction Electrician Helper

Full-time (40 hours per week). Day shift, Monday through Friday, 7:00 A.M.– 3:30 P.M. Prior experience needed: 1 year. Education: High school diploma or equivalent. Must have driver's license and transportation. Must pass drug test. Must achieve satisfactory score on pre-employment test. The test includes questions on the *National Electric Code*, safety, basic math, and electrical theory. Duties include assisting journeyman electricians; wiring luminaires, receptacles, and switches; and running MC cable, pulling wire, and assisting in terminations. Out-of-town work is available. **How to apply:** Press APPLY button, then submit résumé, and complete employer's job application.

Many entry-level job postings are more general, however. They might be written as follows:

Electrician's Helper

Full-time (80 hours per pay period)

Must have: some mechanical experience, own tools, transportation, clean driving record. Ability to listen to instructions and willing to learn. Must be goal oriented!

Your task in evaluating a series of job listings, then, is to determine whether you meet the minimum qualifications for the position and whether you are interested in the types of work described in the ads.

Once you have created a list of jobs that you want to apply for, you have to gather the information the job application forms require and create a résumé. Tips for gathering your work history, filling out application forms, and creating a résumé appear later in this chapter.

Seek Out Apprenticeship Programs

The benefits of going through an approved apprenticeship program far outweigh the disadvantages.

Apprenticeship programs are developed by employers in the electrical trades to train students according to industry standards through a combination of on-the-job experience and classroom instruction. Apprentices who complete registered apprenticeship programs are accepted by the industry as journeymen. Thus, if you complete an apprenticeship program, you are likely to be a more productive employee who does higher quality work than the person who tries to enter the field without undergoing a training program.

The apprenticeship program gives you the tools to find the information you need and the ability to utilize appropriate resources so that you perform effectively on the job; thus, you will be more successful and better paid.

Follow the Procedures of the Particular Program

Different apprenticeship programs have different requirements. In general, to enter an apprenticeship program, you must meet the following requirements:

- Be 18 years old.
- Have a high school diploma or GED equivalent upon entry to the program. (Some programs require the diploma or GED by the time you complete the apprenticeship, however.)
- Have passed an algebra class or completed a math evaluation.
- Be physically able to perform the on-the-job tasks of apprenticeship training.

Identify and Meet Certification Requirements

Certification requirements vary from state to state and municipality to municipality. Contact your local building department or your state building department to determine what certification requirements you have to meet.

Find Job Placement and Sign an Indenture Agreement

Candidates must apply for registered apprentice positions. Admission requirements and eligibility vary by program because program sponsors define them according to their specific training needs. However, federal law defines minimum requirements and mandates that selection criteria be job related.

Indenture agreement a document between an apprentice and an apprenticeship sponsor (training institute) that sets forth the obligations and responsibilities of both parties with respect to the apprentice's employment and training; sometimes called an apprenticeship agreement.

Once you are employed, you will be asked to enter into an **indenture agreement** (sometimes called an apprenticeship agreement). The indenture agreement is a written document between the apprentice and the sponsor (i.e., the training institute). The agreement sets forth the obligations and responsibilities of both parties with respect to the apprentice's employment and training. These agreements are registered with the entity with which the apprenticeship program is registered—either the state or the federal Department of Labor.

LIFE SKILLS

It is important that you become familiar with the jobs you will be applying for. Search the locations suggested in this section, and find three or four positions that interest you. How do you stack up with their requirements? How will you *after* your apprenticeship?

Self-Check 1

1. What are the advantages of enrolling in an apprenticeship program (rather than undertaking a traditional job search) to seek a job in the electrical trades?

2. List some Web sites a job seeker can visit to find ads for job openings.

3. What is an indenture agreement?

16-2 Preparing for the Job Search

Your job search will go smoothly if you first take the time to gather the information you need to complete a job application form or write a résumé. It is also helpful to research each organization you are applying to so you can speak knowledgably about the company when you are called in for an interview.

Meet Basic Qualifications

Before you begin gathering the kinds of information you need to provide to potential employers, you should make sure that you are able to meet the basic qualifications for employment in the field. Managers seeking to hire new employees expect you to meet these criteria:

- Have a clean driving record (no major citations).
- Be able to pass a drug test.

Have a Clean Driving Record

Make sure that your driver's license is current as you are getting ready to begin your job. Be aware of your driving history—know what, if any, violations appear when your license is checked. Be prepared to explain the circumstances under which you were cited for any violations.

In some states, you may be able to order a copy of your driving record through the Bureau (or Registry) of Motor Vehicles. You can go online to find out how to obtain this record (sometimes referred to as an "abstract" of your driving history).

Employers want to know that you can consistently report to work each day: Be prepared to demonstrate that you have reliable transportation to and from jobsites. If your vehicle is in need of major repairs, you should, if possible, have them completed before you start to apply for positions.

Be Able to Pass a Drug Test

As you prepare to undertake a job search, be aware that employers today often test job applicants for drug usage. You should assume that each company you are applying to will administer such a test, and abstain from taking part in activities that would cause you to fail a drug test.

Drug testing can be done at any time in the application process. Some employers do this testing later in the application process, with only the final candidates for a position. Most employers, however, conduct the drug test post-hire, making passing the drug test a condition of employment.

Results of drug tests are considered medical records and must be maintained in confidential files by the employer.

An employer that conducts drug testing should give you a copy of the organization's drug and alcohol policy. You will likely be asked to sign a document

FIGURE 16-2 Drug use can derail your career.

stating that you have received the policy in writing and that you understand any restrictions it places on your conduct. Signing this document may be a requirement of being hired. Make sure you understand what you are agreeing to by signing the document, and ask questions if there are statements you do not understand.

Gather Your Occupational History

Once you have determined that you are able to meet the basic qualifications for employment in the field, you should create a series of lists that enable you to fill out an application form completely and correctly.

Create and save your lists on a computer if possible. You can then revise or update the lists quickly and produce easy-to-read documents that you can use when completing a job application form or writing your résumé. You can even hand one or more of these lists to prospective employers, when appropriate.

The lists you should prepare include these:

- Educational history
- Work history
- Professional references list
- Skills inventory
- Questions about the company and the position for which you are applying

Create an Educational History and a Work History

Résumé a document that provides a summary of a job seeker's educational history and work history. It is used by potential employers to evaluate job candidates' qualification for a position.

To fill out an application form or create a **résumé**, you must to be able to represent your educational background. Begin your educational history by listing the full names of any schools from which you have earned diplomas or degrees. When appropriate, list the full titles of the specific programs in which you were enrolled or the subject(s) on which you focused your studies there. Beside each name, write the year in which you began attending the school and the year in which you completed your work there.

FIGURE 16-3 Creating an educational and work history.

Also make a list of any certificates or licenses you have earned. Write down the full name of the organization issuing the certificate or license, the full title of the program completed, and the dates on which you began and completed each program.

Next, you should make a list of all your previous employers and information about jobs you have held that allows you to create a thorough work history. For each job that you have held, write down this information:

- The name of the company
- The title of the position you held
- The dates on which you began and ended your employment at the company
- The name and job title of your supervisor
- The company's mailing address
- Your supervisor's phone number and e-mail address

Also list the duties for which you were responsible in each job, using specific language to describe your activities.

Finally, make note of the salary you earned at each job and the reason that you left each position.

Create a List of Professional References

On a separate document, compile a **professional references** list—a roster of three to five persons with whom you have worked and who can speak about your on-the-job performances in positive terms. This list may include some of the supervisors you name in your employment history; it might also list coworkers with whom you have

Professional reference a person who is willing to recommend a job seeker for a new position. Ideally, this person has specific knowledge of the job seeker's performance in work settings.

worked closely and effectively or teachers with whom you have had an especially good relationship during a training program.

On your references list, include each person's full name, job title, the name of the company where he or she works, and his or her mailing address, e-mail address, and phone number.

As a courtesy, let your recommenders know that you are including their name and contact information on your references list. This gesture allows the person to expect calls from your potential employers and to prepare remarks about your skills and strengths as a worker and an employee.

Prepare a Skills Inventory

When you apply for jobs in the field, prospective employers want to know what tools you are most familiar with. So that you can readily answers questions about your competence with the necessary equipment, you should prepare a skills inventory—a list of the tools you have worked with and are comfortable using.

Put in writing a list of the tools with which you are most familiar and capable. You should be able to demonstrate that you know how to use 8 to 10 hand tools.

You may want also to create a list of tools already in your possession if you own a significant amount of equipment.

Collect Needed Documents

After you have created your educational and work histories, your list of professional references, and your skills inventory, print several copies of each document. Keep one set of the documents at home, where you can easily find it whenever you need to complete an online job application, and keep a few sets of these documents in a file that you store in your vehicle. You can then give copies to potential employers, when appropriate, or use copies to complete application forms given to you on-site.

Locate your Social Security card and your birth certificate, and keep them at home where you can easily find them. Take these documents with you to job interviews, as potential employers must verify your legal status for Form I-9, a federal document that certifies you as eligible to work for wages in the United States.

Form I-9 a federal document that certifies a person as eligible to work for wages in the United States.

Cover letter a brief document that accompanies a résumé in a job application. In the cover letter, the job seeker states why he or she is a strong candidate for an advertised position and highlights the most important information on his or her résumé.

Research the Company

In addition to assembling important data about yourself as you prepare to search for jobs, you should research the companies to which you are planning to apply. Reading about a company's operations, values, and history helps you to determine whether the organization is a place where your skills, talents, and interests can best be put to use.

When you are ready to complete a job application form or write a cover letter and résumé, you can again draw on your research: The information you gather allows you to write in specific terms about each organization and how you might fit into it.

When hiring managers review job candidates' application materials, they appreciate the fact that you have taken the time to learn about their business and think of ways in which you are suited to the position they are seeking to fill; your knowledge of their company shows them you are sincerely interested in the job you are being considered for.

Collect Information

Probably the most efficient way to find information on an organization for which you would like to work is to search online. Knowing that the average person now turns first to the Internet to find information, most companies maintain a presence on the Web. Generally, you need only to type the name of the business into a search engine to locate the address of its Web site.

Typically, a company's Web site pages:

- Describe kinds of projects the business undertakes.
- List the range of services provided by the organization to its clients.
- State the company's goals and values.
- Recount the business's history or development.

Some sites also include photographs of particular projects completed by the company to indicate the scope and quality of the work done by its employees.

Thus, in just a few minutes' time you can learn a wealth of useful information about a potential employer.

If the company you are applying to does not have a Web site (if it is a small operation with only a few employees, for example), you might instead look for articles about the company in the archives of the local newspaper. (Most newspapers maintain an archive that is searchable via the Web.) When possible, you might also speak to current or past employees of the company about its history, goals, and services.

Prepare a List of Questions about the Company and the Position

The hiring manager who interviews you is likely to ask whether you have any questions about the company or the position you are being considered for. You should be aware that most interviewers want you to ask questions. When you ask questions, you demonstrate your genuine interest in the job. You also reveal to the interviewer something of your personality, level of experience and skill, and/or critical thinking abilities.

If you have researched the business you are applying to and familiarized yourself with the nature of its work, you can prepare a list of intelligent questions. It is better to come up with these questions in advance than to try and formulate them spontaneously at the interview, when you are likely to feel anxious about making a good impression.

Once you have identified the companies you will apply to and have studied the information available about them on the Internet or through other sources, you should prepare a list of three to five questions to ask when you are called in for interviews.

At a first interview, it is a good idea to ask more general questions about the business and/or the position for which you are being considered:

- What are the qualities of the ideal employee at this company?
- What skills are most needed to succeed in this job?
- What are the day-to-day responsibilities (or week-to-week responsibilities) of this job?

FIGURE 16-4 Asking questions shows the interviewer that you're really interested.

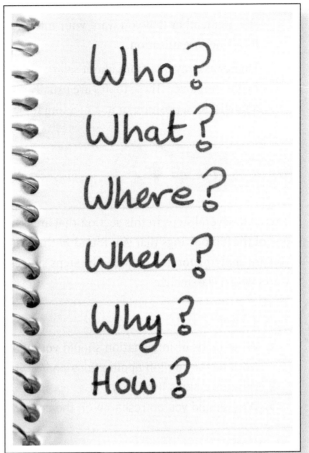

© Stephen Aaron Rees/www.Shutterstock.com

- How far am I expected to drive to get to a jobsite?
- What are the greatest challenges of this job?

At a second or third interview, you can ask more specific questions about the business and the requirements of the position you are applying for:

- How do team members communicate with one another to carry out their work?
- How will my work be evaluated? Who will review my performance?
- What is ahead for this company in the next five years?
- What opportunities might I have for developing my current skills? For gaining new skills?
- If I am willing to work hard and meet all the requirements, how would the company support me in my education?

Whenever you can, phrase your questions so they are specific to the company you are applying to:

- "I saw on your Web site that this company is especially concerned with following green practices. What are some green procedures I might be using in my position?"
- "I read a newspaper article about your company's involvement in the construction of the X Building. What would my role on a project like this one be?"

Note that there are a few types of questions that you should *not* ask at an interview:

• Do not ask any questions that are easily answered by a visit to the company's Web site. Remember that you want your interviewer to know that you have researched his or her organization.

• Do not ask any questions about salary or benefits until these subjects are addressed by the employer. These topics are usually covered later in the process, once you are a finalist for a position or at a meeting in which you are being offered the job.

There are several steps in this section that are related to applying for a position. Take one of the job postings that you found in the first section in this chapter and pretend you are applying for it. Complete the steps suggested in the chapter, and then compare yours with a classmate's.

Self-Check 2

1. What kinds of information should you gather in preparing to represent your work history on job applications and/or your résumé?
2. What is a skills inventory?
3. Why should you do research on the companies where you plan to apply for jobs?

16-3 Completing the Job Application

An employer seeking to fill a job opening asks each job candidate either to fill out an application form or to submit a résumé and cover letter. Following are some guidelines for effectively completing each type of application.

Know How to Fill Out an Application Form

On a typical application form, you are asked to fill in boxes that ask for your educational history, your work history, and contact information for people who are willing to recommend you to your potential employer.

At this point, the lists you compiled in preparing for the job search can be put to use. You can simply transfer the details on where and when you attended school and what diplomas or certificates you have earned into the appropriate spaces on the application form. Likewise, you can fill in the details on where and when you have been employed in the past.

Most application forms ask you to list the names of the companies where you have worked and the dates that you worked there. Usually, they also ask you to list the main responsibilities you fulfilled in your past jobs. You do not have a great deal of space for creating this list of responsibilities, so concentrate on representing the most important duties that you performed.

Some job applications ask you to indicate why you left each of your past jobs. Be honest and concise in explaining why you left each place of employment. Try to use positive (or neutral) language in describing your reasons for leaving a job: "Work was seasonal," "Offered work with more responsibility," "Moved," or "Left to attend school." Avoid negative terms such as "Quit," Fired," "Let go because of lateness," and "Absences."

If the application form asks you to say what sort of position you are looking for, don't write "Any." List a specific position or positions. Use your research on the company to determine what types of jobs are available with the employer.

If the application form asks you to list a desired salary, write "Open" or "Negotiable." You don't want to name a figure that is too high or too low for the position.

If the form asks for your availability to begin work, the best answer is "Immediately."

Work carefully as you are filling out the job application. Write neatly if you are completing a hard copy of the form—those reading the form may reject the application if it is sloppily written, littered with crossed-out phrases, or crumpled from erasing mistakes.

Whether you are completing a paper or an electronic form, read it over carefully as you prepare to submit it; forms that are missing information may also be rejected.

Know How to Prepare an Electronic Résumé

Some employers do not ask for a completed application form, but instead require a cover letter and résumé. The résumé lists information on your work and educational backgrounds, and the cover letter supplies additional information on your history and explains why you believe you are the person the employer should hire for the open position.

FIGURE 16-5   A good résumé can separate you from the crowd.

© alexskopje/www.Shutterstock.com

Begin by creating your résumé. This document follows a fairly standard set of organizational "rules," but there is some room for variety in some sections of the document.

Section 1: Current Contact Information

At the top of the résumé, list the following items of information in the following order:

- Your name
- Your current address (street address, city, and zip code)
- Your current phone number
- Your e-mail address

Note that the e-mail address you use for job applications should be straightforward—your name and, if necessary, a unique combination of numerals. Employers prefer to communicate with johndoe12@yahoo.com than with PartyDude101@gmail.com. Set up an e-mail account that you use only for job applications, if necessary.

Section 2: Career Objective

Career objective
a statement that appears on a résumé to indicate the job seeker's goals concerning employment.

After listing your current contact information at the top of the page, state a career objective. Write one or two sentences about your current employment goals, using specific language:

For example:

Objective: To obtain a position as an electrician's helper that will enable me to use and develop the skills I have gained thus far in my electrical pre-apprenticeship program.

Sections 3 and 4: Education and Work History

After stating your career objective, supply your work history and your educational history—or your educational history and your work history:

- If your work history is limited, begin with your educational information, and then represent your past job experiences.
- If you have held jobs where you have developed skills relevant to the position you are applying for, begin with your work history and then represent your educational experiences.

For each educational program that you have completed (or are currently completing), list the following items of information in the following order:

- The complete name of the school or program and the city where the school is located
- The dates on which you started and finished each program
- The complete title of the degree earned or the certificate earned

For example:

Riverdale High School, Riverdale, NJ
August 2008–June 2011
Diploma

List first the program you most recently completed (or the program in which you are currently enrolled); then go back through your previous programs in reverse chronological order.

For each job that you have held (or currently hold), list the following items of information in the following order:

- Your job title
- The name of the company and the city in which it is located
- The dates on which you began and ended your employment with the company
- The types of work for which you were responsible in the position

List first the job you most recently held (or the job in which you are currently working); then go back through your previous jobs in reverse chronological order.

In describing your job duties, name specific activities, using strong action verbs and concise phrases.

For example:

Electrician's Helper
Pickens Construction and Maintenance, Riverdale, NJ
June 2010–May 2011
Loaded, transported, and unloaded materials and equipment for journeyman electrician. Assisted in lifting, positioning, and securing materials during installation. Performed routine cleaning tasks.

Section 5: Additional Skills and/or Accomplishments

Beneath your educational and work histories, highlight particular skills and/or accomplishments that are not reflected in your work history.

For example:

Additional Skills: Able to read an ohm meter and electrical diagram. Able to bend and install wiring conduit. Proficient in the use of hand tools.

Section 6: References

You should not list your references' contact information on the résumé itself. Instead, make the following sentence the final statement on the résumé: "References available upon request."

On a separate document, have complete contact information for each person who has agreed to recommend you for jobs Take a copy of the list with you to the job interview so you can give it to the interviewer if he or she asks for it.

Format the Résumé

You may be asked to mail a hard copy (paper copy) of your résumé to your prospective employer; if you do, it will likely be scanned into a computer file. More likely, you will be asked to transmit your résumé by e-mail to the company so it can be saved as an electronic file.

So that your résumé reads well when viewed on different computers or printed by different programs, follow these guidelines:

- Place only your name in the first line of the document.
- Begin each line of text at the left margin.
- Use only one font size: 10-, 12-, or 14-point.
- Use a clean, easy-to-read font. Times New Roman, Helvetica, and Arial are the best choices.
- Do not use underlining, bolding, shading, or italics; do not include graphics, vertical lines, or horizontal lines.
- Use hyphens or asterisks instead of bullets in any lists that you write.

Know How to Write a Cover Letter

A cover letter is a brief document that accompanies a résumé in a job application. The cover letter guides the potential employer in reading the job applicant's résumé.

If you are asked to submit a cover letter with your résumé in a job application, you should write a three- or four-paragraph letter, following these guidelines:

FIGURE 16-6 Writing a cover letter.

© Diego Cervo/www.Shutterstock.com

Section 1 (one paragraph): Begin by stating the name of the position you are applying for, and indicate where you learned about the job opening (in the want ads or from a listing provided to the students in your apprenticeship program, for example). Close the first paragraph with a sentence that summarizes why you believe you are the ideal candidate for the position for which you want to be considered.

Sample Section 1

To whom it may concern:

I am writing regarding your company's recent job board posting for an electrician's helper. With a natural inclination toward mechanical work and a strong desire to learn, as well as experience doing maintenance work, I feel that I could be an ideal fit for the position.

Section 2 (one or two paragraphs): In your second paragraph, the body of the letter, list the kinds of experience or training you possess that most significantly make you a strong candidate for the position you are applying for. Explain where and when you gained this knowledge or expertise, mentioning specific projects you have worked on if you can.

If you want to highlight several kinds of knowledge or types of accomplishments, split this section in half, and deal with one or two ideas per paragraph.

Sample Section 2

Throughout my education and my career, I have developed strength in several of the areas that your company is seeking for its electrician's helper candidates. In my previous job as a maintenance technician, I was promoted to a trainer position because of my ability to both give and follow instructions and to quickly learn new processes and procedures. Additionally, I have my own transportation, a clean driving record, and an ever-growing set of tools.

Section 3 (one paragraph): In new language (language different from that used in Section 1 of the cover letter), state your strong interest in the position you are applying for, and indicate why you are suited to the job. Conclude by saying that you look forward to the chance to interview with representatives of the company at their earliest convenience.

Sample Section 3

In addition to my tangible skills and experience, I bring to your organization a relentless work ethic and an eagerness to learn that I believe make me an outstanding Electrician's Helper. I would value the opportunity to speak to you further about my candidacy.

I can be reached by phone at 555-321-7890 should you have any preliminary questions. I appreciate your time in reviewing my qualifications and look forward to speaking with you soon.

Sincerely,
John Doe

Proofread your cover letter and your résumé closely before you submit them. Have several people read the documents if you can. Spelling errors, typing mistakes, and/or missing words can make a negative impression on hiring personnel. You want to imply that you always pay attention to details and that you are always concerned with communicating clearly and correctly. Error-free application documents send exactly these messages.

Create your own electronic résumé. Let your professor take a look at it and give feedback.

Self-Check 3

1. Why is it important to fill out job application forms neatly and carefully?
2. In the third and fourth sections on you résumé, you should list information on your educational background and your work history. Which should you discuss first, education or work experiences?
3. What should you try to explain in the body (second section) of a cover letter?

16-4 Interviewing Successfully

Having a thorough understanding of the kinds of work performed in your field is essential to landing the job that your education has prepared you for. But in addition to reflecting your occupational training and experience, you must show your prospective employers that you know how to behave responsibly and professionally. Effective nonverbal communications and body language are keys to indicating your professionalism at a job interview.

Exhibit Professional Behavior

Listed below are some professional behaviors that help you demonstrate your professionalism.

Write Down the Date, Time, and Location of the Interview

To avoid misunderstandings and missed opportunities, write down when and where your meeting is to take place when you are contacted by an employer to schedule an interview. Take down the address to which you should report, even if you think you already know where the business is located. Also record the name of the person(s) you are to meet with and any instructions you are given as to where you should check in when you arrive for the interview.

Follow the Directions to the Interview Location

Once you have the address at which the interview will take place, be sure you know how to find the correct building or jobsite.

The person who schedules your interview may provide you with directions. If you are unsure of the route you need to travel, you can also chart a route from your home location to the interview site, using an online search engine such as Google Maps, MapQuest, or Yahoo! Maps.

If the interview location is truly unfamiliar to you, you may want to make a practice trip to the site a day or two before the actual interview, at the same time of day that the interview will take place, to verify that your directions are correct and to gauge how much travel time you need to allow yourself on the day of the meeting.

Arrive on Time

It is essential that you arrive on time, or even a few minutes early, for your interview. Your promptness tells your interviewers that you are seriously interested in the job you are applying for and shows your respect for the people who are taking the time to meet with you. You are also signaling that you know the importance of arriving on time for work should you be hired.

Greet Interviewers Politely

Introduce yourself to the person who greets you when you arrive, and state that you are scheduled for an interview. If the person whom you first meet is not your interviewer, introduce yourself again when you meet the person(s) who will conduct the interview.

Shake hands firmly (but not *too* tightly) when you meet your interviewer(s), and make eye contact as you are shaking hands. Maintain good eye contact throughout the

FIGURE 16-7   Always be on time!

© unkreativ/www.Shutterstock.com

interview. Both of these gestures indicate that you are confident in yourself and your abilities and that you have strong social skills.

Reflect a Positive Attitude

Once the interview begins, reflect a positive attitude. Be energetic as you answer (and later ask) questions to show your enthusiasm for the type of work for which you want to be hired.

Phrase your answers to the questions you are asked in the affirmative rather than the negative, whenever possible. That is, say what you can do rather than what you cannot; how you prefer to approach a task rather than what you dislike on a job, and so on. (Be especially careful not to use the interview as an opportunity to complain about former places of employment, former supervisors, or former coworkers.)

Listen Carefully

Listen carefully as an interviewer asks you questions so you understand what the employer is looking for.

Answer Questions Briefly

Answer each of your interviewers' questions fully but concisely. Address directly and specifically the subjects you are invited to discuss as you begin an answer—list the points you want to make in response to the question. Then go on to elaborate a bit on each point if you are encouraged to do so—by nods from the interviewer(s) as you are speaking, for example, or a pause after your initial answer.

Show What You Know

You can't make a second first impression. The following sections cover how to dress with confidence for an interview and how to share your knowledge of the field.

Dress Appropriately

Dress appropriately for the jobsite to let the employer know you have the boots and clothes that are expected. Appropriate clothing that protects the employee in the work environment is called personal protective equipment, or PPE. PPE in the electrical trades generally includes long pants or jeans and boots that go up to the ankle. Whether you should wear a short- or a long-sleeved shirt is dictated by the individual employer. **Figure 16-8** shows an example of some job-specific clothing.

Once you are on the job, standard PPE may also include hard hat, gloves, and safety glasses. (You need not wear these items to job interviews.)

Demonstrate Job-Specific Knowledge

Some employers may ask you to demonstrate knowledge and skill specific to the type of job you are applying for. You should be ready to do the following:

FIGURE 16-8 Dress to impress, but don't overdress.

© auremar/www.Shutterstock.com

- Provide examples of knowledge gained in your apprenticeship program.
- Demonstrate particular skills required of the position.
- Take a written or a computer-generated aptitude exam (an exam that measures how well suited you are for particular types of tasks).

Although it is difficult to "study" for these sorts of tests, you might benefit from reviewing your notes and exams from relevant courses you are taking or have taken at the time of the job interview. You could also practice demonstrating your familiarity with the tools you have listed on your skills inventory.

LIFE SKILLS

There are countless Web sites that list potential job interview questions. Research some common questions, and then hold a mock interview with a partner. Take turns asking and answering the questions.

Self-Check 4

1. Why should you shake hands firmly and make good eye contact with a potential employer when you first arrive at a job interview?
2. What are some ways to show a positive attitude while at a job interview?
3. Why should you dress for the jobsite when you go to a job interview?

Summary

- Attaining a position in the electrical trades is typically accomplished either by employing traditional job hunting techniques, such as searching through the want ads, or through an apprenticeship program.

- Completing an apprenticeship program in order to enter the electrical trades has many advantages, including receiving training according to industry standards, through a combination of instruction methods, and being accepted by the industry as a journeyman upon completion of a registered apprenticeship program.

- Before seeking work in the electrical trades, you must be able to meet the basic qualifications for employment in the field, including having a clean driving record and being able to pass a drug test.

- Create an educational history and an occupational history to fill out an application form or compose a résumé.

- Develop a professional references list that includes individuals with whom you have worked who can provide positive insight about your on-the-job performance.

- Prepare a skills inventory that highlights tools you have worked with and are comfortable using.

- Bring your Social Security card and birth certificate to job interviews. Potential employers need these documents to verify your legal status and to complete the I-9 form.

- Conduct research on the companies where you plan to apply so that you can develop a list of questions to ask at a job interview.

- During an interview, whenever possible, try to phrase questions so they are specific to the company you are applying to.

- Use your educational history, occupational history, and skills inventory lists to assist in successfully filling out a job application.

- Your résumé should include your current contact information, career objective, education and work history, and additional skills and/or accomplishments.

- Do not list your references' contact information on your résumé. Instead, make the following sentence the final statement on the résumé: "References available upon request."

- A cover letter is a brief document that accompanies a résumé in a job application. The cover letter guides the potential employer in reading the job applicant's résumé.

- Follow the formatting guidelines for creating a résumé and cover letter so you can submit either a hard or electronic copy to potential employers.
- At job interviews, exhibit professional behaviors that make a positive impression on potential employers, including arriving on time, greeting interviewers politely, reflecting a positive attitude, listening carefully, answering questions briefly, dressing appropriately, and demonstrating job-specific knowledge.

Review Questions

True/False

1. Today, the best place to look for help-wanted ads is in your local newspaper.   (True, False)

2. The benefits of going through an electrical apprenticeship program far outweigh the disadvantages.   (True, False)

3. It is appropriate to come to an interview prepared to ask three to five questions based on information you learned about the company on the Internet or from other resources.   (True, False)

4. To meet the basic requirements for most jobs in the electrical trades, you must have reliable transportation, a clean driving record, and the ability to pass a drug test.   (True, False)

5. It is not necessary to let your former employers or coworkers know that you are listing them as professional references on your job applications.   (True, False)

6. To represent your educational history on a job application or a résumé, you should be brief, listing only the names of the schools you attended and the degrees you received from them.   (True, False)

7. If a job application asks you to identify what type of position you are seeking, you should write "Any" to indicate your flexibility and eagerness to work.   (True, False)

8. If a job application form asks for your availability to begin work, the best answer is "Immediately."   (True, False)

9. To make your résumé easy to read, start each section with a boldface heading in 16-point type.   (True, False)

10. At a job interview, you should answer each question you are asked as thoroughly and completely as you can; you want the interviewer to hear everything you know on the subjects he or she brings up.   (True, False)

Multiple Choice

11. You should be able to demonstrate to prospective employers that you are able to use at least how many hand tools?
   a. 3 to 5
   b. 6 to 8
   c. 8 to 10
   d. 10 to 12

12. Potential employers must verify your legal status for Form I-9, which certifies you as eligible to work for wages in the United States. Which of the following documents can you use at job interviews to prove your legal status?
   a. Credit card
   b. Driver's license
   c. Proof of medical or automobile insurance
   d. Social Security card

13. In filling out a job application, you are asked to explain why you left your most recent job. Which of the phrases listed below is the best option for answering the question?
   a. "Asked to leave."
   b. "Fired."
   c. "Left to attend a training program."
   d. "Quit."

14. After listing your contact information at the top of your résumé, you should write a one- or two-sentence statement about your employment goals. This statement is called a _____.
   a. Career objective
   b. Career overview
   c. Position statement
   d. Prospective position

15. When representing your work history on your résumé, the job you should list first is the _____.
   a. Job in which you earned the highest wage
   b. Job you first held when you began working
   c. Job you held most recently
   d. Job you liked most

16. The cover letter that is sometimes written to accompany a résumé is typically how long?
   a. Three to four sentences
   b. One to two paragraphs
   c. Three to four paragraphs
   d. Five to six paragraphs

17. In the final paragraph of the cover letter, you should _____.
   a. Describe specific projects in which you played a role at your previous places of employment
   b. Explain in neutral terms why you left your previous places of employment
   c. Say where you read about the job you are applying for
   d. Summarize why you are well suited to the job you are applying for

18. Arriving a little early for a job interview sends all of the following messages *except:*
   a. "I am seriously interested in the job I have applied for."
   b. "I don't know how to manage my time."
   c. "I respect the people who are taking the time to meet with me."
   d. "I will come to work on time when I am hired for the job."

19. Which of the following questions could you appropriately ask at your first interview with a potential employer?
   a. "What are the daily responsibilities of the person who works in this position?"
   b. "What is the salary for this position?"
   c. "What kind of medical insurance will I have if I get this job?"
   d. "What types of services does this company offer to clients?"

20. Personal protective equipment (PPE) in the electrical trades generally includes all of the following items *except:*
   a. Boots that go up to the knee
   b. Boots that go up to the ankle
   c. Long pants or jeans
   d. Safety glasses

ACTIVITY 16-1   Mock Interview

When you go on an interview, you should provide a résumé and be prepared to answer questions similar to those following, identifying why you should be considered for entry into the Electrical Apprenticeship Program.

1. Describe the reasons why you are applying for entry into the Electrical Apprenticeship Program.

2. Describe your knowledge of the electrical trade.

3. What are some ways that you show that you are responsible and can follow through on the job?

4. The electrical program requires a high school knowledge of math. Describe some instances when you have used math in everyday life.

5. How can you assure your employer that you will be at the jobsite and apprenticeship school on time?

6. Explain why it is important to complete tasks and what your personal motivation would be to complete the tasks.

7. Why should you be considered for this position?

8. I have asked you several questions and have provided you with information about the electrical apprenticeship. What questions do you have?

9. Where do you expect to be in one year, five years, and ten years?

Glossary

A

absenteeism a habitual absence from work.

active learning involves learning by being engaged in the educational process.

air-core magnet an electromagnet comprised of a coil wound around a nonmagnetic material.

algebra a branch of mathematics that uses letters or other symbols to represent numbers, with rules for manipulating these symbols.

algebraic expression uses signs and symbols to represent numbers or quantities.

ambient temperature the temperature of the air that surrounds an object on all sides.

American Wire Gauge (AWG) the system used to size the conductors used in residential wiring.

ampacity the current, in amperes, that a conductor can carry continuously under the conditions of use without exceeding its temperature rating.

ampere the number of electrons passing a given point in one second. *One ampere* is equal to 1 coulomb of charge moving past a given point in 1 second.

ampere-turns determined by multiplying the number of turns of wire by the current flow.

appliance branch circuit a branch circuit that supplies energy to one or more outlets to which appliances are to be connected and that has no permanently connected luminaires that are not part of an appliance.

apprenticeship a systematic method of training an individual in a trade or industry.

arbor a mandrel, or a tool component, that can be used to grip other moving tool components.

arc-fault circuit interrupter (AFCI) a device intended to provide protection from the effects of arc faults by recognizing characteristics unique to arcing and by functioning to de-energize the circuit when an arc fault is detected.

area the amount of surface of an object or the amount of material required to cover the surface.

armored cable, Type AC a fabricated assembly of insulated conductors in a flexible metallic enclosure.

artificial respiration the act of simulating respiration, which provides for the overall exchanges of gases in the body by pulmonary ventilation, external respiration, and internal respiration.

ASAP means as soon as possible.

atom the smallest particle into which an element can be divided without losing its identity.

atomic number equal to the number of protons in the nucleus.

attendance the act of being present.

authority having jurisdiction (AHJ) an organization, office, or individual responsible for enforcing the requirements of a code or standard and for enforcing the use of listed materials that have been approved by an NRTL for the specific installation.

awl a small tool for marking surfaces or for punching small holes.

B

ballast a component in a fluorescent luminaire that controls the voltage and current flow to the lamp.

bathroom branch circuit a circuit that supplies electrical power to a bathroom in a residential wiring application.

bevel refers to any edge cut at an angle to a flat surface.

biomass organic matter used as a fuel, especially in a power station for the generation of electricity.

bonded (bonding) connected to establish electrical continuity and conductivity.

bonding the connection of two or more conductive objects to establish electrical continuity and conductivity.

bonding conductor or jumper a reliable conductor to ensure the required electrical conductivity between metal parts required to be electrically connected.

bonding jumper, equipment the connection between two or more portions of the equipment grounding conductor.

bonding jumper, main the connection between the grounded circuit conductor and the equipment grounding conductor at the service.

bonding jumper, supply-side a conductor installed on the supply side of a service or within a service equipment enclosure, or for a separately derived system, that encloses the required electrical conductivity between metal parts required to be electrically connected.

bonding jumper, system the connection between the grounded circuit conductor and the supply-side bonding jumper, or the equipment grounding conductor, or both, at a separately derived system.

branch circuit the circuit conductors between the final overcurrent device (fuse or circuit breaker) and the power and/or lighting outlets.

buddy system a procedure in which two people, the buddies, operate together as a single unit so they are able to monitor and help each other.

C

cable a factory assembly of two or more insulated conductors that have an outer sheathing that holds everything together; the outside sheathing can be metallic or nonmetallic.

cable layout a drawing that shows the number of conductors in each cable, how cables are run, and what devices are connected together in a specific area or room.

cardiopulmonary resuscitation (CPR) an emergency procedure for people in cardiac arrest or, in some circumstances, respiratory arrest.

career objective a statement that appears on a résumé to indicate the job seeker's goals concerning employment.

chassis ground a point used as a common connection from other parts of a circuit.

chisel an edge tool with a flat steel blade with a cutting edge.

chuck the part of the drill that holds the drill bit securely in place.

circuit breaker an automatic overcurrent device that trips where an overcurrent is detected into an open position and stops the current flow in an electrical circuit.

coaxial cable, or coax an electrical cable with an inner conductor surrounded by a flexible, tubular insulating layer, surrounded by a tubular conducting shield.

code-making panel (CMP) one of 19 groups responsible for producing sections of the *NEC*. CMPs consist of elected volunteers from a variety of electrical backgrounds.

coefficient a multiplier of a term in a formula.

combination circuit a circuit configured (wired) with both series and parallel connections.

combination switch a device that consists of more than one switch type on the same strap or yoke.

compound exists when two kinds of atoms combine chemically.

conductivity a measure of the ability of a substance to conduct electricity.

conductor a material that readily conducts electricity.

conduit a protective passageway (pipe or tubing) for cables.

constants letters representing numbers that do not vary.

construction electrician basic responsibilities include installing and repairing telephone systems and high- and low-voltage electrical power distribution networks, both overhead and underground.

continuing education the acquisition or improvement of work-related skills, generally referring to classes and seminars that focus on job-related skills and knowledge.

coulomb the practical unit of electrical charge. One coulomb of charge is the total charge on 6.25×10^{18} electrons.

cover letter a brief document that accompanies a résumé in a job application. In the cover letter, the job seeker states why he or she is a strong candidate for an advertised position and highlights the most important information on his or her résumé.

cram a slang term for last-minute studying.

current the rate of flow of electrons through a conductor.

D

de-energized when electrical power is removed from a circuit.

detailed drawing a separate large-scale drawing that shows very specific details of a small part or particular section of a structure.

device box an electrical device that is designed to hold devices such as switches and receptacle.

digit a single character in a numbering system.

dimmer switch a switch type that raises or lowers the lamp brightness of a luminaire.

directly proportional means that changing one factor results in a direct change to another factor in the same direction and by the same magnitude.

distributed generation a variety of small, modular, power-generating technologies that can be combined to improve the operation of the electricity delivery system.

distribution electrician helper basic duties include assisting with electrical equipment, materials, and cables/conductors, and responding to trouble calls.

dividend a number to be divided.

divisor a number that divides.

doff to remove an article of clothing.

don to put on or dress yourself in.

double-insulated a form of electrical protection featuring two separate insulation systems to help protect against electrical shock from internal malfunctions.

double-pole switch a switch type used to control two separate 120-volt circuits or one 240-volt circuit from one location.

drag the longitudinal retarding force exerted by air or other fluid surrounding a moving object.

duplex receptacle the most common receptacle type used in residential wiring; it has two receptacles on one strap; each receptacle is capable of providing power to a cord-and-plug-connected electrical load.

E

earth ground a ground point made by physically driving a pipe or rod into the ground.

effective ground-fault current path an intentionally constructed, low-impedance, electrically conductive path designed and intended to carry current under ground-fault conditions from the point of a ground fault on a wiring system to the electrical supply source and that facilitates the operation of the overcurrent protective device or ground-fault detectors on high-impedance grounded systems.

electric power the amount of electric energy converted to another form of energy in a given length of time.

electrical blueprint a document containing all the necessary information needed to create an electric circuit.

electrical continuity a complete, continuous, and connected path that keeps the electrical potential difference at zero.

electrical current the flow of electrons.

electrical distribution the final stage in the delivery of electricity to end users.

electrical distributor a worker, often considered a *solutions provider,* who distributes materials to end users and contractors.

electrical drawing a part of the building plan that shows information about the electrical supply and distribution for the structure's electrical system.

electrical inspector examines the installation of electrical systems and equipment to ensure that they function properly and comply with electrical codes, standards, and regulations.

electrical product distribution the provision of electrical products to end users and electrical contractors.

electrical products manufacturer produces those items that are needed by individuals working in the electrical trade.

electrical shock a reflex response to the passage of electric current through the body.

electrical supervisor responsibilities include supervising electricians and trades helpers in the repair, maintenance, and installation of electrical systems.

electricity the theorized flow of electrons that cannot be seen, although its effects can be seen and measured.

electromagnet an object that acts like a magnet, but its magnetic force is created and controlled by electricity.

electron an elementary particle carrying one unit of negative electrical charge.

element a group of identical atoms.

elevation drawing a one-dimensional (flat) projection that shows the side of a structure or house as seen "head on" and facing a certain direction.

EMT sometimes called thinwall, is commonly used instead of galvanized rigid conduit (GRC), as it is less costly and lighter.

energized when electrical power is connected to a circuit.

energy audit an assessment of the energy needs and efficiency of a building or buildings.

energy carrier moves and delivers energy in a usable form to consumers.

energy conservation achieved through efficient energy use, whereby energy use is decreased while achieving a similar outcome, or by reduced consumption of energy services.

equation a mathematical statement that asserts the equality of two expressions.

estimator a person who often performs in a supervisory capacity for a general construction company and accurately assesses the amount and cost of electricity required for a building or process.

even number a number exactly divisible by 2.

F

factor any of the numbers (or symbols) that form a product when multiplied together.

fiber-optic cable a thin, flexible, transparent fiber that acts as a wave guide or "light pipe," to transmit light between the two ends of the fiber.

fire-rated wall made of fireproof material to prevent the spread of a fire from one part of a dwelling to another; also called a fire wall.

flashcards cards with words or numbers on either side, which can be used in classroom drills or in private study.

floor plan a diagram, or drawing, usually to scale, that shows the relationships among rooms, spaces, and other physical features at one level of a structure.

foreman a tradesman who oversees and manages the building, maintenance, and troubleshooting of electrical and electronic systems.

Form I-9 a federal document that certifies a person as eligible to work for wages in the United States.

formula a rule expressed in letters, symbols, and constant terms.

four-way switch a switch type that, where used in conjunction with two 3-way switches, allows control of a 120-volt lighting load from more than two locations.

fraction a number that can represent part of a whole.

free electrons valence electrons that have been temporarily separated from an atom.

fuse an overcurrent protection device that opens a circuit where the fusible link is melted away by the extreme heat caused by an overcurrent.

G

general-purpose branch circuit a branch circuit that supplies two or more receptacles or lighting outlets, or a combination of both.

geothermal energy thermal energy generated and stored in the earth.

geothermal heat pump a central heating and/or cooling system that pumps heat to or from the ground. It uses the earth as a heat source (in the winter) or a heat sink (in the summer); also called ground source heat pump (GSHP).

graduations linear measure markings on a tape measure.

green technology the application of environmental science and green chemistry to conserve the natural environment and resources, and to curb the negative impacts of human involvement.

greenhouse gas a gas that contributes to the greenhouse effect by absorbing infrared radiation, for example, carbon dioxide and chlorofluorocarbons.

ground the earth.

ground fault an unintentional, electrically conducting connection between an ungrounded conductor of an electrical circuit and the normally non-current-carrying conductors, metallic enclosures, metallic raceways, metallic equipment, or earth.

ground-fault circuit interrupter (GFCI) a device intended for the protection of personnel that functions to de-energize a circuit or portion thereof within an established period of time when current to ground exceeds the values established for a Class A device.

ground-fault current path an electrically conductive path from the point of a ground fault on a wiring system through normally non-current-carrying conductors, equipment, or the earth to the electrical supply source.

grounded conductor a system or circuit conductor that is intentionally grounded.

grounded, solidly connected to ground without inserting any resistor or impedance device.

grounding is the act of connecting an electrical device to the ground (earth), generally by means of a conductor such as a wire or rod, so that the device has zero electrical potential.

grounding electrode a conducting object through which a direct connection to earth is established.

grounding electrode conductor a system or circuit conductor that is intentionally grounded.

H

habitable floor area floor area or rooms in a residential occupancy used for living, sleeping, cooking, and eating, but excluding bath, storage and service areas, and corridors.

hexagonal refers to six sides or edges.

home runs the branch-circuit wiring from the panel to the first outlet or junction box in that circuit.

hydrocarbon a compound of hydrogen and carbon, such as any of those that are the chief components of petroleum and natural gas.

hydroelectric power the force, or energy, of moving water that is captured and turned into electricity.

hydrogen fuel cell an electrochemical cell in which the energy of a reaction between a fuel (liquid hydrogen) and an oxidant (liquid oxygen) is converted directly and continuously into electrical energy.

I

impedance the total opposition to current flow.

indenture agreement a document between an apprentice and an apprenticeship sponsor (training institute) that sets forth the obligations and responsibilities of both parties with respect to the apprentice's employment and training; sometimes called an apprenticeship agreement.

individual branch circuit a circuit that provides electrical power to only one piece of electrical equipment.

industrial electrician basic duties are targeting and troubleshooting electrical maintenance issues that arise in an industrial setting, and making necessary repairs.

interrupting rating the highest current rated voltage that a device is intended to interrupt under standard test conditions.

inversely proportional means that increasing one factor results in a decrease in another factor by the same magnitude, or a decrease in one factor results in an increase of the same magnitude in another factor.

ions atoms that have more or less than their normal complement of electrons.

iron-core magnet an electromagnet made by wrapping a coil of wire around an iron core.

J

journeyman a journeyman has the required skills and knowledge and has met the requirements of time in the field.

journeyman electrician an experienced electrician who may have completed training through an electrical apprenticeship program and is able to work independently without supervision.

junction box a box whose purpose is to provide a protected place for splicing electrical conductors.

K

Kirchoff's current law (KCL) states that the sum of currents flowing into a junction equals the sum of currents flowing away from the junction.

Kirchoff's voltage law (KVL) states that the sum of all the voltage drops across all circuit resistors must equal the voltage source.

knob-and-tube wiring an early standardized method of electrical wiring in structures that used knobs as insulators and ceramic tubes to isolate the wiring from neighboring wood.

L

label an identifying mark of a specific nationally recognized testing laboratory (NRTL).

laundry branch circuit the type of branch circuit used in residential wiring applications that supplies electrical power to laundry areas.

law of charges opposite charges attract and like charges repel.

legend that part of a building plan that describes the various symbols and abbreviations used on the plan.

lift an upward force that counteracts the force of gravity, produced by changing the direction and speed of a moving stream of air.

listed indicates that a device has met the testing and other requirements set forth by a listing agency. To be listed, a material or device has to meet the testing and other requirements set by a Nationally Recognized Testing Laboratory (NRTL).

lockout/tagout (LOTO) specific practices and procedures to safeguard you and your coworkers from unexpected energization or startup of machinery and equipment, or the release of hazardous energy during service or maintenance activities.

luminaires lighting fixtures

M

magnetic field the result of the force of magnetism.

magnetism a force field that acts on some materials, but not on other materials.

main idea words that tell the reader what the paragraph is mostly about.

maintenance electrician basic responsibilities include but are not limited to repairing, installing, replacing, and testing electrical circuits, equipment, and appliances.

master electrician an electrician who can display extensive job knowledge and has been tested to have an extensive understanding of the electrical safety code.

material safety data sheets (MSDS) a detailed document that provides product users and emergency personnel with information and procedures needed for handling and working with chemicals.

metal clad cable, Type MC a factory assembly of one or more insulated circuit conductors enclosed in an armor of interlocking metal tape or a smooth or corrugated metallic sheath.

metallic tubing or conduit is an electrical piping system used for protection and routing of electrical wiring.

metric system an international decimalized system of weights and measures.

microamperes (µA) a unit of current equal to one-millionth (10^{-6}) of an ampere. (Ampere is the base unit used to define the rate of current flow.)

milliamperes (mA) a unit of current equal to one thousandth (10^{-3}) of an ampere.

mnemonic a memory or learning aid.

molecule consists of two or more atoms.

multiple-load circuit a circuit that contains two or more loads.

multiwire circuit a circuit in residential wiring that consists of two ungrounded conductors that have 240 volts between them and a grounded conductor that has 120 volts between it and each ungrounded conductor.

N

National Electrical Code (NEC) a United States standard for the safe installation of electrical wiring and equipment.

National Fire Protection Association (NFPA) an international standards-making organization dedicated to the protection of people from the ravages of fire and electric shock.

Nationally Recognized Testing Laboratory (NRTL) a listing organization that has passed the OSHA recognition process.

neutral the conductor connected to the neutral point of a system that is intended to carry current under normal conditions.

neutron a small particle that possesses no electrical charge and is typically found within an atom's nucleus.

node refers to any point on a circuit where two or more circuit elements meet.

non-time-delay fuse a fuse with one fusible element, or link. Where an overcurrent condition exists, the link melts, offering protection to connected circuits.

nonmetallic-sheathed cable, Type NMC a factory assembly of two or more insulated conductors having an outer sheathing of moisture-resistant, flame-retardant, nonmetallic material.

normal loading where the current flowing is within the capability of the circuit and/or the connected equipment.

nucleus the very dense region consisting of protons and neutrons at the center of an atom.

O

Occupational Safety and Health Administration (OSHA) an agency of the US Department of Labor whose mission is to prevent work-related injuries, illnesses, and occupational fatalities by issuing and enforcing standards for workplace safety.

odd number a number that is not evenly divisible by 2.

ohm the amount of resistance that allows 1 ampere of current to flow when the applied voltage is 1 volt.

on-the-job training (OJT) the acquisition of skills within the work environment under normal working conditions.

open circuit a condition where the circuit is *not closed* somewhere in the circuit.

outlet the point on the wiring system at which current is taken to supply electrical equipment.

outside lineman typically installs and repairs underground power lines, switch gears, transformers, and other hardware on the primary distribution system.

overload a condition where the current flowing is more than the circuit and/or connected equipment is designed to safely carry.

P

paraphrasing stopping at the end of a section to check comprehension by summarizing and restating the information and ideas in the text.

passive learning acquiring knowledge without active effort, such as listening to lectures and watching experienced professionals in the working environment.

percent hundredths, or number per 100.

percentage a way of expressing a number or value as a fraction of 100.

permanent magnet a magnet that does not require power or force to maintain its field.

permeability a term used to describe the measure of a material's ability to become magnetized.

personal protective equipment (PPE) equipment worn to minimize exposure to a variety of hazards (e.g. gloves, foot and eye protection, earplugs, hard hats).

photovoltaic system a system that uses solar cells to convert light into electricity.

plant electrician basic duties include installing, troubleshooting, and maintaining lighting, electrical equipment, and power distribution systems.

plot plan an architectural drawing that is generally the first sheet in a set of building plans. It shows all the major features and structures on a piece of property and is drawn as if you were looking down on the property from a considerable height.

plumb perfectly vertical; the surface of the item being leveled is at a right angle to the floor or platform the electrician is working from.

polarity the type (positive or negative) of charge.

polarized plug a type of plug that has one prong longer than the other so it can only be inserted into an outlet in one way.

poles the two areas (ends) of a magnet where the magnetic force is the greatest.

power a mathematical notation indicating the number of times a quantity is multiplied by itself.

preventive maintenance the care and service provided for the purpose of maintaining equipment and facilities in satisfactory operating condition by providing for systematic inspection, detection, and correction of failures before they occur.

prime number a number that has no factors except itself and 1.

product the result of multiplying two or more numbers together.

professional reference A person who is willing to recommend a job seeker for a new position. Ideally, this person has specific knowledge of the job seeker's performance in work settings.

project manager a person who works on site and interprets technical statements of work and design documentation as it relates to project planning, budgeting, procurements, implementation, testing, training, and project completion.

proton found in the nucleus and has a positive electrical charge.

punctuality being on time.

Q

quotient the result of dividing one number by another.

R

raceway an enclosed channel of metal or nonmetallic materials designed expressly for holding wires or cables.

reading comprehension understanding the information that you read.

receptacle a device installed in an electrical box that allows an electrician to access current from the wiring system and deliver it through a cord and attachment plug to a piece of equipment.

reciprocating moving back and forth.

reluctance resistance to magnetism.

renewable energy generated from natural resources, such as sunlight, wind, rain, tides, and geothermal heat, which are regenerative (i.e., naturally replenished).

residual magnetism the amount of magnetism left in a material after the magnetizing force has been removed.

resistance the opposition to the flow of electrons (current).

résumé A document that provides a summary of a job seeker's educational history and work history. It is used by potential employers to evaluate job candidates' qualification for a position.

retrofitting the fitting in or adding of new technology or features to an older structure or building.

rotor the rotating armature of a motor or generator.

rough-in stage in an electrical installation, the time when the raceways, cable, boxes, and other electrical equipment are installed; this electrical work must be completed before any construction work can be done that covers wall and ceiling surfaces.

S

saturation the state reached when an increase in applied external magnetizing field cannot increase the magnetization of the material further.

scaffolding a temporary structure used to support people and material.

schedule a table used on building plans that provide information about specific equipment or materials used in the construction of a structure or house.

schematic diagram an electrical drawing that uses symbols to represent physical components.

schematic symbol an electrical drawing used to represent a physical component.

section line a broad line consisting of long dashes followed by two short dashes; at each end of the line are arrows that show the direction in which the cross section is being viewed.

sectional drawing a part of the structure that shows a cross-sectional view of a specific part of the dwelling. It is a view that enables you to see the inside of a structure or house.

senior construction electrician basic responsibilities include maintaining building electrical infrastructure such as low and medium ac voltage, low dc voltage, LAN network and communication cabling, and fire alarm, security, and PLC control systems.

series circuit a circuit that has two or more loads but only one path for current to flow, from the voltage source, through circuit loads, and back to the source.

service the conductors and equipment for delivering electric energy from the serving electric utility to the wiring system of the premises served.

service cables conductors that are made up in the form of a cable to supply the electrical power from the utility company to the building's electrical system.

service entrance that section of the wiring system where electrical power is supplied to a dwelling or building from the electric utility.

service mast a piece of rigid metal conduit or intermediate metal conduit, usually 2 or 2½ inches in diameter, that provides service conductor protection and the proper height requirements for service drops.

service equipment the necessary equipment connected to the load end of the service conductors supplying a building and intended to be the main control and cutoff site.

sheathed cable a cable protected by a nonconductive covering, such as vinyl.

sheathing a protective covering that wraps or surrounds something.

short circuit a condition where two or more conductors come in contact with one another, resulting in a current flow that bypasses the connected load.

signs of operation in algebra, the same signs and meanings as used in arithmetic (+, −, ×, and ÷).

single receptacle a single contact device with no other contact device on the same yoke.

solar energy radiant energy from the sun that is converted into thermal or electrical energy.

snap switch a manually operated switch used in interior electric wiring; usually used for the control of lighting or small motors.

specifications the part of a building plan that provides more *specific* details about the construction of a building.

split-wired receptacle a duplex receptacle wired so that the top outlet is "hot" all of the time and the bottom outlet is switch controlled.

static electricity the accumulation of an electric charge on an insulated body.

subscript a small letter or number written or printed to the right of and slightly below a letter symbol. Subscripts may be used to identify different variables that are similar enough to each other to have the same main letter symbol.

substance abuse a maladaptive pattern of use of a substance (or drug) that is not considered dependent.

superintendent a person who oversees the implementation, maintenance, and safety of an electrical system.

supplemental grounding electrode a grounding electrode used to back up a metal water pipe grounding electrode.

T

tardiness the quality or habit of not adhering to a correct or usual or expected time.

technology the application of knowledge for practical purposes.

temporary magnet acts as a magnet only as long as it is in the magnetic field produced by a permanent magnet or an electric current.

term the part of an algebraic expression not separated by a plus or minus sign.

three-way switch a switch type used to control a 120-volt lighting load from two locations.

time management a range of skills, tools, and techniques used to manage time when accomplishing tasks, projects, and goals.

time-delay fuse a fuse with a time-delay configuration, or time-current characteristic, that is designed with a heat sink next to the fusible element.

tradesman a manual worker in a particular skill or craft.

transformer an electrical device by which alternating current of one voltage is changed to another voltage.

T®-Stripper Wire Stripper a wire stripper designed to strip insulation from several different wire sizes without having to be adjusted for each size

U

underground feeder cable, Type UF a listed factory assembly of one or more insulated conductors with an integral or an overall covering of nonmetallic material suitable for direct burial in the earth.

Underwriters Laboratories (UL) perhaps the most well-known Nationally Recognized Testing Laboratory (NRTL). UL wrote many of the standards for safety devices.

V

valence shell the outermost shell, or ring, of an atom.

variables those letters or literal symbols that can be assigned different values.

volt the base unit for voltage.

voltage the electrical pressure, or the potential force, or the difference in electrical charge between two points.

voltage drop the voltage or potential energy difference across a resistor.

voltmeter an instrument used to measure the electrical potential difference between two points in an electric circuit.

W

want ad an advertisement printed in a newspaper or posted on job-search Web sites by an employer seeking to fill a job opening. A want ad lists the kinds of work required of the position and the qualifications a potential employee for the position must meet; sometimes called a classified ad or a help-wanted ad.

watt the base unit for power.

wind turbine a machine that captures the energy of wind and transfers the motion to a generator shaft to produce electricity.

wiring diagram a schematic drawing that shows the physical relationship of all components and the information needed to hardwire the circuit.

Z

zero tolerance the strict policy of enforcing all the laws of a state, or the rules of an institution, and allowing no toleration or compromise for first-time offenders or petty violations.

Index

Page locations in *italics* indicate figures and tables

12-piece electrician's tool set, 241–242
13-piece electrician's tool set, 242

A

Aboveground voltage, 145
Absenteeism, 34
ac (current), 119, 120
AC cable, 312
Account manager, 17
Active class participation, 35–37
Active learning, 36
Active reading strategies, 40
Active workplace participation, 28–29
Addition of whole numbers, 78–79
Aden, Henry, 74
Adjustable wrenches, 216
AFCI, 299–300, 395
Air-core magnets, 124
Alcohol in the workplace, 33–34
Algebra
  algebraic expressions, 89–91
  area, 92
  basics, 88
  defined, 88
  grouping symbols, 89
  introduction to, 87–88
  numbers, 89
  powers and exponents, 90
  signs of operation, 89
  solving equations, 91–92
  terms, 90

Algebraic expressions, 89–91
Allen key, 238
Allen wrench set, *238*, 238–239
Alternating current (ac), 119, 120
Ambient temperature, 295
American Heart Association (AHA), 55
American National Standards Institute (ANSI), 64
American Red Cross, 55
American Wire Gauge (AWG), 367
Ammeter, 133
Ampacity of a conductor, 354–355, *355*
Ampere, 133
Ampere-turns, 124
Amps, 133
Anderson, Greg, 308
Angle of declination, 122
Annex C, 357
Annexes to the *NEC*, 172
ANSI, 64
Appliance branch circuits, 350, *351*
Apprentice, *30*, *32*
Apprentice-journeyman interactions, 32–33
Apprenticeship, 5
Apprenticeship agreement, 425
Apprenticeship programs, 424–425
Arbor, 230
Arc-fault circuit interrupter (AFCI), 295, 299–300

Arcing, 299
Area calculations, 93
Area foreman, 20
Armored cable, Type AC, 312
Articles of the *NEC*, 169
Artificial respiration, 54
ASAP, defined, 34
Atomic number, 113
Attendance, 34
Attics, 385–386
Auger bit, 227, *228*
Authority having jurisdiction (AHJ), 174, 265, 312
Autotransformer dimmers, 290, *290*
AWG (American Wire Gauge), 367
Awls, 222

B

Backstabs, 284
Ballast, 291
Bandsaw, portable, 231–232, *232*
Basement receptacles, 384
Basements, 385–386
Basic algebra, 88
Basic electrical circuit, 137
Bathroom branch circuits, 352, *353*
Bathroom receptacles, 383, *383*
Bedrooms, master
  cable layout for, 388–391, *390*
  outlet count, *390*

Behavior, professional. *See* Professional behavior in the workplace
Belowground voltage, 145
Bevel, 217
Binomial, 90
Biofuels, 409
Biomass combustion, *409*
Biomass energy, 408–409
Biopower, 409
Bioproducts, 409
Blades, cutting, 231
Blade-type cartridge fuses, 271, *271*
Body harness, 66
Bonding
  defined, 184, 185
  introduction to, 186–187
  to maintain continuity, *203*
  metal boxes, *186*
  for residence service installation, 192
  a single-family dwelling, 195–197
  wiring devices to outlet boxes, 202–206
Bonding conductor or jumper, 187
Bonding jumper
  equipment, 187
  installation, 205–206
  main, 187, *187*, *192*
  screw as, 192
  supply-side, 187
  system, 187
Bonding receptacles, 204
Borrow "1" method, 80–81
Borrowing, 79–81
Box types. *See* Electrical boxes
Brack, Patricia, 212
Branch circuits
  bathroom, 352, *353*
  conduit fill calculations, 356–358
  connections, *347*
  defined, 346
  general lighting, 347–350, *348*
  general-purpose, 347
  individual, 352–353, *354*
  laundry, *351*, 351–352
  requirements, *369*
  small-appliance, 350, *351*
  types of, 346
Branch/feeder AFCI, 299
Bubble level, 236–237

Buddy system, 53–54
Building plans, 248

C

Cable, defined, 310
Cable cutter, *238*, 238
Cable layout, 388
Cable ripper, 219, *220*
Cabling diagrams, 259, *264*
Cardiopulmonary resuscitation (CPR), 54
Career objective, 434
Career paths
  construction electrician, 12
  construction industry, 7–13
  cost engineer, 18
  distribution electrician helper, 16
  electrical inspector, 20
  electrical maintenance, 13–14
  electrical manufacturer, 17–18
  electrical power distribution, 14–16
  electrical supervisor, 21
  electro-mechanical assembly technician, 18
  estimator, 20
  foreman, 20
  industrial electrician, 14
  journeyman electrician, 19–20
  master electrician, 21
  outside lineman, 15–16
  plant electrician, 13–14
  power distribution electrician, 14–16
  product line sales manager, 18
  products distribution, 16–17
  project manager, 20
  senior construction electrician, 12
  superintendent, 20
  tradesman, 18–19
  vertical mobility, 19–21
Carlyle, Larry, 376
Cartridge fuses, *270*, 270–271, *292*, 293
Certification requirements, 425
Chassis ground, 144, *146*
Chemicals, working with, 60
Chisels, 222, 237
Chitwood, Keith, 420
Chuck, 227

Circle, formula used to find area of, *93*
Circuits
  circuit-breaker panel, *11*
  circuit breakers, *271*, 271–272, *294*, 294–295, *298*, 387
  circuit essentials, 137–139
  conditions, types of, 336–340
  GFCI circuit breakers, 387
  ground fault, 339, *339*
  layout of, 378–380, *379*
  normal circuit, *337*
  open circuit, 339, *340*
  overloaded circuit, *337*
  receptacle ratings for, *369*
  recessed cans, 379
  short circuit, *338*, 338–339
  symbols and diagrams, 138
Circular saw, *230*, 230
Class participation, 35–37
Classroom behaviors, 36
CMP. *See* Code-making panels (CMP)
"CO/ALR" conductor, 284
Coaxial cable, *10*, 10
Code-making panels (CMP), 163, 165
Coefficients, 90
Cold chisels, *237*, 237
Combination AFCI, 299
Combination switch, *322*, 322
Commercial and industrial facilities, grounding, 197–200
Compass, 121
Compounds, 112
Comprehension strategies, 40–41
Concrete-encased electrodes, 197–198
Conductivity, 366
Conductor material markings, 284
Conductors
  ampacity of, 354–355, *355*, 368
  AWG system, 367
  defined, 116, 137
  grounded, 280–281
  identification markings, *367*
  NEC requirements, 367–370
  residential applications, *368*
  size of, 284
  types of, 366
Conduit, 7, 8, 233. *See also* Raceways

Confined spaces, 58–59, *59*
Constants, 88
Construction electrician, *12*, 12
Construction site, *31*
Continuing education, 6
Control device, 137
Cordless drills, 222, 229, *229*
Cordless power tools, 225
Cordless screwdrivers, *221*, 221–222
Cost engineer (electrical), 18
Coulomb, 130
Countertop receptacles, 380
Cover letter, 429, 436–438
CPR, 54
"cradle to cradle" design, 396
Cram, 41
Crescent wrench, 216, *217*
Critical thinking skills, 36
CSA International, 266–267
"Cu-Clad" conductor, 284
"CU" conductor, 284
Current, 130–133, 147–148
Cutting blades, 231

D

dc current, 120
Deal, Jim, 394
Decimal fractions, 85–86
De-energized, 53
Definite numbers, 89
Denominator, 84
Dependability, 35
Detailed drawings, 252, *253*
Device box, 358, *359*, *360*, *361*
Diagonal cutting pliers, 218, *218*
Digit, defined, 76
Dimmer control devices, 289–291, *290*, 321, *321*
Dining room receptacles, 382–383
Direct current (dc), 120
Directly proportional, 140
Distributed generation, 402
Distribution electrician helper, 16
Dividend, 78, 84
Division of whole numbers, 82–83
Divisor, 78, 84
Doff, 53
Don, 53
Double-insulated, 225
Double-pole switch, 286, *287*, 319, *320*
Drag, 404

Draw tape, 234–236
Draw wire, 234–236
Drill bits, 227, *228*
Driving record, 426
Drugs
    testing for, 426–427
    use of, 33, *427*
    in the workplace, 33–34
Duplex receptacle, *283*, 283–284, *284*

E

Earmuffs, 64
Earplugs, 64
Earth ground, 144, *145*
Earth's poles, *121*
Edison-base plug fuses, 269
Education, 5–6, 40. *See also* Personal success; Training
Effective ground-fault current path, 186
Electrical blueprint, 138
Electrical boxes
    allowances to be considered, *363*
    box fill and sizing, 361–365
    metal boxes, *362*
    nonmetallic box, 362
    selection guide, *365*
    square box, *364*
    types of, 358–360, *359*, *360*, *361*
    volume allowance required per conductor, *362*
Electrical burn, 55
Electrical circuit, 137
Electrical continuity, 202
Electrical current, 117
Electrical distribution, 14–15
Electrical distributors, 17
Electrical drawings, 252
Electrical hardware, 17–18
Electrical industry, 5–6
Electrical inspector, 20
Electrical installations. *See also* Wiring methods
    arc-fault circuit interrupter (AFCI), 295, 299–300
    building plans, 248
    circuit breakers, 271–272, 294–295
    detailed drawings, 252, *253*
    dimmer control devices, 289–291, *290*
    electrical drawings, 252

elevation drawings, 250, *251*
    floor plans, 249, *249*, *250*, *254*
    fuses, 268–271, 293–295
    ground-fault circuit interrupter (GFCI), 295, *296*, 296–298, *297*, *298*
    interrupting rating, 272–273
    plot plans, 249
    schedules, 252–253
    sectional drawings, *251*, 251–252
    specifications, 253–255
    specifications used in, 248–258
    switches, 285–289
    symbols and notations used in, 259–264
Electrical maintenance, 13–14
Electrical metallic tubing (EMT), *7*, *318*, 318, 356, *357*
Electrical metallic tubing (EMT) bender, *236*, 236
Electrical nonmetallic tubing (ENT), 317, *317*
Electrical power distribution electrician, 14–16
Electrical power grid, *14*
Electrical products distribution, 16–17
Electrical products manufacturers, 17–18
Electrical quantities
    current, 130–133
    power, 135–136
    resistance, 134
    voltage, 133–134
Electrical shock, 53, 55
Electrical supervisor, 21
Electrical switch symbols, *262*
Electrical symbols, *263*
Electrical theory
    circuit essentials, 137–139
    Ohm's Law, 139–141, *141*
    parallel circuits, 146–152
    quantities and units, 130–137
    series circuits, 141–146
    series-parallel circuits, 152–156
Electrical wiring symbols, *264*
Electric charge, 114–115
Electric current, affects on the body, 55–56, *56*
Electricians
    with analog multimeter, *9*
    defined, 4
    licensed, 6

and other trades workers, 18–19
    at work, *9*
Electrician's hammer, 222, *223*
Electrician's knife, 222, *223*
Electricity
    atoms and, 111–114
    defined, 4
    electric charge, 114–115
    electric sources, 120–121
    electrons in motion, 116–119
    ions, 117
    magnetism, 121–125
    static electricity, 117–119
    valence electrons, 116–117
Electric power, 135–136
Electric shock, effects of, *208*
Electric sources, 120–121
Electrode conductor, 193
Electrode system, 193
Electromagnet, 123
Electromagnetism, 123–125
Electro-mechanical assembly
    technician, 18
Electron flow, 117
Electronic dimmers, 289
Electronic résumé, 433–435.
    See also Résumé
Electron orbits, *114*
Electrons
    defined, 112–113
    electric charges, *115*
    electron flow, 117
    free electrons, 117
    lines of force, *113*
Element, 111
Elevation drawings, 250, *251*
Employer responsibilities to safety,
    57–58
EMT. *See* Electrical metallic
    tubing (EMT)
EMT (electrical metallic tubing)
    bender, *236*, 236
Energized, defined, 53
Energy audits, 398, 411
Energy carrier, 410
Energy conservation, 397
Energy efficiency, 397–398
Energy production, 398
English area measure, *92*
English-Metric equivalents length
    measure, *92*
English system length
    relationships, *96*

Equations, 91–92
Equipment bonding jumper, *205*,
    205–206
Equipment spaces, 385–386
Estimator, 20
Even number, 78
Exceptions to the *NEC*, 171
Extractions from the *NEC*, 171
Eye and face injury
    protection, 64

F

Face injury protection, 64
Factor, 78
Fall arrest system, 66, *66*
"Fall Protection," 66
Faraday, Michael, 110
Ferrule-type cartridge fuses,
    270, 270
Fiberglass fish tape, *235*
Fiber-optic cable, *10*, 10
Figures in the *NEC*, 170
Fire-rated wall, 379
Fire wall, 379
First aid, 54–55
Fish tape, 234–236, *235*
Flashcards, 42
Flat-bladed spade bit, 227
Flexible metal conduit (FMC),
    316, *316*
Floor plans, *249*, 249, *250*, *254*
Fluorescent lamp dimming, 291
Flux, lines of, *122*, 123
Folding rule, *224*
Foot and leg injury
    protection, 63
Foreman, 20, 29
Form I-9, 429
Formula, defined, 93
Four-way switch
    installation, 330–333
    operation of, 288, *289*, *331*
    wiring methods, 320, *321*
Fraction, 84
Fraction-to-decimal conversion,
    84–85
Franklin, Ben, 110
Free electrons, 117
Fuse puller, 239, *239*
Fuses, *11*, 268–271, *269*, *270*,
    293–295
Fuse sockets, *11*

G

Gas emssions, 408
General Duty Clause, 57–58
General foreman, 20
General lighting loads by
    occupancy, *349*
General numbers, 89
General-purpose branch
    circuit, 347
Geothermal energy, 405–407
Geothermal heat pump
    diagram, *406*
Geothermal heat pumps, 406
GFCI. *See* Ground-fault circuit
    interrupter (GFCI)
Gloves, 64, *65*
Goggles, *63*, 64
Graduations, 98
Greenhouse gas, 397
Greenhouse gas emissions, 408
Greenjobsearch.org, 415
Green technology
    biomass energy, 408–409
    defined, 396
    employment opportunities for
        electricians, 412–413
    energy conservation, 397
    energy efficiency, 397–398
    energy production, 398
    energy solutions, 397–398
    geothermal energy, 405–407
    hydroelectric power, 407–408
    hydrogen energy, 409–410
    objectives of, 396–397
    renewable energy, 398–399
    solar energy, 399–402
    wind energy, 402–405
Grosser, Jeremy, 344
Ground rods, 198
Ground
    defined, 185
    as a reference, 144–145
Ground clamps, 198, *199*, *200*
Grounded, solidly, 186
Grounded conductor, 186, 280
Grounded neutral
    conductor, 281
Ground fault, 185, *207*, 339, *339*
Ground-fault circuit interrupter
    (GFCI)
    circuit breakers, 387
    defined, 206

Ground-fault circuit interrupter (GFCI) (*Continued*)
installation and operations of, *296*, 296–298, *297*, *298*
NEC requirements for, 386–387
objective of, 206–208
receptacles, 387
Ground-fault current path, 185
Grounding
commercial and industrial facilities, 197–200
concrete-encased electrodes, 197–198
defined, 184, 185
electrode conductor connection, 193
electrode system, 193
incorrect grounding, 200–201
introduction to, 185–186
lack of, 201–202
for residence service installation, 191–192
self-grounding receptacle, *204*
service, 188–200
a single-family dwelling, *195*, 195–197
supplemental grounding electrode, 190, *196*, 196
UFER ground, 196
Grounding electrode, 186
Grounding electrode conductor, 186, 193, *194*, *199*
Grounding receptacle, *204*
Grounding screw terminal, 285
Ground rods, 198
Ground source heat pump (GSHP), 406
Ground symbols, 144, 145
Grouping symbols, 89

H

Habital floor area, 348
Hacksaw, 222, *223*
Hammer, 222, *223*
Hammer drill, *228*, 228–229
Hand tools. *See also* Power tools; Tools
basic, 215–225
caring for, 214–215
fish tape, 234–236, *235*
folding rule, *224*, 224
hacksaw, 222, *223*
hammer, 222, *223*

keyhole saw, 233–234, *235*
knife, 222, *223*
pliers, 217–218
ruler, *224*, 224
screwdrivers, 219–222
tape measure, 97–100, *98*, *99*, 224, *224*
wire strippers, 219
wrenches, 216
Hand Tools Institute, 240
Hard hats, *63*, 63
Hard magnetic material, 122
Hawkbill knife, 222, *223*
Head injury protection, 63
Hearing loss protection, 64
Helium, 112
Helium atom, *112*
Helmets, 64
Hexagonal, 238
Hex key, 238
Hex key set, 238–239
High, Sarah E., 182
Hiring process, 422–425
Hockaday-Bey, George, 130
Home run symbol, 259
Hoover Dam, *407*
Hydraulic knockout set, 233, *234*
Hydrocarbons, 409
Hydroelectric power, 407–408, *408*
Hydrogen energy and fuel cells, 409–410
Hydrogen fuel cells, 410

I

I-9 form, 429
Indenture agreement, 425
Index to the *NEC*, 172
Individual branch circuits, 352–353, *354*
Industrial electrician, 14
Industrial facilities, grounding, 197–200
Informational notes to the *NEC*, 171
Inline ammeter, *77*
Innovation, 397
Installations. *See* Electrical installations
Insulator, 137
Intermediate metal conduit (IMC), 315, *316*, 356
International Association of Electrical Inspectors, 175

International System (SI), 94–96
Interrupting rating, 272–273
Intertek Testing Services (ITS), 267
Interview, 438–441
Inversely proportional, 140
Ions, 117
Iron-core magnets, 124

J

Job application
application form, 430–433
career objective, 434
contact information, 434
cover letter, 429, 436–438
education and work history, 427–429, *428*, 434–435
electronic résumé, 433–435
formatting résumé, 436
references, 435
Job interview, 438–441
Job search
application form (*see* Job application)
apprenticeship programs, 424–425
certification requirements, 425
company research, 429–432
cover letter, 429, 436–438
hiring process, 422–425
interview, 438–441
online job posting example, 424
preparing for, 426–432
professional references, 428–429
questions, preparing, 430–432
traditional job hunting, 422–424
want ads, 422–423
Web sites, 415, 423
work history and, 427–429, *428*, 434–435
Johns, Edmund T. "Ned," 2
Johnston, Trenton, 246
Journeyman, 6, 19–20, *30*, *32*
Journeyman-apprentice interactions, 32–33
Junction box, 358, *360*

K

Kevlar gloves, 64, *65*
Keyhole saw, 233–234, *235*

Keystone-tip screwdrivers, 220, *220*
Kirchoff's voltage law (KVL), 143, 148
Kitchen receptacles, 382–383
Knife, 222, *223*
Knob-and-tube, 333
Knockout punch, 233

L

Label, 265
Ladders, 58, 69–70
Ladder safety, 70
Laundry room
   branch circuits, *351*, 351–352
   receptacles, 383
Law of charges, 113, 114–115, *115*
Leadership in Energy and Environmental Design (LEED) Green Building Rating System, 410–412
Lead foreman, 20
Leather gloves, *65*
LEED rating system, 410–412
Left-hand rule, 124, *124*
Legend, 259
Leg injury protection, 63
Level, 236–237
Licensed electricians, 6
Lift, 404
Lighting branch circuits, 347–350, *348*
Lighting outlets
   locating, *384*, 384–386
   symbols, *260*
   wiring, *380*
Lineman. *See* Outside lineman
Lineman pliers, *217*, 217–218, 238
Lines of flux, *122*, 123
Liquidtight flexible metal conduit (LFMC), 317, *317*
Liquidtight flexible nonmetallic conduit (LFNC), 317, *317*
Listed, 166
Load, 137
Lockout/tagout (LOTO), 60–62
Lockout/tagout devices, *61*
Lodestone, 121
Long-nose pliers, 218, *218*
Long-winded comments, 36

Luminaires, 167
Luminaire schedule, *255*

M

Magma, 406
Magnetic devices, 125
Magnetic field, 122, *123*
Magnetic lines of flux, *122*, 123
Magnetic poles, *121*, 123
Magnetism, 121–125
Main bonding jumper, *187*, 192, *192*
Main idea, 40
Maintenance electrician, *13*, 13–14
Manual knockout set, 233, *234*
Masonry bits, 227, 229
Master bedrooms, 388–391, *390*
Master electrician, 21
Material Safety Data Sheet (MSDS), 60
Mathematics
   addition, 78–79
   borrowing, 79–81
   defined, 76
   division, 82–83
   fractions and decimal conversion, 84–87
   metric system, 94–96
   multiplication, 81–82
   subtraction, 79–81
   tape measure, 97–100
   terms, 77–78
   units of measure, 94–96
   whole number operations, 76–83
MC cable, 312–313
Measurement, with tape measure, 99
Metal clad cable, Type MC, 312–313
Metallic tubing, 231
Metric area measure, *92*
Metric system, 92, 94–96
Metric system length relationships, *97*
Microamperes (mA), 56
Miller, James W., 26
Milliamperes (mA), 56
Mnemonic, 42
Molecule, 111
Monomial, 90

Mounting ears, 285
Multiple-choice questions, 43
Multiple-load circuit, 142
Multiplication of whole numbers, 81–82
Multiwire circuits, 280
Municode Library, 175

N

Nameplate, 352–353
National Electrical Code (NEC). See also Wiring methods; Wiring requirements
   bonding wiring devices, 202–206
   branch-circuit sizing and loading, 346–356
   chapter organization of, 167–169, *168*
   conductors and, 367–370
   defined, 7
   EMT installation, 315–318
   form for proposal, 164
   four-way switches, 322–323
   history of, 162–163
   layout of, 166–173, *167*
   MC cable installation, 312–313
   NMC installation, 310–311
   online version, 163
   overhead service, 255
   purpose of, 165–166
   replacing receptacles, 333–336
   requirements for ground-fault circuit interrupters, 208
   scope of, 172
   standards and local authorities, 173–175
   structure of, 166–167
   three-way switches, 322–323
   UF cable installation, 314
   underground service, 256–258
   updating, 163–165
   using the, 176–177
National Electrical Manufacturers Association (NEMA), 267–268
National Fire Protection Association (NFPA), 162, 163
Nationally Recognized Testing Laboratory (NRTL), 166, 265–267
NEC. See National Electrical Code (NEC)
Neutral, 281

Neutron, 112
NFPA. *See* National Fire Protection Association (NFPA)
NFPA 70E, 57
NM-B cable, 311
NMC-B cable, 311
NMS-B cable, 311
Node, 153
Nonmetallic-sheathed cable, Type NMC, 310–311, *311*
Non-time-delay fuse, 269, 293
Normal loading, 336
Normally loaded circuit, *337*
North Pole, *121*
Nucleus, 112
Numbers, 89
Numerator, 84

O

"Occupational Noise Exposure," 64
Occupational Safety and Health Administration (OSHA), 52, 57–58, 240
Odd number, 78
Oersted, Hans, 110
Ohm, George, 139
Ohmmeter, 134
Ohms, 134
Ohm's Law, 139–141, *141*
On-the-job training, 29–31
Open circuits, 339, *340*
Open-end wrenches, *216*, 216
OSHA. *See* Occupational Safety and Health Administration (OSHA)
Outdoor entrances, 385
Outlet, 281–282
Outlet circuit AFCI, 299
Outside lineman, *15*, 15–16
Outside sales, 17
Overhead service entrance, *189*, *190*, 255, *256*
Overload, 336
Overloaded circuit, *337*

P

Parallel circuits. *See also* Series-parallel circuits
  calculating values, 149–151, *150*
  current, *147*, 147–148
  resistance, 148–149

voltage, 146–147
Paraphrasing, 40, 58
Participation monopolizers, 36
Parts of the *NEC*, 169
Passive learning, 36
Percent and percentages, 86–87
Permanent magnets, 122
Permeability, 124
Personal fall arrest system, 66, *66*
Personal protective equipment (PPE). *See also* Tools
  defined, 30, 53, 62
  eye and face injury protection, 64
  foot and leg injury protection, 63
  hand injury protection, 64–65
  head injury protection, 63
  hearing loss protection, 64
  safety harness, 66
  whole body proection, 65
Personal success. *See also* Education; Training
  active class participation, 35–37
  active learning, 36
  classroom behaviors, 36–37
  comprehension strategies, 40–41
  critical thinking skills, 36
  passive learning, 36
  study techniques, 38–40
  test-taking strategies, 41–43
  time management strategies, 37–38
Pew Charitable Trusts, 412
Phillips-tip screwdrivers, 220, *221*
Photovoltaic (PV) systems, 399–401
Pistol-grip drill, 227, *227*
Plant electrician, 13–14
Pliers, *217*, 217–218, 238
Plot plans, 249
Plug fuses, 268–270, *269*, *270*, *292*, 292, *293*, *293*
Plumb, 236
Polarity, 114
Polarized plug, 226
Poles, Earth's, *121*
Polynomial, 90
Polyvinyl chloride conduit (PVC), 315, *316*, 356
Portable AFCI, 300
Portable bandsaw, 231–232, *232*
Potential difference, 133
Power (algebra), 90

Power, electric, 135–136
Power distribution electrician, 14–16
Power drills, 226–229
Power saws, 229–233
Power source, 137
Power tools. *See also* Hand tools; Tools
  cordless power tools, 225
  power drills, 226–229
  power saws, 229–233
PPE. *See* Personal protective equipment (PPE)
Pre-reading strategies, 40
Preventive maintenance, 13
Prime number, 78
Product (mathematics), 78
Product line sales manager, 18
Products distribution, 16–17
Professional behavior in the workplace
  active workplace participation, 28–29
  alcohol and drugs, 33–34. *See also* Drugs
  apprentice-journeyman interactions, 32–33
  dependability, 35
  on-the-job training, 29–31
  punctuality and attendance, 34
Professional references, 428–429, 435
Project manager, 20
Protection device, 137
Protons, 112, *113*, *115*
Pump storage plant, 408
Punctuality, 34
Push-in terminals, 284
PV arrays, 400–401, *401*
PVC. *See* Polyvinyl chloride conduit (PVC)
PV systems, 399–401

Q

Quotations specialist, 17
Quotient, 78

R

Raceways, 8, 8, 310, *315*, 315–317, *316*, 356. *See also* Conduit
Reading comprehension, 40
Receptable outlet symbols, *261*

Receptacle outlets, *282*, 282–284
Receptacles
  in basements, 384
  bathroom, *383*, 383
  countertop, 380, 382–383
  defined, 281
  in dining room, 382–383
  duplex receptacle, *283*, 283–284, *284*
  GFCI, 387
  in kitchens, 382–383
  laundry room, 383
  locations for, 381–384
  mounting, *383*
  placement, *381*, *382*
  replacing, 333–336, *334*, *335*
  single receptacle, 283
  split-wire receptacles, 283–284
  wall, 383
Recessed cans, 379
Reciprocal formula, 149
Reciprocating saw, 230, *231*
Rectangle, formula used to find area of, *93*
References, professional, 428–429, 435
Reforming, 409
Reluctance, 124
Renewable energy
  benefits of, 398–399
  defined, 398
  geothermal energy, 405–407
  solar energy, 399–402
  wind energy, 402–405
Repetitive responses, 36
Residential floor plan, *389*
Residential service entrance, grounding, *191*, 191–192
Residential service installation, *194*
Residual magnetism, 125
Resistance, 134, 143, 148–149
Résumé, 427, 433–435, *436*
Retrofitting, 411
Rigid metal conduit (RMC), 315, *315*, 356
Rigid nonmetallic polyvinyl chloride conduit (PVC), 315, *316*, 356
Romex, 310–311
Ross, Chris, 50
Rotary BX cutter, 239, *239*
Rotor, 403

Rough-in stage, 381
Ruler, *224*

S

Saber saw, 230
Safety
  basic rules of, 53–55
  confined spaces, 58–59
  drugs and, 33
  employer responsibilities, 57–58
  how body is affected by electric current, 55–56
  ladders, 69–70
  lockout/tagout, 60–62
  Material Safety Data Sheet, 60
  personal protective equipment (PPE), 62–66
  proper work procedures, 57–62
  scaffolding, 67–68
  trenches, 58
Safety glasses, *64*
Safety harness, 66, *67*
Safety watch, 59
Saturation, 124
Sawzall, 230, *231*
Scaffolding, 67–68, *68*
Schedules, 252–253
Schematic diagram, 138
Schematic symbols, *138*, 138, *139*, 144, 145
Scope of the *NEC*, 172
Screw, as main bonding jumper, 192
Screwdrivers, 219–222, *220*
Screw terminals, 285
Sectional drawings, *251*, 251–252
Section line, 252
Sections of the *NEC*, 170
Self-grounding receptacle, *204*
Senior construction electrician, 12
SER (service cable), 257, *258*
Series circuits, 141–146, *142*, *143*. *See also* Series-parallel circuits
Series circuits values, 143–144
Series-parallel circuits, *152*, 152–156, *153*, *155*
Service, 188
Service cables, 256–257
Service entrance, 188, *189*, 255–257
Service-entrance cables, *257*, 258

Service equipment, 188–190
Service grounding, 188–200. *See also* Grounding
Service mast, 315
SEU (service cable), *257*, 257
Sheathed cable, 219
Sheathing, 219
Shorkey, Ray, 108
Short circuits, *338*, 338–339
SI base units, *96*
Side-cutter pliers, 217–218
Signs of operation, 89
Single-family dwelling, grounding, *195*, 195–197
Single-pole circuit breaker, *298*
Single-pole switch
  installation of, 323–326, *324*, *325*
  operations of, 285–286, *286*
  wiring methods, 319, *319*
Single receptacle, 283
Skil, 230
Skills inventory, 429
Skilsaw, 230
Slemp, Jimmie, 160
Small-appliance branch circuits, 350, *351*
Snap switches, 323
Soft magnetic material, 122
Solar cell array, 399–401
Solar cells, 399–402, *400*
Solar energy, 399–402
Solar panels, *404*
Source reduction, 397
South Pole, *121*
Spaces, confined, 58–59
Specifications, 253–255
Spectacles, 64
Split-circuit receptacle outlets, 389
Split-wire receptacles, 283–284
S plug fuse, 269, *270*
Spray painting, 118–119, *119*
The SQR3 Study Reading System, *41*
Square, formula used to find area of, *93*
Stairways, 385
Standard engineering units, *95*
Standard units of metric measure, *95*
Static electricity, 117–119
Steel fish tape, *235*
Stepladder, *69*
Storage rooms, 385–386

Straight ladder, 69
Strip gauge, 284
Stubby screwdrivers, 220, *221*
Study techniques, 38–40
Sturgeon, William, 110
Subscript, 142
Substance abuse, 33
Subtraction of whole numbers, 79–81
Superintendent, 20
Supplemental grounding electrode, 190, 196, *196*
Sustainability, 396
Switch box, *12*
Switches
  combination, 322, *322*
  double-pole, 319, *320*
  four-way, 320, *321*
  four-way installation, 330–333
  installation, 323–333
  operations of, 285–289
  single-pole, *319*, 319
  single-pole installation, 323–326
  snap, 323
  three-way, 319, *320*
  three-way installation, 326–330, *327*
  types of, 319–322
Switching circuit, *329*, *332*
Switch toggle, 285
Symbols, *254*
Symbols used in notations
  electrical switch, *262*
  electrical wiring, *264*
  lighting outlet, *260*
  miscellaneous electrical, *263*
  overview, 259–264
  receptacle outlet, *261*
Système International, 94–96

T

Table of Contents to the *NEC*, 171
Table of elements, *111*
Tables in the *NEC*, 170
Tape measure, 97–100, *98*, *99*, *224*, 224
Tardiness, 34
Taylor, Jim, III, 278
Technology, defined, 396
Temporary magnets, 122
Term (algebra), 90
Terms and definitions to the *NEC*, 172

Testing laboratories, 265–268
Test-taking strategies, 41–43
Theory. *See* Electrical theory
Thermal-magnetic circuit breakers, 294–295
Three-way switch
  installation, 326–330, *327*
  operations of, 286–288, *287*
  wiring methods, 319, *320*
Time-delay fuse, 269, 293
Time management, defined, 37
Time management strategies, 37–38
Tool kits, 240–242
Tools. *See also* Personal protective equipment (PPE)
  caring for, 214–215
  hand tools, 215–225
  power tools, 225–233
  specialty tools, 233–240
  tool kits, 240–242
Tool sheds, 385
Torpedo level, 236–237, *237*
Total energy management (TEM), 411
Tradesman, 18–19
Training, 5–6, 29–31. *See also* Personal success
Transformer, 120
Trapezoid, formula used to find area of, *93*
Traveler terminals, 288
Trenches, 58
Triangle, formula used to find area of, *94*
Trinomial, 90
T®-Stripper Wire Stripper, 219, *219*
Twist bit, 227
Type AC cable, 312, 313
Type MC cable, 312–313
Type NM-B cable, 311
Type NMC-B cable, 311
Type NMS-B cable, 311
Type S plug fuse, 269, *270*
Type UF cable, 314

U

UF cable, 314
Ufer, Herbert G., 196
UFER ground, 196
Underground conductors, 281
Underground feeder cable, Type UF, 314

Underground service entrance, *189*, 190, 256–258
Underwriters Laboratories (UL), 166, 265–266
Units of measure, 94–96, *95*
U.S. Bureau of Labor, 5, 7
USE (service cable), 257, *258*
U.S. Energy Information Agency, 411
U.S. Green Building Council (USGBC), 410–412
Utility knife, *223*

V

Valence electrons, 116–117
Valence shell, 116
Variables, 88
Ventilation, 59
Viability, 397
Voltage, 55–56, 133–134, *134*, 145, 146–147
Voltage drops, 142, *143*
Voltmeter, 62, 134
Volts (V), 133

W

Wall receptacles, 383
Want ads, 422–423, *423*
Waterpower, 407–408
Watt, 135
Whole body protection, 65
Whole number operations
  addition, 78–79
  borrowing, 79–81
  division, 82–83
  mathematical terms, 77–78
  multiplication, 81–82
  numbers, 77
  subtraction, 79–81
  symbols, 76–77
Wind energy, 402–405
Windmill, *402*, 402. *See also* Wind turbines
Wind turbines, 402–405, *403*, *404*, *405*
Wiring devices. *See* Electrical installations
Wiring diagram, 138, 259, *264*, 388
Wiring methods
  electrical metallic tubing (EMT), 318, *318*

metal clad cable, Type MC, 312–313
nonmetallic-sheathed cable, Type NMC, 310–311, *311*
raceways, 315–317
replacing receptacles, 333–336, *334, 335*
switches. *See also* Switches
underground feeder cable, Type UF, 314

Wiring requirements
arc-fault circuit interrupter (AFCI), 380
circuit layouts, 378–380
countertop receptacles, 380
locating lighting outlets, 384–386
NEC requirements for GFCI protection locations, 386–387
receptacle locations, 381–384
Wood chisels, 237, *237*

Work history, 427–429, *428,* 434–435
Wrenches, *216,* 216, *217*

Z

Zero tolerance, 33